Moisture Storage and Transport in Concrete

Moisture Storage and Transport in Concrete

Experimental Investigations and Computational Modeling

Lutz H. Franke

WILEY-VCH

Author

Prof. Dr. Lutz H. Franke
Hamburg University of Technology (TUHH)
Institute of Building Materials
Building Physics and Building Chemistry
Hamburg, Germany

Cover Image: © Westend61/Getty Images

Library of Congress Card No.: applied for

British Library Cataloguing-in-Publication Data
A catalogue record for this book is available from the British Library.

Bibliographic information published by the Deutsche Nationalbibliothek
The Deutsche Nationalbibliothek lists this publication in the Deutsche Nationalbibliografie; detailed bibliographic data are available on the Internet at <http://dnb.d-nb.de>.

Print ISBN: 978-3-527-35378-1
ePDF ISBN: 978-3-527-84686-3
ePub ISBN: 978-3-527-84685-6
oBook ISBN: 978-3-527-84687-0

Typesetting: Straive, Chennai, India
Printing and Binding: CPI Group (UK) Ltd, Croydon, CR0 4YY

C9783527353781_260424

Contents

Preface

This book deals with the modeling and computational **simulation of moisture absorption, moisture storage, and moisture distribution** mainly in concretes or cement mortars, based on our own experimental investigations and numerous national and international publications on the subject. New aspects of moisture transport and its modeling are presented. For comparison, transferable aspects of moisture transport in inert materials such as ceramic bricks and materials with comparable porosity are considered.

One aim of the book is to present the research results as comprehensibly as possible in such a way that the presented content does justice to the title of the book. In doing so, it is attempted to reproduce the presentations as simply as possible, but in a physically and mathematically correct manner.

The results of targeted experiments carried out at the TUHH (Hamburg University of Technology) on moisture storage and transport serve as an important basis. The materials used were cement mortars and concretes, mainly made of Portland cement. The computational simulations were carried out with the help of a separate computational program additionally developed for this problem area. The models used and, where appropriate, mathematical formulations of interest or required for understanding and comparison with previous approaches are given.

Investigation results on the various moisture transport mechanisms such as vapor transport, capillary transport, and effusion are presented. Water-molecule adsorption in the pore system is considered in more detail and the influences on the transport coefficients, sorption isotherms, and moisture storage functions are highlighted. Special effects such as hysteretic behavior with scanning isotherms, water content distribution, and self-sealing are considered and modeled, and the influence is highlighted.

Furthermore, not only an alternative modeling of drying, especially of concrete, is presented, but also the drying of ceramic bricks and materials of similar pore size distribution is considered for comparison.

Capillary pressure (and vapor pressure) is used as the driving potential for moisture transport, since this mechanism allows a broader consideration of influences, especially the hysteresis of the moisture storage function. Other authors also suggest using preferably capillary pressure as driving potential for the modeling of transport processes in concrete materials.

The simulation program developed and used in the present project to calculate and verify the results is written in Fortran source code. A number of additional calculations were performed using the program MathCad. The Windows interface to the program was developed by Dr. G. Deckelmann at the TUHH and exists at the institute.

January 2024

Prof. Dr. Lutz H. Franke
Emeritus Head of the Institute of Building Materials
Building Physics and Building Chemistry
Hamburg University of Technology (TUHH)
Germany

1

Surface Energetic Principles for Moisture Storage in Porous Materials

1.1 Introduction

Most natural mineral materials, with the exception of crystals, have a pore system whose pores can range from very fine nanometer (nm)-sized pores to the millimeter (mm) range. A recognized **classification of pore sizes** has been made by International Union of Pure and Applied Chemistry (IUPAC). In the 2015 update of the 1985 report [1]. In this paper, the pores are classified into macropores, mesopores, and micropores.

Building materials such as natural stone, brick and especially concrete cover the full pore size range mentioned. Concrete materials usually contain a substantial concentration of particularly fine pores, which are classified in the group of nanopores.

These porous materials can therefore store liquids in the pore system, especially water, which can be carried in vapor form or in liquid form via surface forces.

The capillary absorbed liquid content is measured in [kg/m^3] or in [m^3/m^3], the velocity usually with good approximation as $W = ww \cdot \sqrt{t}$ in [kg/m^2] with the material coefficient ww in [kg/(m$^2 \cdot$ s$^{0.5}$)] when constant fluid supply is ensured.

To describe the storage of liquid from vapor uptake, the resulting water content is presented in the form of **sorption isotherms** as a function of the external relative humidity φ or after converting the relative humidity to the corresponding capillary pressure in the material.

In a number of (building) materials, such as brick products, a portion of the pores is so large that it is no longer filled by vapor adsorption, even at about 100% relative humidity. These pores can then only be filled capillary by external liquid-water supply. This water fraction is called the superhygroscopic range of total water uptake. In such cases, the total moisture storage is described by the so-called **moisture storage function**.

As indicated in the schematic moisture storage function in **Figure 1.1a**, many authors allow the hygroscopic range of the moisture storage function to extend only to about 98% relative humidity, when in fact it must be defined to about 100% RH. The reason for this is the difficulty of precisely setting the moisture and measuring it accurately in this 100%-near range. If the material also contains large pores that cannot be filled by capillary action – for example, air pores – the associated pore volume,

Moisture Storage and Transport in Concrete: Experimental Investigations and Computational Modeling, First Edition. Lutz H. Franke.

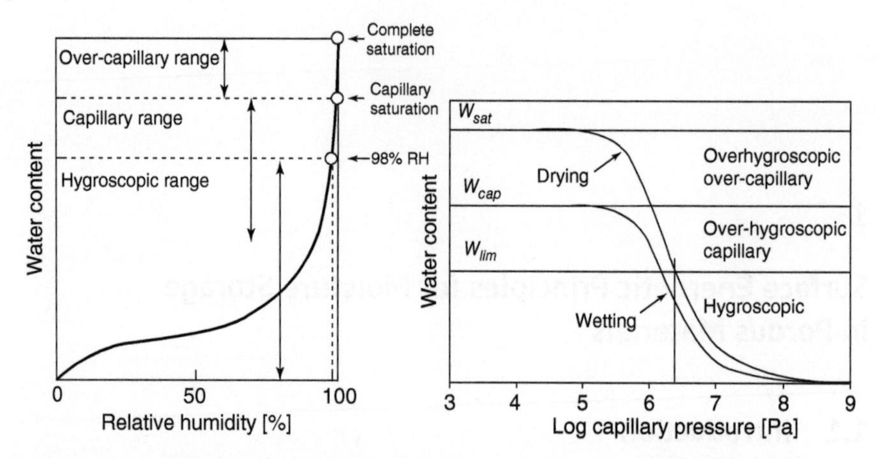

Figure 1.1 (a) Model of the water-storage function for cement-bound material. Source: Adapted from Fagerlund [2] and Eriksson et al. [3]. (b) Sample moisture adsorption and desorption storage functions for building materials as a function of capillary pressure p_C. Source: Carmeliet and Roels [4]/Sage Publications.

which can usually only be filled under pressure, is assigned to the **overhygroscopic range**.

The curve region above 98% RH can be determined using the pressure plate experiment [5, 6], and Espinosa–Franke [7] as a so-called suction stress curve depending on the applied capillary pressure. The mutual conversion of φ *in* p_k is done with Eq. (1.1). In this way, moisture-storage functions can also be represented completely as a function of p_k instead φ as, for example, by Carmeliet in **Figure 1.1b**. At 98% relative humidity, the associated capillary pressure is $p_k = 2.7 \cdot 10^6$ [Pa].

This means that in **Figure 1.1a,b** only the lower section of the curves (in the hygroscopic region) was determined by sorption measurements. The over-hygroscopic ranges thus concern additional capillary water absorption as well as further water absorption under pressure also with (partial) filling of **the processing-related or artificially inserted air pores.**

Using the Eq. (1.39) explained in more detail in Section 1.3.2, the vapor pressure dependence of the adsorption and desorption curves can be converted to the corresponding dependence on the associated capillary pressure in [Pa] as follows:

$$p_k = R_D \cdot T \cdot \rho_W \cdot |\ln(\varphi)| \tag{1.1}$$

Sorption tests on building materials, in particular cement-bound materials, yield desorption curves that deviate significantly from the adsorption curves or moisture storage functions during water absorption. The reason for this behavior will be discussed in more detail in Section 4.6. Measurements by, for example, Feldman and Serada [8] or Ahlgren [9] have already made this clear in 1968 and 1972, see **Figure 1.2a,b**. The measurements also show that a transition between an absorption and desorption curve, or vice versa, occurs on a "short path," **referred to as scanning loops or scanning isotherms**. The main focus of Chapters 2, 3,

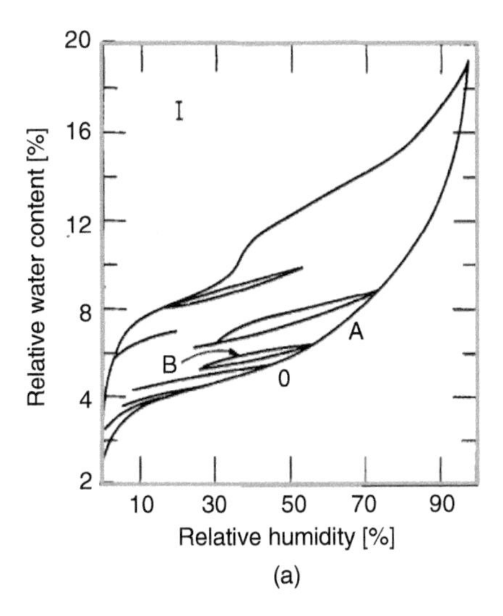

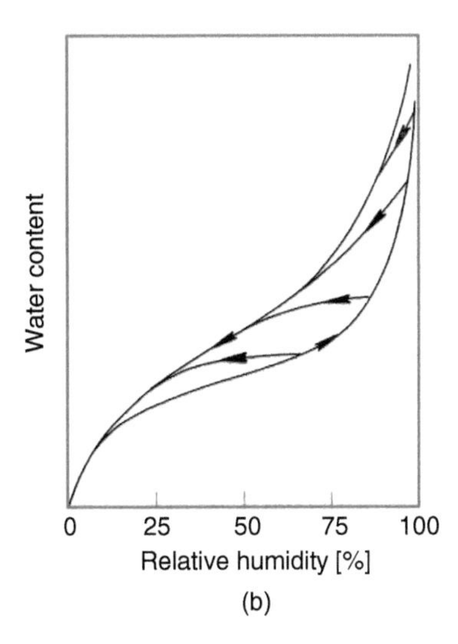

Figure 1.2 (a) Adsorption and desorption isotherms and scanning loops measured on HCP of Portland cement WZ = 0.80. Source: Feldman and Sereda [8]/Springer Nature. (b) Principle course of moisture storage functions including scanning isotherms of building materials. Source: Ahlgren [9]/Lund Institute of Technology/CC BY 4.0.

and 4 will be to show the effects of this behavior on the moisture transport and the moisture household of corresponding material bodies.

It will be shown that the individual moisture storage functions may have a fundamental importance for the moisture balance of porous materials and the modeling of moisture transport.

1.2 Surface Energy and Spreading of Liquids on Solid Surfaces

Since pore water within moist porous bodies is transported by capillary pressure (and vapor pressure) in the presence of sufficiently fine pores, and therefore capillary pressure is a crucial quantity with respect to moisture transport, the origin of capillary pressure within the pore system is first addressed. This first requires **explanations of the role of the surface energy** of the substances involved.

1.2.1 Explanations on Surface Energy and Surface Tension

The molecular arrangement on a water surface surrounded by air is shown schematically in **Figure 1.3**. In contrast to the interior of water, where the molecules are surrounded by similar molecules in all spatial directions and therefore force effects between the molecules cancel each other out in the summation, the surface lacks balancing molecular partners on the air side.

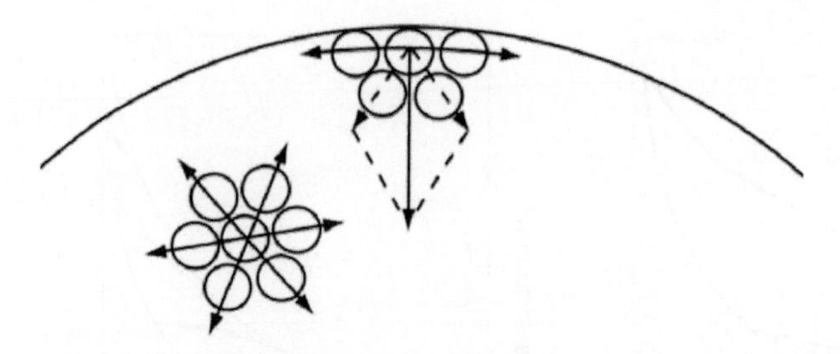

Figure 1.3 Orientation of water molecules and schematic representation of the attractive forces at the liquid surface, by D. Drummer, Erlangen-Nürnberg, Germany.

Therefore, a molecular arrangement is formed at the surface, which leads to inwardly directed cohesive forces (hydrogen bonds) and force effects in the surface plane.

To increase the liquid surface area, work must be done to overcome the cohesive forces of a considered amount of water while the volume remains unchanged. The work to be done per unit area to increase this surface area A is called the surface energy γ_{LV} [LV means liquid versus air], **here abbreviated as** γ_L, corresponding to Eq. (1.2) in differential formulation:

$$\gamma_L = \frac{dW_{surface(L)}}{dA} \quad \left[\frac{\text{Nm}}{\text{m}^2} = \frac{\text{N}}{\text{m}}\right] \tag{1.2}$$

$$F = 2 \cdot b \cdot \gamma_L \quad [\text{N}] \quad \Rightarrow \quad \gamma_L = \frac{F}{2 \cdot b} \quad \left[\frac{\text{N}}{\text{m}}\right] \tag{1.3}$$

Using Figure 1.4, it can be shown how surface energy can be determined by surface enlargement in a model experiment:

A water membrane (producible by addition of surfactant) of dimension $b \cdot s$ is stretched by δs with force F. The surface (front and back) increases by $2 \cdot b \cdot \Delta s$. The work done to increase the surface $\Delta W_{surface}(L)$ is given in **Figure 1.4** using Eq. (1.2).

The displacement work is $\delta s \cdot F$ (δs and F measured).

The formulations of the work $\Delta W_{surface(L)}$ *and* $\Delta W_{boundary}$ **describe the same change in the sample and therefore must be equal in magnitude.** Thus,

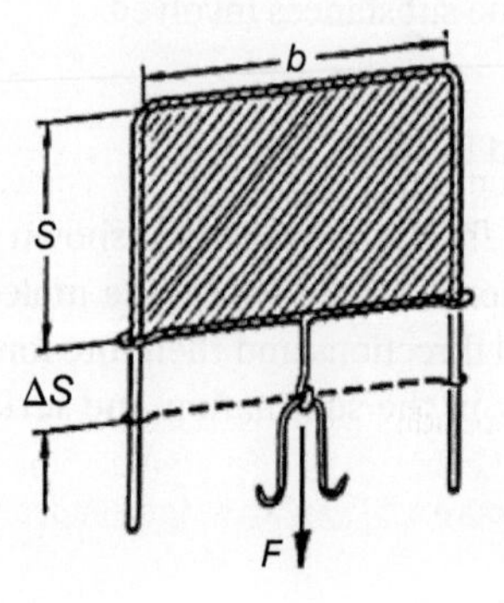

Figure 1.4 Testing Model : Measuring the surface energy and surface tension by the work of displacement Δs and the boundary force F on a (soap)-watermembrane.

$\Delta W_{surface(L)} = 2 \cdot (\Delta s \cdot b \cdot \gamma_L)$ [Nm]
$\Delta W_{boundary} = \Delta s \cdot F$ [m · N]
$\Delta W_{surface(L)} = \Delta W_{boundary}$.

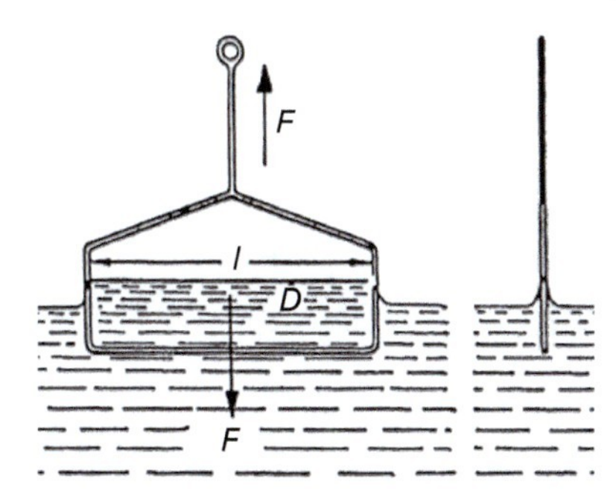

Figure 1.5 Measuring of the surface tension by the bracket-method $\eta_L = \frac{F}{2 \cdot l}\left[\frac{\text{N}}{\text{m}}\right]$.

Eq. (1.3) yields the magnitude of the (specific) surface energy of the surfactant-added water. It can be seen that the special molecular orientation or the resulting surface cohesion in surface plane of the water membrane can introduce an edge force leading to an increase of the surface area, which is called the surface stress. From Eq. (1.3), it can be derived at the same time that the boundary force F related to the unit of the boundary length in [m] corresponds to the surface energy γ_L of the liquid.

The true value γ_L for non-surfactant water can be determined fairly accurately with the experiment shown in **Figure 1.5,** in which a wire stretched in a stirrup structure is lifted out of a water surface via a precision balance. The water surface around the wire is lifted until it breaks off when the maximum force F is reached. The surface energy of water is 0.07275 [Nm/m^2 = N/m] at 20 °C, correspondingly 72.75 [mN/m].

The examples of surface stress measurements shown in **Figure 1.6** illustrate the shapes that water surfaces can attain largely due to surface stress alone. Instead of the bracket test, the Noüy method or the Wilhelmy method are predominantly used; compare **Figure 1.6** and Welcome to DataPhysics-Instruments [10].

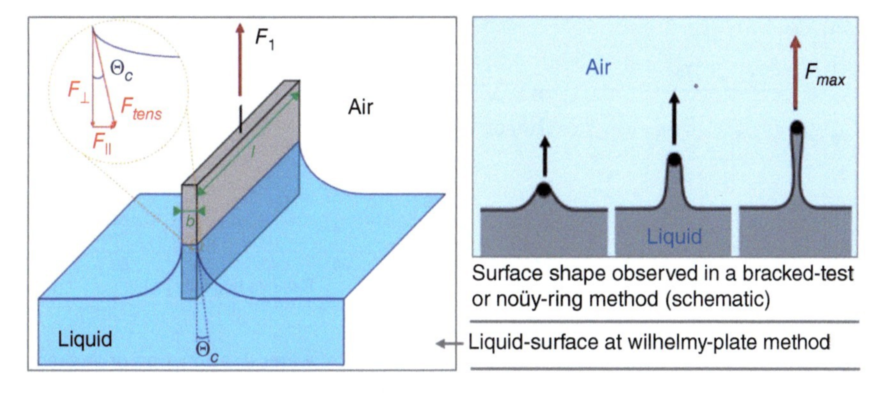

Figure 1.6 Measuring surface tension and corresponding liquid surface shapes by two different methods [10]. **Left**: Situation during a Wilhelmy-plate test. **Right**: Three states of the film surface during a bracket test (compare **Figure 1.5**) or with a ring-shaped wire during a Noüy test.

In the Noüy method, a ring-shaped wire is immersed and then drawn. A high-speed camera shows that in this process, the water surface takes on the shapes sketched on the right in **Figure 1.6** as the pulling force F increases. From the maximum tensile force F_{max}, the surface energy of the liquid is then calculated according to Eq. (1.3), corresponding to the Wilhelmy method.

1.2.2 Dependence of Surface Energy of Water on Temperature, on Relative Humidity of Air, and for Aqueous Salt Solutions

Results on the dependence of the surface energy of water on the relative humidity of the surrounding air were apparently first reliably determined in 2012 by Pérez-Díaz et al. [11], supplemented in 2017 by investigation results by Portuguez et al. [12]. Up to now, obviously, no corresponding results could be determined with capillary suction tests or on drops on flat material surfaces due to the mutual influence of liquid and solid or the resulting influences by simultaneous water evaporation.

As a new measuring method, Pérez-Díaz et al. and Portuguez et al. developed the **method of hanging drops** in climatic chambers combined with precise drop shape measurements via microscopy and image analysis.

The following **Figure 1.7a** shows the variation of the surface energy γ_L at 100% RH as a function of temperature in comparison to the already known behavior as a confirmation of the measurement methodology used. **Figure 1.7b** contains the results for different relative humidities at different temperatures. This shows, for example, that at 20 °C and 20% RH, the value of γ_L is about 5% larger than at 100% RH.

The dependence of surface energy on temperature at 100% RH can be calculated by the following Eq. (1.4). At 60 °C, γ_w= 0.067 [Nm/m^2].

$$\gamma_w(\vartheta) = 0.07275 \cdot (1 - 0.002 \cdot (T_\vartheta - 293)) \tag{1.4}$$

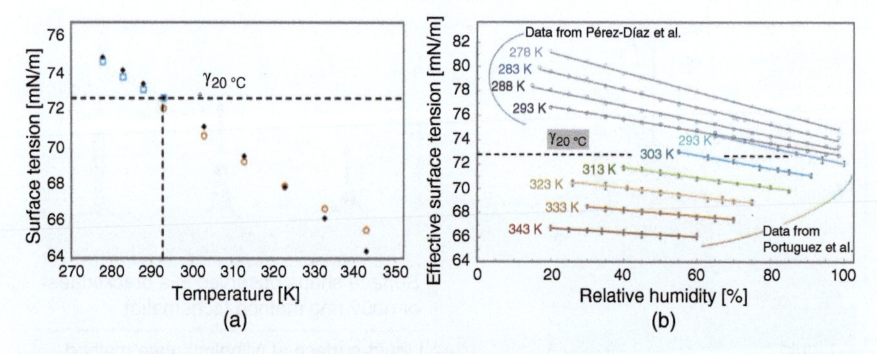

Figure 1.7 Measurement results for the surface energy of water as a function of ambient relative humidity and temperature, basic diagrams from Portuguez et al. [12] with data from Pérez-Díaz et al. [11] (a) Dependence of the surface energy at 100% RH as a function of temperature, comparison of the measured values of Potuguez (red circles) and of Pérez-Díaz (blue circles) with previous tabulated measured values (black signs). (b) Dependence of surface energy on relative humidity and temperature.

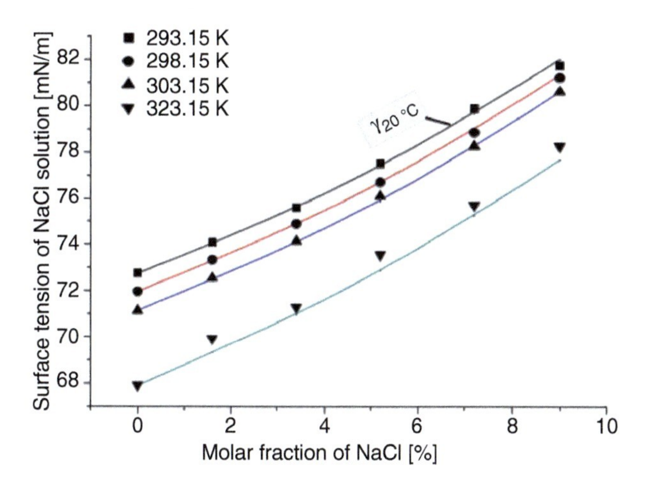

Figure 1.8 Surface tension of aqueous sodium chloride solution at 15–35 °C. Source: Chen et al. [13]/with permission of Elsevier.

According to American National Institute of Standards and Technology (NIST), γ_w at 100% RH **and 100 °C:** $\gamma_w \approx 0.059$ and at 200 °C: $\gamma_w \approx 0.037$ [Nm/m^2]. The surface energy becomes zero at the critical point of water at 373 °C.

Furthermore, it is necessary to ask to what extent the surface energies change for aqueous salt solutions or electrolytes. For limited-concentration solutions, H. Chen et al. [13] presents a "Gibbs phenomenological surface-phase" method for numerous salts. **Figure 1.8** shows results from this publication for sodium chloride and sodium sulfate solutions. The relatively limited influence up to salt contents of 1 [mol/kg] can be seen.

1.2.3 Spreading of Liquids on a Solid Surface

The ability of a liquid to wet a solid depends largely on the surface energies of the substances involved. Therefore, **the surface energies of solids** are also important.

These surface energies can only be measured indirectly at room temperature using drops of test liquids whose surface energies are known. Contact angle measurement is usually used, in which the angle of inclination or the edge angle to the solid surface is determined and the surface energy is derived from this, compare for example Kinloch [14]. Also, so-called test inks allow an approximate determination of the energetic surface states.

For the evaluation of the results, the following considerations are important:

Is defined first

γ_L = **specific surface energy of the liquid** in the environment air plus vapor.

γ_{SV} = **specific surface energy of the solid** in the environment air plus vapor, as well as

γ_{SL} = **specific interfacial energy**, which can exert a separating effect in the interface between the liquid in contact and the solid due to the different molecular nature.

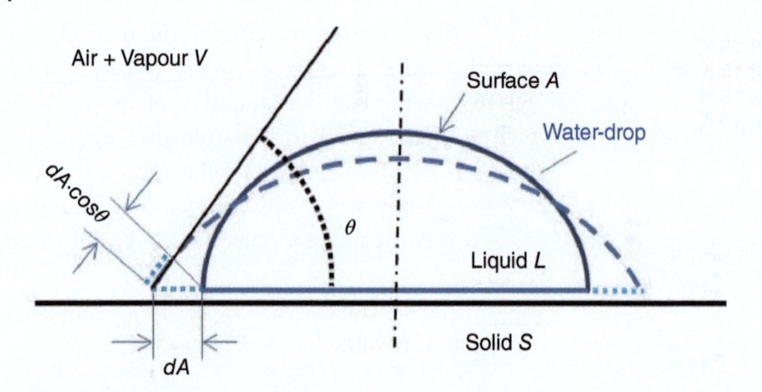

Figure 1.9 Incremental progress dA of a water droplet (L) on a solid surface (S) shortly before reaching the energetic balance of the surface energies involved.

For an understanding of the following relationships, we refer to **Figure 1.9** of a partially contacting droplet with total surface area of A and a given volume.

The spreading of the droplet (L) on the solid surface (S) occurs, if the preconditions are met, due to the effect of attractive forces between the liquid and the solid, which are able to overcome the surface tension of the liquid when the contact area is increased.

The spreading comes to a stop when a minimum of the total energy of the ongoing process is reached. **Figure 1.9** represents the situation just before this standstill: The already existing contact area between (L) and (S) still increases by the fraction dA. On the air side of the contacting droplet, the surface area is thus approximately increased by the value $dA \cdot cos\theta$. In total, the droplet surface area thus increases by $dA + dA \cdot cos\theta$. For this, the following work must be done on the drop side:

$$dW_{surface}(L) = \gamma_L \cdot (dA + dA \cdot cos(\theta)) \tag{1.5}$$

At the same time, adhesion energy is released in the area of contact area increasing dA:

$$dW_{Adh}(SL) = -dA \cdot (\gamma_L + \gamma_{SV} - \gamma_{SL}) \tag{1.6}$$

Here γ_{SL} expresses that there is not a full saturation of the surface energy in contact but is reduced by the fraction $\gamma_{SL} \cdot dA$. At equilibrium:

$$dW_{surface}(L) + dW_{Adh}(SL) = 0 \tag{1.7}$$

It follows

$$(1 + cos(\theta)) \cdot \gamma_L + \gamma_{SL} - \gamma_L - \gamma_{SV} = 0 \tag{1.8}$$

and for the boundary angle θ the well-known Young equation follows after the transformation:

$$cos(\theta) = \frac{\gamma_{SV} - \gamma_{SL}}{\gamma_L} \tag{1.9}$$

At $cos\theta = 1$ *or* $\theta = 0$ complete wetting or dissolution of the drop takes place.

The surface energies present now allow **statements about the dispersion behavior of different liquids on different surfaces**. This also determines whether a liquid can penetrate a pore system. The measurement of the contact angle in the drop test provides at least approximate information on this. If the contact angle is about 90°, $\gamma_{SL} \approx 0$. This results is a "neutral" behavior of the given liquid on the given solid surface. No capillary water absorption takes place in this case. In order that capillary takes place, the angle θ must be appreciably less than 90°.

According to the following equation by van Honschoten et al. [15], spreading occurs for a solid–liquid combination when the value S is positive:

$$S = U_{dry}^{substrate} - U_{wet}^{substrate} = \gamma_{SV} - (\gamma_{SL} + \gamma_{LV}) \tag{1.10}$$

1.2.4 Determining the Surface Energies of Solid Surfaces

For mineral materials used in construction, the contact angle is usually less than 90°, unless the surface has been modified, for example, by a hydrophobizing measure. Kaolin has a surface energy of about 500–600 [mN/m]; HPC concretes with aggregate from granodiorite and from granite with a compressive strength of about 130 [MPa] have, according to Barnat-Hunek γ_S = 1000 to 1800 [mN/m] [16] and a contact angle of about 10° to 30° when wetted with water at 22 °C, resulting in capillary water absorption into a pore system.

Calcium carbonate has a surface energy of only γ_S = 75–80 [mN/m], so that largely standing water droplets can be expected on such surfaces.

Plastics, paints, and waxes have a γ_S of 25–40 [mN/m], so that no spreading of water can take place on these surfaces, which is very well seen, for example, on waxed car bodies. Teflon exhibits a γ_S of about 20 [mN/m].

In contrast, ethanol or isopropanol, for example, with surface energies of 22 or 23 [mN/m] on mineral surfaces always show a contact angle close to 0°, i.e. complete spreading and a strong readiness for capillary penetration into capillary pores.

Metal alloys also usually have very high surface energy, γ_{SV}-values of metals can be taken from Kumikov and Khokonov [17]. Nevertheless, with respect to adhesion, adequate surface pretreatment is especially important for metals.

If, in the case of "unknown" solid surfaces, their surface energy γ_S is needed, however, it cannot be determined only by determining a boundary angle with a known liquid. A more accurate determination of the surface energy can be made by at least two edge angle measurements with two different test liquids with different, known **polar interaction fractions** $\gamma_L{}^p$ (for example, from hydrogen bonding) **and dispersive fractions** $\gamma_L{}^d$ (from van der Waals forces).

The surface energy γ_{SV} of the solid as well as the interfacial energy γ_{SL} can then be determined from the results of the boundary angle measurements according to the accepted OWRK method of Owens, Wendt, Rabel, and Kaelble, compare Yuan and Lee [18], Lauth and Kowalczyk [19], and Barnat-Smarzewski and Smarzewski [16].

In Literature [20], measurement and calculation results from two institutes are reported by Cwikel et al., in which the performance of five different computational models for determining the surface energy of solids is investigated. For this purpose, the contact angles of selected test liquids on 42 different solid surfaces are measured and compared with the predictions of the computational models.

1.3 Basic Equations for Liquid Absorption in Material Pores

1.3.1 Liquid Absorption in Pores by Effect of Surface Energies

Liquid is also drawn into the inner surface of material pores by the effect of surface tension. The finer the pores, the greater the depth of penetration or rise relative to gravity. In such pores, the liquid is transported as in a tube. This is traditionally shown by the liquid rise height in a cylindrical capillary pore.

1.3.1.1 Derivation of the Capillary Rise via the Adhesion Works and the Potential Energy in Capillaries

The adhesion works in a standing cylindrical capillary are for the rise height h:

$$W_{surface}(h) = +2 \cdot \pi \cdot r_{pore} \cdot \gamma_L \cdot h + W_{surface}(0) \tag{1.11}$$

$$W_{Adh}(h) = -(\gamma_L + \gamma_{SV} - \gamma_{SL}) \cdot 2\pi \cdot r_{pore} \cdot h - W_{Adh}(0) \tag{1.12}$$

The potential energy to be overcome is:

$$E_{pot}(h) = +r_{pore}{}^2 \cdot \pi \cdot \varrho_L \cdot g \cdot \frac{h^2}{2} + E_{pot}(0) \tag{1.13}$$

At equilibrium is

$$\frac{dW_{surface}(h)}{dh} + \frac{W_{Adh}(h)}{dh} + \frac{E_{pot}(h)}{dh} = 0 \tag{1.14}$$

From this follows, shortened by $2 \cdot \pi \cdot r_{pore}$:

$$-(\gamma_{SV} - \gamma_{SL}) + r_{pore} \cdot \varrho_L \cdot g \cdot \frac{h}{2} = 0 \tag{1.15}$$

Using Eq. (1.9), this gives the relationship for the achievable height due to adhesion work in a cylindrical capillary:

$$h_{max} = \frac{2 \cdot \cos(\theta) \cdot \gamma_L}{r_{pore} \cdot g \cdot \varrho_L} \tag{1.16}$$

If instead slit pores with a constant spacing of the pore surfaces of $d = 2 \cdot r_{pore}$ are present, the result for the rise height is

$$h_{Slit,max} = \frac{2 \cdot \cos(\theta) \cdot \gamma_L}{d \cdot g \cdot \varrho_L} \tag{1.17}$$

If $d = 2 \cdot r_{pore}$ gives $h_{slot} = 1/2 \cdot h_{cap}$. At this point, we refer the reader to Section 1.3.1.3, where a more general derivation of the fluid uptake in pores due to adhesion work is described.

1.3.1.2 Capillary Pressure in Cylindrical Pores and in Slit Pores

The transport in capillary pores or slit pores caused by the surface energy creates a tensile stress below the menisci in the pores, which is called capillary pressure. As a result of the previous explanations, the resulting capillary pressure can be simply represented according to **Figure 1.10**. The water column of the pores resp. pore filling is further pulled by the effect of the surface energies resp. the corresponding surface tension at the wall of the pores. In dependence of the existing edge angle θ from Eq. (1.9), the force component f_{cap} in [N/m] is now generated there in the pore direction.

$$f_{cap} = \gamma_L \cdot cos(\theta) \tag{1.18}$$

For a cylindrical capillary pore with a given perimeter, the resulting total force follows to

$$F_{cap} = f_{cap} \cdot 2\pi \cdot r_{pore} \tag{1.19}$$

The resulting capillary pressure p_{cap} cannot be exceeded by the surface energy alone because of the limitation of the surface tension to γ_L. The capillary pressure is then (when the surrounding air pressure is not taken into account)

$$p_{cap}(r) = \frac{F_{cap}}{\pi \cdot r_{pore}^{2}} = \frac{2 \cdot \gamma_L \cdot cos(\theta)}{r_{pore}} \quad [\text{Pa}] \tag{1.20}$$

In the presence of slit pores with d = surface distance of the pore walls, the following results instead

$$p_{cap}(d) = \frac{2 \cdot \gamma_L \cdot cos(\theta)}{d} \quad [\text{Pa}] \tag{1.21}$$

When $d = 2 \cdot r_{pore}$, $p_{cap}(d)$ is only half as large as for a cylinder pore, according to Eq. (1.20).

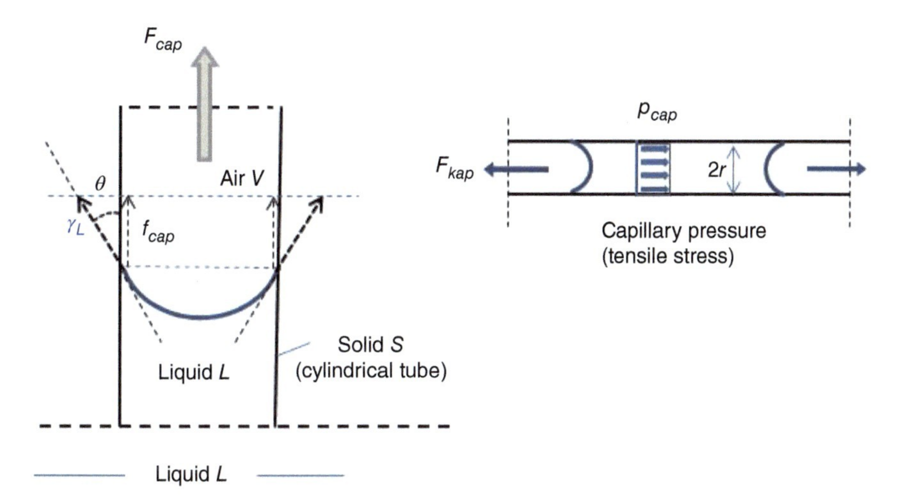

Figure 1.10 Formation and definition of capillary pressure in a cylindrical pore during water absorption.

The influence of changing temperatures and possible salinity on γ_L has been given previously.

1.3.1.3 Capillary Pressure as a Cause of Fluid Transport and Rise Height in a Capillary Pore

From Extrand and Moon [21], using simple water absorption experiments on glass capillaries, it is shown that the derivation of the rise height in capillary pores in terms of surface work and potential energy presented in Section 1.3.1.1 leads to an unjustified formal restriction on the applicability of the Eq. (1.16).

According to this equation, varying ratios of surface energies along the capillary pore and a varying pore cross section, for example, a larger pore radius in the lower part of the capillary, should also lead to the same riser height (1.16). Extrand and Moon [21] concludes that the capillary pressure at the head of the water column in the capillary is responsible for the increase, independent of the other parameters. In fact, this can be shown and somewhat specified in the following way:

Instead of (1.11) and (1.12), let be written only the local increase of the surface energy fractions or work fractions of the liquid surface $\Delta W_{surface}$ and the solid surface ΔW_{Adh} in the region of the pore radius r_{pore}

$$\Delta W_{surface} = +2 \cdot \pi \cdot r_{pore} \cdot \gamma_L \cdot \Delta h \tag{1.22}$$

$$\Delta W_{Adh} = -(\gamma_L + \gamma_{SV} - \gamma_{SL}) \cdot 2 \cdot \pi \cdot r_{pore} \cdot \Delta h \tag{1.23}$$

The resulting energy sum does the work of raising the corresponding liquid level. The corresponding mechanical work portion force · displacement results at the considered radius from the product of the pore cross-sectional area and the hydraulic stress acting in the cross-section times the displacement Δh. For example, the surface of the advancing fluid has the shape of a meniscus. The associated mechanical work is then, with the hydraulic (tensile) stress p_{cap}

$$\Delta W_{meniscus} = r_{pore}{}^2 \cdot \pi \cdot p_{cap} \cdot \Delta h \tag{1.24}$$

At equilibrium is

$$\Delta W_{surface} + \Delta W_{Adh} + \Delta W_{meniscus} = 0 \tag{1.25}$$

From this follows, shortened by δh and $2 \cdot \pi \cdot r_{pore}$, with $\gamma_{SV} - \gamma_{SL} = cos(\theta) \cdot \gamma_L$ the relation

$$p_{cap} = \frac{2 \cdot \gamma_L \cdot cos(\theta)}{r_{pore}} \quad [\text{Pa}] \tag{1.26}$$

With the suction stress p_{cap}, the fluid bulk density ϱ_L, and the acceleration due to gravity g, the hydraulic potential or the fluid pressure head is given by

$$h = \frac{p_{cap}}{\varrho_L \cdot g} \tag{1.27}$$

Figure 1.11 shall schematically illustrate the relationship between the effect of capillary pressure and gravity in different capillaries. h_1 is the rise height in capillary 1 with the associated capillary pore radius $r1$ according to Eq. (1.16) resp. (1.27).

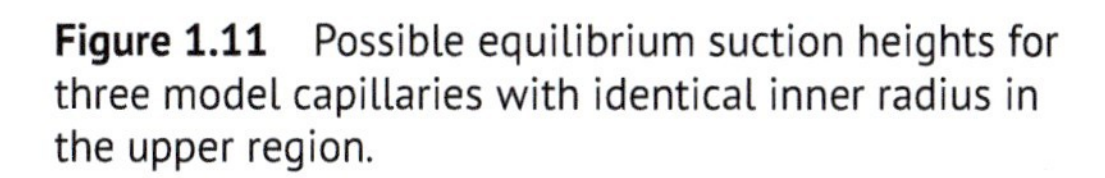

Figure 1.11 Possible equilibrium suction heights for three model capillaries with identical inner radius in the upper region.

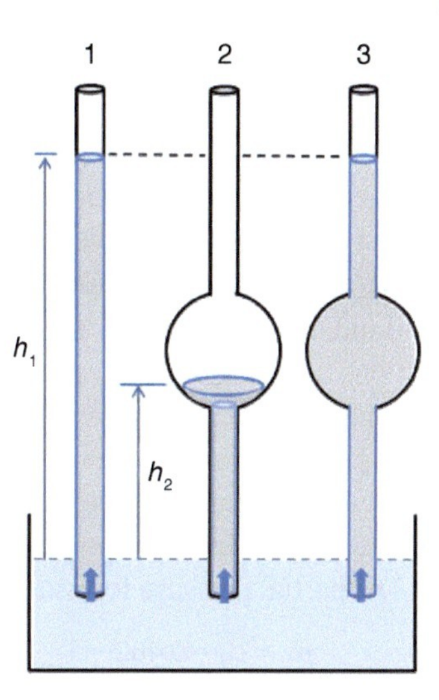

If the capillary has a radius extension, for example, in the form of a spherical pore, the liquid uptake stops when entering the spherical pore according to the associated radius extension, resp. the lower capillary pressure (capillary 2 in 1.11). If, on the other hand, the sphere pore is filled with water by some action in the sense of a continuous pressure connection to the surface, the rise height reaches the suction height h1 for a capillary radius $r_3 = r_1$.

It is to be noted that only the adhesion work in the area of the front meniscus and the capillary pressure there (as suction force) lead to the progress of the water column.

If **capillary condensation** takes place at the front of the capillary pore due to the boundary conditions (pore radius and vapor pressure, compare Sections 1.3.2 and 1.4.5), liquid is added there accordingly, but the capillary head $h1$ is not increased.

When the capillaries are horizontal or there is no gravity (but water contact at the base), the capillary is completely filled by the adhesion work up to the head, regardless of its length.

1.3.2 Pore Filling by Capillary Condensation

In an empty pore system of a dried material, if the external vapor pressure is increased from very small pressures to saturation pressure p_{sat}, the vapor diffusing into the pores leads to sorption layer thicknesses at the pore walls corresponding to the associated "base isotherm." For water vapor, this is the Eq. (1.50) resp. **Figure 1.15a**. These sorption liquid layers are also called liquid films.

Practical experience confirms that fine-porous materials have moisture contents that correlate with the moisture content of the surrounding air. In the following, it will be shown again why this must be so.

As a reminder, the general gas equation for ideal gases is written first.

$$p = \frac{n}{V_{ges}} \cdot R \cdot T \tag{1.28}$$

There is p = Gas pressure in [Pa], $R = 8.31$ [Nm/(K·mol)] = General gas constant, and n = Number of moles of gas in V_{ges}.

If $n = 1$ follows $V_{ges} = \overline{V}$ = Molar volume = 24.46 [l/mol] under thermodynamic standard conditions 25 °C and 101.3 [kPa] and 22.414 [l/mol] under standard conditions 0 °C and 101.3 [kPa].

Equation (1.28) is also valid for the partial pressures of individual gas components in gas mixtures, for example also for the water vapor fraction in air.

Let the saturation concentration for water vapor in air be c_S. At 20 °C, c_S = 17.3 [g/m³]. In contrast, air weighs about 1170 [g/m³] under standard conditions. The partial pressure for water vapor in air is also:

$$p_S = c_S \cdot 0.462 \cdot T \ \text{[Pa]} \tag{1.29}$$

Herein, 0.462 is the specific gas constant (independent of boundary conditions) for water vapor in [Pa·m³/(g · K)] and c_S in [g/m³]. Let p_S be the saturation partial pressure of water vapor in air. The temperature dependence is $\delta \geq 0$ °C.

$$p_{S(\delta)} = 288.58 \cdot \left(1.098 + \frac{\delta}{100}\right)^{8.02} \tag{1.30}$$

The vapor concentration is c_S in equilibrium with a (flat) water surface A_0 at a given temperature T_0.

If the water surface is curved concavely, this leads to a higher relative concentration c_S with respect to the curved water surface A_1. Expressed as relative concentration c_S/A is now $c_S/A_1 > c_S/A_0$, which is especially clear when imaging the inner surface of a sphere.

To maintain thermodynamic equilibrium, the liquid surface must absorb vapor molecules from the air in an effort c_S to reduce. In the initial plane state, the equilibrium between the phases of water and vapor in contact in confined space is as follows:

$$d\overline{G} = \mu_L \cdot dn_L + \mu_D \cdot dn_D = 0.$$

Since $dn_L = dn_D$ must be, it follows (at unchanged temperature and pressure) $\mu_L = \mu_D$.

When there is a curvature of the fluid surface, there is (due to the relationship between surface curvature and pressure) a change in pressure in the system. Here, first of all, the fluid pressure is meant.

$$d\mu = -S \cdot dT + V \cdot dp = V \cdot dp \quad \text{at } T = const.$$

On the part of the liquid, the change in chemical potential (or free enthalpy) to be compensated is compared to the initial state at a flat surface and at unchanged temperature, **with $\overline{V}$ = molar volume of the liquid or water** and p_S the corresponding saturation pressure is

$$d\overline{G}_L = \overline{V} \cdot dp, \text{ respectively :} \tag{1.31}$$

$$\Delta\overline{G}_L = \int_{p1}^{p_S} \overline{V}_L \cdot dp = -\overline{V}_L \cdot (p - p_S) \tag{1.32}$$

The **fluid pressure** can be $p \leq p_S$ *or* $p > p_S$. This depends on whether the pressure generated in the fluid is negative or positive. In the case of a cylindrical pore with a progressing meniscus of fluid filling in **Figure 1.10**, the fluid pressure should be assumed to be negative as a tensile stress.

In Eq. (1.20), r_{pore} is the radius of the cylindrical pore. In a circular– capillary, the pressure-transmitting meniscus has two main radii of curvature

$$r_{meniscus} = \frac{r_{pore}}{cos(\theta)} \tag{1.33}$$

Inserted into Eq. (1.20), the Young–Laplace relation for a circular–cylindrical pore follows.

$$p_{cap(meniscus)} = \frac{2 \cdot \gamma_L}{r_{meniscus}} \quad [\text{Pa}] \tag{1.34}$$

In the plane initial state, $r_{meniscus} = \infty$ and thus $p_S = 0$ in Eq. (1.32). In the present case, p and hence δp is negative. From this follows for the fluid in the region of the meniscus

$$\Delta\overline{G}_L = \overline{V}_L \cdot \Delta p = -\overline{V}_L \cdot \frac{2 \cdot \gamma_L}{r_{meniscus}} = -\overline{V}_L \cdot \frac{2 \cdot \gamma_L \cdot cos(\theta)}{r_{pore}} \quad [\text{J/mol}] \tag{1.35}$$

The enthalpy change of the liquid is followed by the corresponding **reaction of the vapor phase *V***: From the saturation initial state c_S *or* p_S as the saturation pressure of the plane surface.

$$\Delta\overline{G}_V = \int_{p1}^{p_S} \frac{R \cdot T}{p} dp = -R \cdot T \cdot ln\left(\frac{p1}{p_S}\right) \quad [\text{J/mol}] \tag{1.36}$$

where $p1$ is the equilibrium saturation vapor pressure in the concave surface region.

From the requirement $\Delta\overline{G}_V = \Delta\overline{G}_L$ follows with Eq. (1.35) and $\overline{V}_L$= mole volume of condensed liquid (18 [cm^3] or 18/10^6 [m^3/mol] for water), the **Kelvin equation** given by $p1 \leq p_S$ in the following formulation for concave surfaces:

$$r_{pore} = \frac{-\overline{V}_L \cdot 2\gamma_L \cdot cos(\theta)}{R \cdot T \cdot \, ln\left(\frac{p1}{p_S}\right)} = \frac{M_\gamma}{ln(\varphi)} \quad [\text{m}] \tag{1.37}$$

Therein, $p1/p_S = \varphi_{ext}$ **is the external relative humidity** φ. The summary term M_γ (constant for given T) corresponds to Eq. (1.68).

Equation (1.37) indicates up to which pore radius, at a given vapor partial pressure p_1 a cylindrical pore is completely filled by condensation from the vapor entering the pore, denoted **capillary condensation**. Here, **for cylindrical pores with constant cross-section**, it is assumed that a meniscus supposed to be stationary is already present due to an interrupted water supply at the base of the capillaries, or in the case of an equilibrium situation as indicated in **Figure 1.10**. When exposed to an external relative vapor pressure of at least $p1$, in the curvature region of this meniscus, the external vapor pressure becomes the saturation vapor pressure, with the consequence of condensation of the vapor present and pore filling.

If, for a given radius r_{pore}, the critical relative vapor pressure $p1$ or the external relative humidity above which capillary condensation takes place is sought, the result is (Eq. (1.37) transformed):

$$\varphi_{ext} = e^{-\frac{\overline{V}_L \cdot 2\gamma_L \cdot \cos(\theta)}{R \cdot T \cdot r_{pore}}} \tag{1.38}$$

For materials with pore size distributions "small to large," the entire material is filled only up to the pore size $r_1 = r_{pore}$ according to Eq. (1.37) by the vapor diffusing in from the outside.

The Kelvin Eq. (1.37) or the resulting Eq. (1.38) are generally valid also for curved concave liquid surfaces characterized by different principal radii of curvature, as well as **for pores with noncircular cross section.**

Combining the relations (1.37) and (1.20), we immediately obtain the relation between p_{cap} and $\varphi = p1/p_S$

$$p_{cap} = -\frac{R \cdot T}{\overline{V}_L} \cdot ln(\varphi) = R_D \cdot T \cdot \rho_W \cdot |ln(\varphi)| \quad [\text{Pa}] \tag{1.39}$$

1.3.2.1 Extent of Validity of the Kelvin Equation

An important question is up to which lower pore radius the Kelvin equation is valid. The applicability is considerably limited by the charge distribution and the relation of the pore radius to the molecule size of the fluid under consideration. Numerous authors have commented on this issue in the past. **Matsuoka et al. [22]** perform atomic force microscopy (AFM) studies on liquid films between curved muscovite-mica surfaces at different relative humidities. They give a lower pore radius of $r = 1.5$ [nm] for (the polar) water, e.g. $r = 0.5$ [nm] for cyclopentanes.

Fifteen years later, **Kim et al. [23]** remark, based on measurements with more advanced atomic force microscope (AFM) technology, that a much lower boundary pore radius $r = 0.5$ [nm] for water can be assumed. They measure in the AFM apparatus at the curved contact surface of the samples capillary condensation down to the mentioned radius and evaluate the results based on the accepted Kelvin–Tolman theory, which gives a correction to Laplace's equation for very small pore radii, compare also [24].

They emphasize explicitly that this result (also based on AFM studies) refers to the classical Kelvin equation according to Eq. (1.37).

If one includes the relation of Eq. (1.79) according to the representation in **Figure 1.22** or Section 1.4.5 and **Figure 1.15a**, then yields the corresponding real **radius $r_R = 0.7$ [nm], which could then be called as the lower physically detectable pore radius for the applicability of the Kelvin relation.**

1.3.3 Saturation Vapor Pressure at the Surface of Convex Shapes

In contrast to concave liquid surfaces, for example, at a meniscus in cylindrical pores, the saturation vapor pressure at convex external surfaces is increased by $p > p_S$, so that condensation of vapor occurs only at relative humidities $\varphi_1 > \varphi_S$.

It is then valid as liquefaction vapor pressure over an external surface (with the main curvature radii r). Instead of Eq. (1.38), the corresponding equation with reversed sign is then valid.

Since the relative vapor pressure is lower on such surfaces, faster evaporation or drying also takes place there.

In Section 1.4.6.2, the layer thicknesses of water molecules that can form by adsorption on concave or convex surfaces as a function of relative humidity are investigated.

1.3.4 Explanations of the Young–Laplace Equation for Stress on Curved Fluid Surfaces

The relation presented at the same time by Young and Laplace is as follows:

$$\Delta p = \gamma_L \cdot \left(\frac{1}{r1} + \frac{1}{r2}\right) \tag{1.40}$$

Equation (1.40) gives the mechanical relationship between the curvature of a nonplanar membrane-like surface, given by the two principal radii of curvature $r1$ and $r2$, and the stress acting in the membrane plane γ_L in [N/m] and the pressure Δp in [N/m^2] acting on the membrane surface (orthogonal).

This pressure can act from a liquid or a gas. Since a "water-membrane" is formed at the water surface, the surface can be treated with the Young Laplace relation (1.40). Since the membrane stress γ_L acts largely as a constant for water surfaces, Eq. (1.40) provides the relationship between the sum Δp of inside and possibly outside orthogonal pressure components on the surface and the curvature of the surface.

In the area of the concave meniscus in a cylindrical capillary pore the water pressure acts on the water side as a two-dimensional tensile stress, on the air side the air pressure (including vapor partial pressure). However, since the air pressure acts on the entire system, it is also present as a component on the "water side" of the meniscus and can therefore be disregarded (except in special cases).

This also indicates that for cylindrical capillaries with a constant cross-section, the vapor partial pressure does not play a role in meniscus formation, unless the sorption properties of the inner capillary-pore surface are affected by varying vapor pressure, which would manifest itself, for example in a change in the contact angle. There is no evidence for this in glass capillaries, for example. **In contrast, for capillaries with increasing pore radius** or pore systems with medium pore sizes increasing from "small to large" there is a clear relationship between the vapor partial pressure $p1$ and the pore size or meniscus shape and thus the capillary pressure according to the Eqs. (1.37) and (1.34).

With reference to the Eqs. (1.33) and (1.34), results analogously to Eq. (1.40)

$$p_{cap(meniscus)} = \gamma_L \cdot \left(\frac{1}{r_{meniscus}} + \frac{1}{r_{meniscus}}\right) = \frac{2 \cdot \gamma_L}{r_{meniscus}} \tag{1.41}$$

Due to the limitation of the possible tensile stress in the "meniscus membrane" to γ_L, the possible capillary pressure (tension) is limited accordingly. The principal radii of curvature $r1 = r2 = r_{meniscus}$ to be used for a circular–cylindrical capillary pore differ from the pore radius r_{pore} for a noncircular cross section according to Eq. (1.33).

Eslami and Elliott [25] have investigated **the exact shape of menisci under gravity and vapor pressure** in cylindrical capillaries mathematically in more detail by integrating Laplace's equation and taking thermodynamic considerations into account.

Figure 1.12 shows from this work the results on the meniscus shape for the edge angle $\theta = 0$ and for varying edge angles. It can be seen that even the largest capillary pores in cement-bonded materials assumed to have circular cross-sections with radii on the order of 300 [nm] still exhibit pronounced curvature radii of nearly hemispherical shape in the presence of a edge angle of $\theta \approx 0$.

The edge angles on concrete surfaces were measured to be 10°–15°. Even for such edge angles, the deviations from the hemisphere shape are small.

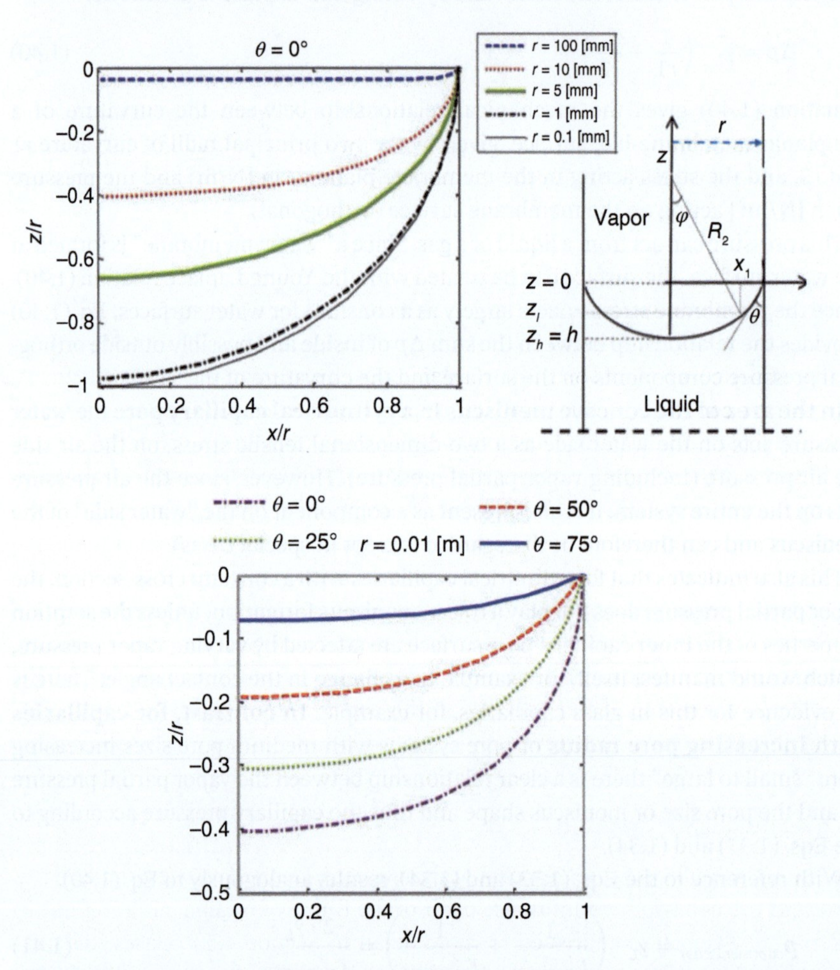

Figure 1.12 Meniscus shape in cylindrical pores as a function of pore radius at constant contact angle of 0° and influence of the contact angle at constant radius 10 [mm] (lower figure). Source: Eslami and Elliott [25]/Springer Nature/CC BY 4.0.

1.3.5 Application of Kelvin Equation to Floating Droplets

Spherically assumed **individual droplets in a vapor-supersaturated air** form different droplet sizes depending on the surrounding relative humidity φ_{ext} resp. degree of supersaturation, which can be determined by means of a fitted Kelvin equation.

In the droplet the internal pressure is positive in contrast to the liquid at a concave meniscus, furthermore at the "surface-membrane" of the droplets there is no more contact to a solid surface, so that the contact angle $\theta = 0$ and $cos(\theta) = 1$ is.

The consequence is that in Eq. (1.32) p becomes $>p_0$ and from this follows instead of Eq. (1.34) for the fluid pressure

$$p_{drop} = \frac{2 \cdot \gamma_L}{r_{drop}} \tag{1.42}$$

and instead of Eq. (1.35) for the change of the chemical potential of the liquid

$$\Delta \overline{G}_L = +\overline{V}_L \cdot \frac{2 \cdot \gamma_L}{r_{drop}} \quad [\text{J/mol}] \tag{1.43}$$

Due to the increase of $\Delta \overline{G}_L$, a positive change of the chemical potential of the gas phase vapor must also result. However, in the Eq. (1.36), which also applies here, $p1/p_S \geq 1.0$ must be used. From this follows the Kelvin equation adapted for free drops:

$$r_{drop} = \frac{+\overline{V}_L \cdot 2 \cdot \gamma_L}{R \cdot T \cdot \, ln\left(\frac{p1}{p_S}\right)} \quad [\text{m}] \tag{1.44}$$

In the absence of condensation nuclei or condensation-enabling surfaces in the vapor-containing air in question, supersaturation $\varphi > 1.0$ can be produced by supplying additional vapor or by cooling the air. Artificially, obviously, several times the saturation vapor pressure can be produced; in nature, normally only a few percent supersaturation is produced in the atmosphere. If the vapor meets condensation nuclei, a spontaneous condensation takes place, which immediately changes into an evaporation of the droplets if the pressure falls below $\varphi = 1.0$. The respective droplet sizes can be derived according to Kelvin from the respective local vapor concentration. **Overall, this is an unstable dynamical process that depends, among other things, on the transport velocity of the vapor molecules in the air**. Extensive research is still ongoing on this topic [see for example B. Waigand, University of Stuttgart, Germany].

1.3.6 Solubility of Gases in Water

The issue of solubility of gases, especially air in water, plays a role, for example, in the context of the behavior and influence of air of the filling of original air pores in the material by water. It is addressed in more detail in Section 4.5.5. Gases dissolve in water, for example, from the water surface under their present partial pressure. The relationship is described by Henry's law. It indicates which dissolved gas concentration c_{W_i} of gas i results in water as a function of water temperature and partial

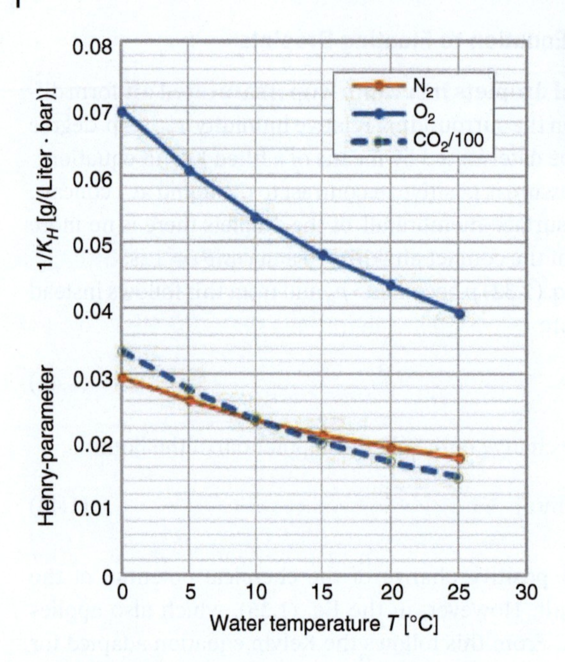

Figure 1.13 Temperature dependence of the Henry parameter $1/K_H$ of O_2 and N_2, and $(1/K_H)/100$ for CO_2, Values from W. Aeschbach-Hertig (Heidelberg University).

pressure p_i or total pressure. According to this

$$c_{W_i} = \frac{1}{K_{H_i}} \cdot p_i \quad [\text{g/(Liter} \cdot \text{bar)}] \tag{1.45}$$

The Henry parameters K_H are determined experimentally. **Figure 1.13** gives for the three selected gases O_2, N_2 and CO_2 the values of the factor $1/K_{H_i}$ as a function of temperature and a gas pressure of 1 bar. The correct value for CO_2 is the diagram value ×100.

For oxygen, at a water temperature of 20 °C, an atmospheric pressure of 1 bar and the partial pressure of 0.21 bar, the concentration of dissolved O_2 is as follows

$$c_{W_i} = 0.043 \cdot 0.21 = 0.00903 \text{ [g/liter]} = 9.03 \text{ [g/m}^3\text{]}$$

Assuming partial pressures of 21% for oxygen and 78% for nitrogen, the following concentration for air $c_{W_{air}}$ at 20 °C is obtained (approximately) using the Henry values of the diagram:

$$c_{W_{air}} = (0.21 \cdot 0.043 + 0.78 \cdot 0.019) \cdot 1000 = 23.9 \text{ [g/m}^3\text{]}$$

The solubility of gases increases significantly with decreasing temperature, but especially with increasing total pressure. In 20 [m] water depth, for example, the solubility of nitrogen due to the pressure increased by 2 bar is instead of 14.8 [g/m^3]

$$c_{W(N_2)} = 0.019 \cdot (0.78 + 2.0) \cdot 1000 = 52.8 \text{ [g/m}^3\text{]}$$

In the presence of appreciable amounts of other ions or very saline waters, the soluble gas concentrations may deviate from the Henry values. Thermodynamic equilibrium calculations are then necessary.

1.3.7 Cavitation in the System Water/Vapor

In the context with the tensile stress of water in capillary pores, the problem area of cavitation in the water/vapor system will first be highlighted.

The phenomenon of cohesion loss due to cavitation (i.e. the spontaneous transition of water into vapor) can occur when a volume of water is subjected to tensile stress, especially when the water contains small vapor bubbles, for example, on external contact surfaces or impurities. **Cavitation can occur, for example, in free water at strong underpressure formation** $\ll p_{atm}$ in the area of fast-rotating ship propellers and lead there to propulsion loss in the long run to material damage, compare for example [26].

The danger of cavitation arises when the pressure in the water approaches the boiling point pressure corresponding to the given temperature. **From the phase diagram of water,** it can be seen that when the external pressure drops from $p_{atm} = 10^5$ [Pa] to 2340 [Pa], the water already boils at 20 °C and thus loses cohesion without active external tensile stress.

The influence of the negative pressure can also be seen in the so-called "geodesic suction head" for suction pumps, where water is pumped from depth by suction at the top. In this case, a maximum suction head is obtained by applying the Eq. (1.27) with $p_{cap} = p_{atm} = 1.013 \cdot 10^5$ $[Pa]$ and $\varrho_L \cdot g = 1000 \cdot 9.81 = 9810$ [Pa] to $\boldsymbol{h} = \mathbf{10.33}$ **[m]**, at which the suction tension at the top corresponds to the atmospheric pressure and the cohesion of the water molecules fails there. The suction heights that can be realized in practice are significantly lower.

As a rule, the occurrence of cavitation is favored by impurities or water vapor bubbles. In this case, possible cohesion failure is closely **related to any vapor bubbles that may be present and to existing underpressure**. According to [27] or [26], a new formation or enlargement of vapor bubbles requires an amount of energy, which can be simply expressed with the fraction from the internal surface enlargement and the fraction of work to increase the volume of the bubble under the given pressure $p - p_{atm}$ as follows:

$$\Delta E = 4 \cdot \pi \cdot r^2 \cdot \gamma_L + \frac{4 \cdot \pi}{3} \cdot r^3 \cdot (p - p_{atm}) \tag{1.46}$$

Here, p is a tensile stress applied on the water surrounding the air bubble. This tensile stress is applied in addition to the atmospheric pressure p_{atm} (operating on all sides) and reduces the total pressure acting on the bubble or air pore.

Putting into the derivative the radius r of this equation equal to zero, we get the critical radius $r^\star$ depending on γ_L and $(p - p_{atm})$, at which the energy demand for bubble formation shows the maximum and is considered as "energy barrier."

$$r^\star = \frac{2 \cdot \gamma_L}{p_{atm} - p} \text{ [m]} \tag{1.47}$$

This result is also consistent with the following consideration of the equilibrium between the pressure within the bubble from surface tension and an imposed negative pressure p [Pa] in the surrounding water:

$$p_{atm} + \frac{2 \cdot \gamma_L}{r^\star} = p + p_{atm} \quad \Rightarrow \quad p = \frac{2 \cdot \gamma_L}{r^\star} \quad \Rightarrow \quad r^\star = \frac{2 \cdot \gamma_L}{p} \tag{1.48}$$

Cohesion is abruptly lost when vapor bubbles larger than $r^\star$ in diameter are present. **In fine capillary pores, negative capillary pressures much larger than $-p_{atm}$ exist** and are therefore much larger than the "geodesic suction heights" previously calculated.

If the water has no impurities and no incipient vapor bubbles that cause heterogeneous nucleation, it can withstand high tensile forces in the "undisturbed state" in laboratory experiments up to the magnitude order of several 100 bar until cohesion failure [26].

1.4 Sorptive Storage on Material Surfaces and on the Inner Surface of Pore Systems

1.4.1 Preliminaries

The saturation of surface free energy on solid surfaces by gas molecules, especially water vapor molecules from air, and the sorption of molecules in contact with liquids, have long been the subject of research. The sorption behavior is usually described by sorption isotherms based on empirical or thermodynamic methods. As is well known, a distinction must be made between chemical and physical sorption.

The following remarks concentrate on the important area of physical sorption, in particular also for internal surfaces, and on the consequences for the "water balance" of the material in the case of water vapor sorption.

The sorption isotherms measurable in physical sorption give the relationship between the ambient gas pressure, for example, the water vapor partial pressure, and the mass of gas molecules V_{ads} bound to the surface under consideration as a liquid layer in equilibrium with the external partial pressure p_0.

Groups of adsorption forces, which are based on polar effects or on dispersion forces or Van der Waals forces, are responsible for the physisorption. The recorded atomic or molecular distances are ≈0.2 to ≈10 [nm]. A large number of researchers have dealt with this. Reference is made to Israelachvili [28], Rouquerol et al. [29], and Lauth and Kowalczyk [19].

In Figure 1.14 are shown the names for the processes and the phases as they are encountered in physisorption of most gases or vapors on nonporous or macro-porous materials.

For the calculation resp. prediction of the mentioned physical adsorption mechanisms, there are a number of models resp. approaches that consider interactions between molecules in different ways and model a mono-molecular to multi-molecular coverage of the surface, compare [19]. These methods allow the prediction of adsorption on pore-free surfaces and on internal surfaces of porous solids **up to relative gas pressures or relative vapor pressures of about 0.40**. Adolphs and Setzer [30] shows a comparison of the ability of a number of the models to simulate the sorption measurement results of N_2 on SiO_2 powder.

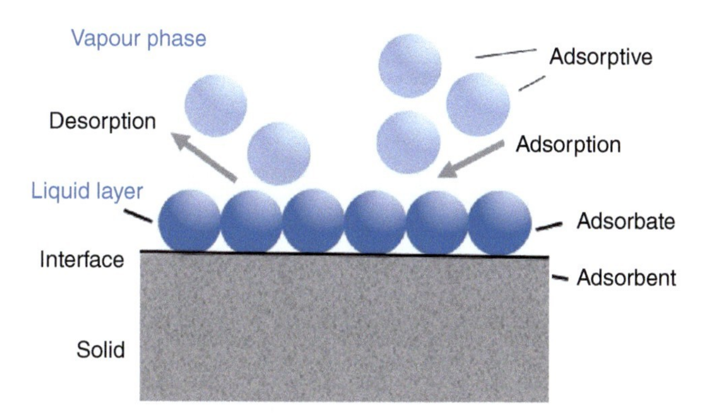

Figure 1.14 Vapor molecule physisorption on solid surfaces. Designations of the phases and components.

The most common of these methods is the BET method of Brunauer/Emmett/Teller-1938 [31]. This can be used to reliably determine the monolayer capacity and total internal surface area. The BET method will be discussed in more detail below.

Another approach to describing sorption isotherms can be based on thermodynamic derivatives, as shown by Adolphs and Setzer [30] and Badmann et al. [32]. Further discussion of this is also given below.

A new approach to the description of adsorption isotherms has been worked out by Zandavi and Ward [33] and Zandavi [34]. They assume (actually existing) cluster formation of vapors of different adsorptive, increasing covering densities as well as chemical potential of clusters of different sizes. The physical parameters required for the theory, however, must be determined from sorption measurements, but are then valid for the entire partial pressure course of the isotherms and can also be led beyond $\varphi = 1.0$ to a certain extent.

1.4.2 Measured Surface Sorption of Water Vapor on Flat Nonporous Surfaces

Sorption isotherms measured on porous mineral materials contain increasing amounts of water from capillary condensation with increasing relative humidity. For a more detailed analysis of sorption processes and possible thermodynamic calculations, however, it is necessary to know the condensate layer thicknesses developing with increasing gas partial pressure **on pore-free surfaces,** especially when approaching saturation partial pressure.

Measurement results from different authors are available for this purpose. The sorption layer thicknesses are determined with the aid of gravimetric fine measurements, in particular TGA measurements, and are usually presented as a mathematical function between layer thickness and relative partial pressure or the relative humidity in the case of water vapor sorption measurements.

Reference is made, for example, to **Badmann et al. [32]**. This reports on water vapor sorption measurements on cement phases and samples from hydrated cements of different compositions and different water–cement ratios. The functional

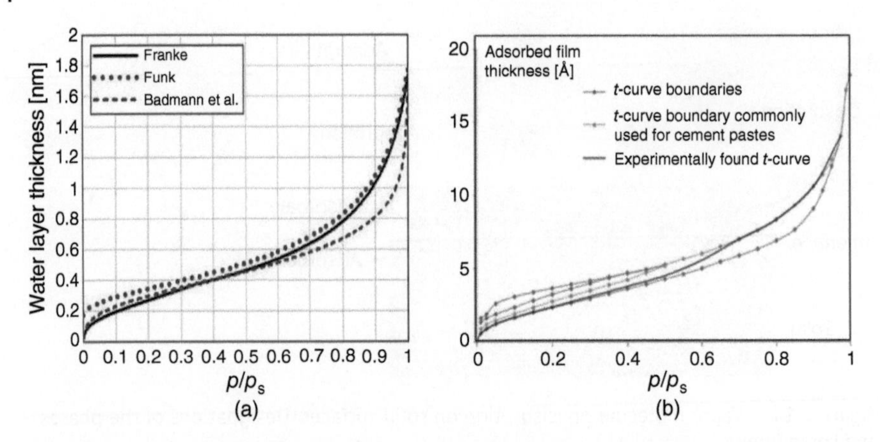

Figure 1.15 Sorption isotherms for water vapor with pore-free surface. (a) From Franke, Eq. (1.50). (b) Source: From Snoeck et al. [37]/with permission of Elsevier.

dependence of the averaged values for water vapor sorption is given as:

$$t(\varphi) = 0.385 - 0.189 \cdot ln(-ln(\varphi)) \ \ [\text{nm}] \tag{1.49}$$

Funk [35] compares $t(\varphi)$-curves of different authors and recommends the function dependence of deBoer et al. [36]. **These curves are plotted in Figure 1.15a** together with a curve of [Franke]. The comparative calculations made by Franke with sorption isotherms lead to a slightly modified dependence of the sorption layer thicknesses and to the proposal of the following function. **Figure 1.15b** shows measurement results from Snoeck et al. 2014 [37] with a curve shape very close to the curve proposal from Franke (Eq. (1.50)):

$$t(\varphi) = \varphi^{0.028} - 1.03 + \left(\frac{0.09}{log\left(\frac{1}{\varphi}\right) + 0.03} \right)^{0.5} \ \ [\text{nm}] \quad (\varphi \geq 0.0115) \tag{1.50}$$

With this function, the monolayer is obtained with $t = 0.30$ [nm] at $\varphi(n_{mono}) = 0.23$ and $t = 1.70$ [nm] at the saturation partial pressure.

In Section 1.4.6.1, we investigate to what extent the basic isotherm for nonporous (hydrophilic) surfaces derived here on the basis of experimental results can also be approximated theoretically on the basis of thermodynamic approaches. In addition, Section 1.4.6 considers issues relating to sorption film formation within material pores as well as on differently curved surfaces.

1.4.3 Use of Water Vapor Sorption Isotherms to Determine the Value of the Internal Surface Area

Knowledge of the **internal surface area of porous material**, particularly cement-bound products, is important for moisture storage and transport considerations. Typically, sorption experiments with nitrogen N_2 or helium H_2 are performed

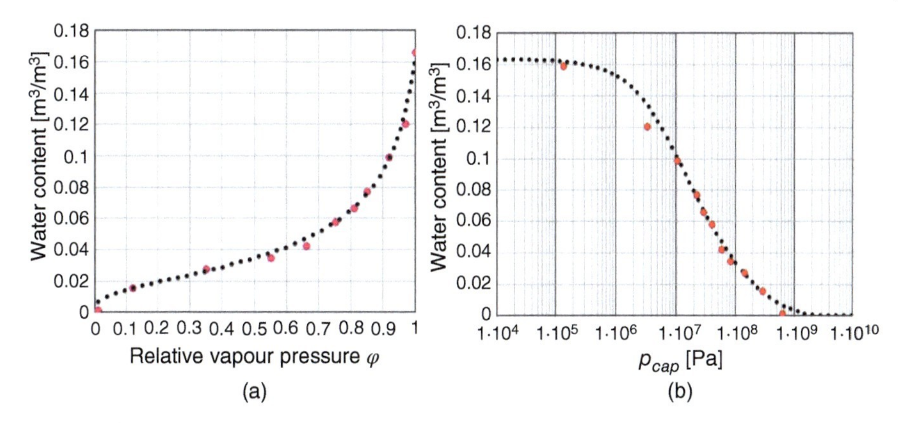

Figure 1.16 (a) Measured points and regression curve of the adsorption isotherm of material REF ($W/C = 0.45$). (b) Isotherm of the material REF, related to the capillary pressure p_{cap}.

to determine these. **It will be shown that the inner surface can be determined with the known evaluation methods on the basis of existing water vapor sorption isotherms**. This is shown below by means of an example based on the BET method and the method of Adolphs and Setzer. **Figure 1.16a** shows the adsorption isotherm of the reference material REF mainly used in experiments in the present project. The function of the measured adsorption isotherm is given by Eq. (1.51). **Figure 1.16b** shows the adsorption isotherm as a function of capillary pressure p_{cap} for comparison. This capillary pressure dependent isotherm or moisture storage function is obtained by converting the output function (1.51) of **Figure 1.16a** using the Eq. (1.1).

1.4.3.1 Total Internal Surface Determined Using BET Method

To determine the internal surface area of the Portland cement on reference mortar REF, the BET method is used.

The derivation of the equations to be used can be taken in detail from Brunauer et al. [31], Rouquerol e al. [29], and, in particular, Lauth and Kowalczyk [19]. The BET method is capable of simulating coverage with multiple layers of the respective gas molecules, in this case, water vapor molecules. However, previous experience shows that if a partial vapor pressure of about 0.40 is exceeded, an overestimation of the sorbed liquid quantity occurs, so that an **application of the BET method should only be made in the partial pressure range of about 0.05 to 0.4**.

In the following example, the starting point of the evaluation is the sorption isotherm already shown in **Figure 1.16a**.

The aim of the evaluation is first to determine the so-called monolayer capacity in order to derive from it the inner surface area using the example of the REF material mainly used in our experiments.

Equation (1.51) is the equation of the adsorption isotherm or moisture storage function in [m^3/m^3] for the REF material. Indicated are the coefficients $A, a, b,$

and c to be used. The inverse function $\varphi_\theta(\theta)$ is given as Eq. (1.52) with the same coefficients.

$$\theta\varphi(\varphi) = a \cdot \frac{1}{\left[\left(\frac{A}{\varphi}\right)^{c} - 1\right]^{\frac{1}{b}}} \tag{1.51}$$

$A = 1.065 \quad a = 0.0197 \quad b = 1.70 \quad c = 0.43$

$$\varphi_\theta(\theta) = A \cdot \left[\left(\frac{a}{\theta} + 1\right)^{b}\right]^{\frac{-1}{c}} \tag{1.52}$$

The Eq. (1.53) is the well-known basic equation of the BET method, derived or explained in the previously given publications. By varying the parameter C, the course of the curve can be adapted to given sorption measurement results. **Figure 1.17** shows the resulting curves for parameters 1–100 as a function of the gas partial pressure. One can see the good adaptability to different measurement results.

$$\frac{n}{n_{mono}} = \frac{C \cdot \varphi}{(1-\varphi) \cdot (1 - \varphi + C \cdot \varphi)} \tag{1.53}$$

The basic Eq. (1.53) can be transformed into the well-known linear form $y(\varphi) = a + b \cdot \varphi$ of the Eq. (1.54) as a function of φ. In the left part of the equation, $\boldsymbol{n}$ corresponds to the sorbed liquid amount, n_{mono} in the right side of the equation corresponds to the monolayer capacity. The left part of the Eq. (1.54) is given as $y_0(\varphi)$ with the measurement results of the moisture storage function (1.51) as Eq. (1.55)

$$\frac{\varphi}{n \cdot (1-\varphi)} = \frac{1}{C \cdot n_{mono}} + \frac{(C-1)}{C \cdot n_{mono}} \cdot \varphi \tag{1.54}$$

$$y_0(\varphi) = \frac{\varphi}{\theta\varphi(\varphi) \cdot (1-\varphi)} \tag{1.55}$$

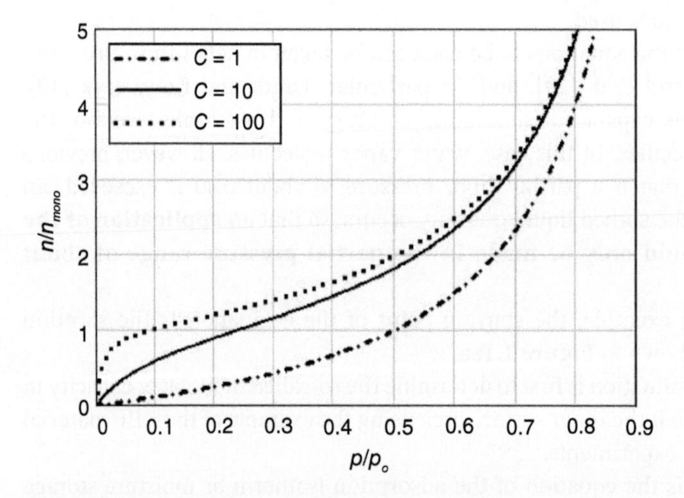

Figure 1.17 BET sorption curves according to Eq. (1.53) as a function of parameter C.

It is generally assumed that the BET Eq. (1.53) is valid up to a relative partial pressure of about 0.40 for surfaces. At higher partial pressures, the process results in overemphasized adsorption.

From the measured value curve resp. sorption isotherms $\theta\varphi(\varphi)$, now two points $\varphi1/y_0(\varphi1)$ and $\varphi2/y_0(\varphi2)$ are taken in the mentioned partial pressure range to determine the axis intercept $y_{00}(\varphi1, \varphi2)$ of the straight line $y_0(\varphi)$. The following values were used:

$$\varphi1 = 0.15 \quad y_0(\varphi1) = 10.56 \qquad \varphi2 = 0.35 \quad y_0(\varphi2) = 20.51$$

$y_{00}(\varphi1, \varphi2)$ results from

$$y_{00}(\varphi1, \varphi2) = y_0(\varphi2) - \frac{y_0(\varphi2) - y_0(\varphi1)}{\varphi2 - \varphi1} \cdot \varphi2 \tag{1.56}$$

From Eq. (1.57), the coefficient b is obtained as the slope of the straight line Eq. (1.54). Using the intercept y_{00}, the curve parameter C is then obtained from Eq. (1.58).

$$b = \frac{y_0(\varphi2) - y_0(\varphi1)}{\varphi2 - \varphi1} \tag{1.57}$$

$$C = \frac{b}{y_{00}(\varphi1, \varphi2)} + 1 \tag{1.58}$$

For the REF material (resp. the curve in Figure 1.16a corresponding to the black curve in Figure 1.19) the following values are obtained:

$$y_{00}(\varphi1, \varphi2) = 3.1 \qquad b = 49.74 \qquad C = 17.0.$$

With C known, n_{mono} is obtained from Eq. (1.59):

$$n_{mono} = \frac{1}{C \cdot y_{00}(\varphi1, \varphi2)} \tag{1.59}$$

For the REF material, $n_{mono} = 0.019$ [m³/m³]. The inner surface area A_{BET} we are looking for follows from Eq. (1.60):

$$A_{BET} = \frac{n_{mono} \cdot N_A \cdot A_{molecule}}{n_{mol}} \tag{1.60}$$

with the Avogadro number $N_A = 6.022 \cdot 10^{23}$, the molar volume for water n_{mol}, and the area occupied by a water molecule (according, for example, to [38]) $A_{molecule} = 11.4 \cdot 10^{-20}$. For the REF material under consideration, the inner surface area of the REF material we are looking for amounts to:

$$A_{BET} = 7.24 \cdot 10^7 \ [\mathrm{m^2/m^3}] = 72.4 \ [\mathrm{m^2/cm^3}] \qquad \varphi_\theta(n_{mono}) = 0.20$$

In Figure 1.18, the resulting curves n/n_{mono} according to Eq. (1.53) and the REF sorption isotherm divided by the value n_{mono} (red curve) are compared. Very good agreement is seen, generally up to a relative partial pressure of about 0.40.

For comparison, the inner surface A_{BET} of the red curve in Figure 1.19 was also determined according to the steps shown previously. The following values were obtained, from which one can see the suitability resp. precision of the BET method:

$$C = 9.1 \quad n_{mono} = 0.020 \quad \varphi_{mono} = 0.25 \quad A_{BET} = 76.4 \ [\mathrm{m^2/cm^3}]$$

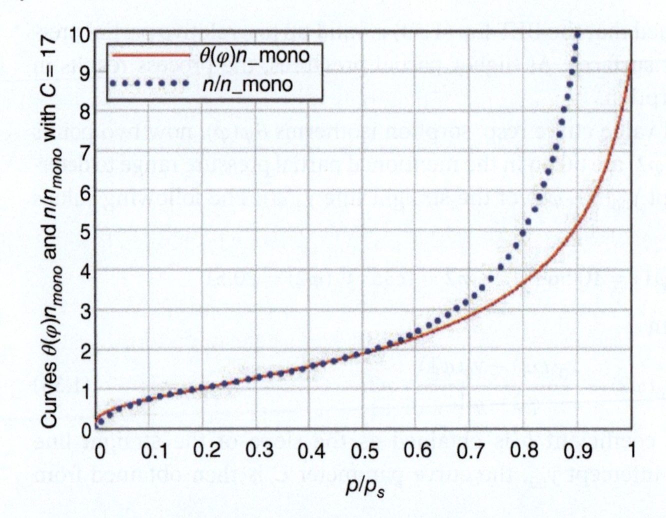

Figure 1.18 Comparison of the BET modeled sorption curve according to Eq. (1.53) with $C = 17.0$ (blue) and the measured n_{mono} related basic (red) curve from REF isotherm (curve **Figure 1.16a** or the black dotted curve in **Figure 1.19**).

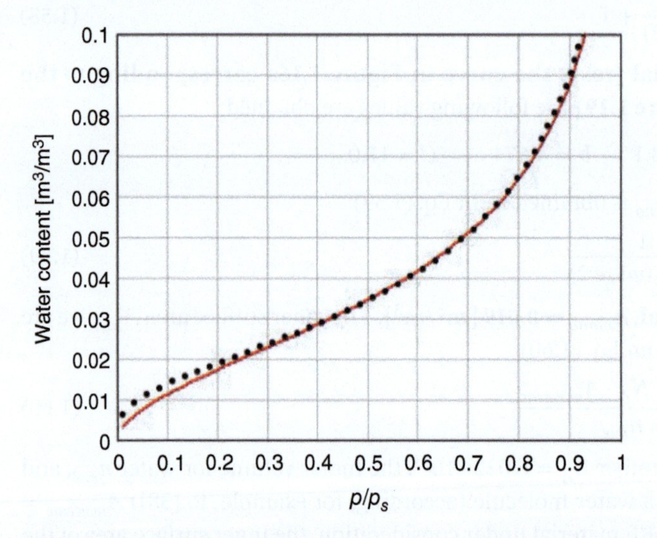

Figure 1.19 Sorption isotherm for the REF material according to **Figure 1.16a** (black dotted) and sorption isotherm (red) with slightly lower water absorption up to $p/p_S = 0.40$ with respect to a computational comparison of the monolayer capacity.

In their 1969 publication [39], Brunauer et al. propose an extension of their BET method in which the vapor partial pressure is multiplied by a **factor** $k < 1.0$ in the linear basis Eq. (1.54). This can be used to extend the agreement to measured values beyond the relative partial pressure of about 0.40. Pavlik et al. [40] apply the modified procedure to comparative modeling of sorption isotherms of different building materials and find improved agreement. As a criticism, there is no physical justification for the extended BET method, and thus this extended equation is merely an empirical equation for modeling sorption isotherms, see also Rouquerol/Sing [29]/section 5. In contrast, the classical BET method is sufficient for determining the internal surface area of materials.

1.4.3.2 Total Internal Surface Determined by the Method of Adolphs and Setzer

The Adolphs/Setzer method for determining the internal surface area of a porous material is described in Adolphs/Setzer [30, 41], and [42]. It can be applied to different gases and different material surfaces.

The following shows what results are obtained with respect to the internal surface area when water vapor sorption isotherms are used for its determination. This also results in a comparison to the BET method.

Adolphs/Setzer assume the following equation:

$$\Phi = n_{ads} \cdot \Delta\mu \tag{1.61}$$

Φ is used to describe a change in surface free energy upon sorption of gas molecules, here specifically water vapor. Φ is defined by Adolphs/Setzer as "**Excess surface work.**" In Eq. (1.61), n_{ads} is the sorbed liquid mass, and $\Delta\mu$ is the change in chemical potential corresponding to Eq. (1.62).

$$\Delta\mu = RT \cdot ln\left(\frac{p}{p_S}\right) \quad [\text{J/mol}] \tag{1.62}$$

After introducing the sorbed water of sorption isotherms $\theta\varphi(\varphi)$, the corresponding excess surface work equation follows from Eq. (1.61).

$$\Phi(\varphi) = RT \cdot ln(\varphi) \cdot \theta\varphi(\varphi) \tag{1.63}$$

According to the method of Adolphs/Setzer, the low point of the curve 1.63 corresponds to the value n_{mono} resp. θ_{mono}.

Accordingly, if we first evaluate the red curve of **Figure 1.20** with respect to the (red) sorption isotherm **of Figure 1.19**, we obtain for $T = 293.15$ the following values:

$$\theta_{mono} = 0.020 \qquad \varphi_\theta(\theta_{mono}) = 0.219$$

These results agree well with the values determined after BET procedure.

From Eq. (1.62), we get then with θ_{mono} the relation for $\Delta\mu_0$:

$$\Delta\mu_0 = RT \cdot |ln(\varphi_\theta(\theta_{mono}))| \tag{1.64}$$

The remarks of Adolfs/Setzer show that $nn(\varphi_\theta) = n_{ads}/n_{mono}$ lead to Eq. (1.65).

$$nn(\varphi_\theta) = 1 - ln\left[\frac{RT \cdot |ln(\varphi_\theta)|}{\Delta\mu_0}\right] \tag{1.65}$$

From (1.65), $nn(\varphi_\theta) = 1$ can be used to derive the value for $\Delta\mu_0$ resp. Eq. (1.64). Substituting the previously calculated value $\varphi_\theta(\theta_{mono}) = 0.219$, we obtain the following values:

$$nn = n_{ads}/n_{mono} \qquad \Delta\mu_0 = -3.703 \cdot 10^3 \qquad \text{nn}(\varphi_\theta(\theta_{mono})) = 1$$

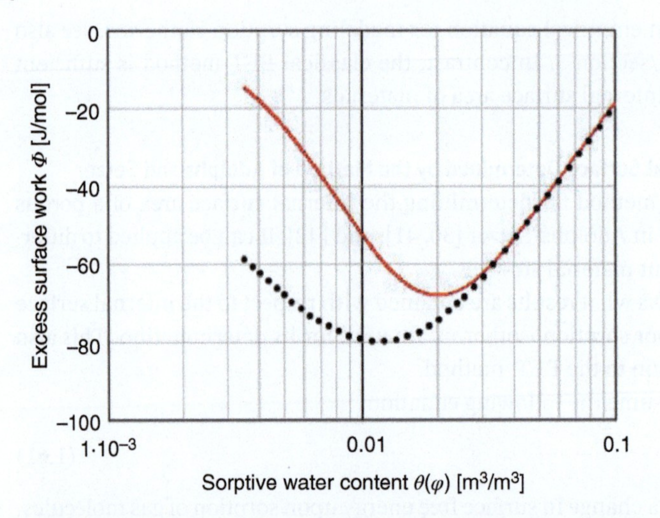

Figure 1.20 Excess surface work curves according to Eq. (1.63) for the adsorption isotherms according to **Figure 1.19**. The low points give the values for n_{mono} and θ_{mono}, respectively.

If, on the other hand, the same calculation is performed for the (black dotted) sorption isotherm according to **Figure 1.19** with its equation $\varphi_\theta(\theta)$ according to Eq. (1.52), the following values result with the black dotted determination curve according to **Figure 1.20**:

$$\theta_{mono} = 0.013 \qquad \varphi_\theta(\theta_{mono}) = 0.081.$$

The corresponding internal surface area for this according to Eq. (1.60) is $A_{Adolphs} = 49.6$ [m^2/cm^3] and is thus significantly lower than the reference value $A_{BET} = 72.2$ [m^2/cm^3] according to BET.

The data of other authors on the internal surface area of mortar products similar to the material considered here amounts to at least 30 [m^2/g] corresponding to at least approx. 65 [m^2/cm^3], so that the calculated approx. 50 [m^2/cm^3] appears too low. The reason for this would require a more detailed analysis.

Figure 1.21 shows the comparison of the resulting curve $nn(\varphi_\theta)$ and the associated sorption isotherms related to n_{mono}. Agreement between the curves is obtained up to a partial pressure of about 0.25.

1.4.4 Calculation of the Pore Size-Dependent Distribution of the Inner Surface Using the Moisture Storage Function

Although the relationships and results presented below are specific to the reference REF material, they are applicable to other porous materials with known moisture storage functions. The explanations show the calculation of the course of the inner surface as a function of the relative partial pressure or the pore-radius distribution, which is also calculated, using the moisture storage function resp. the adsorption isotherm as a starting point.

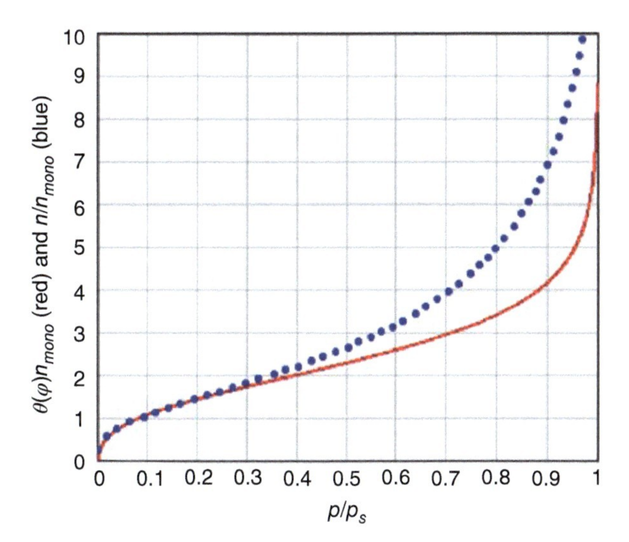

Figure 1.21 Comparison of the curve $nn(\varphi_\theta)$ calculated according to Adolphs/Setzer with the sorption isotherms related to n_{mono}.

Furthermore, the distribution of the different moisture fractions surface sorbate and capillary condensate are also calculated in dependence of the relative partial pressure.

1.4.4.1 Determination of the Net Sorptive Storage Function

Figure 1.22 schematically shows that the original sorptive isotherm resp. moisture storage function, for example, of the material REF in **Figure 1.16a** is not very suitable for a more accurate determination of the pore size distribution or for the determination of the course of the inner surface. This is because the function value $\theta_{Isoth(\varphi 1)}$ associated with a $\varphi 1$ has not only the content of condensed water at $\varphi 1$, but additionally the water adsorbed on the inner surface of the pores in the not yet

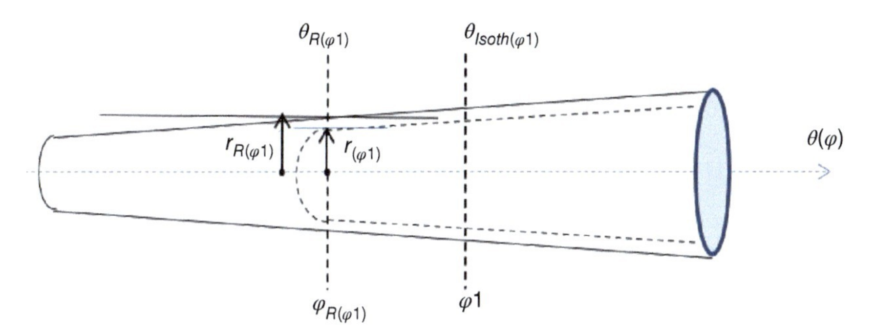

Figure 1.22 Schematic representation of a cylindrical pore with increasing radius as well as the meniscus associated with a selected relative partial pressure $\varphi 1$ and the associated surface film of thickness $t(\varphi 1)$ in the non-water-filled pore region.

filled region before a meniscus. Therefore, the curve without the sorbed fraction must be determined first.

The determination of this **"net storage function"** can be done by first calculating the pore-size distribution from the original isotherms as shown below, and then using the sorbed surface fraction to calculate the net storage function we are looking for. This procedure is described in further detail below in Section 1.4.6.3. First, an approximate equation is used. Comparison calculations show that the following Eq. (1.66) gives the net storage function with a very good approximation:

$$\theta\varphi_N(\varphi) = \theta\varphi(\varphi) \cdot \varphi^{0.5} \tag{1.66}$$

The net storage function is given in **Figure 1.23a** as a function of $\varphi = p/p_S$ as a line $\theta\varphi_N(\varphi)$. In addition, the "exact" curve $[\theta(\varphi) - \Delta Wh(\varphi)]$ is given, resulting from the calculation of the internal surface area and vapor sorption curve still shown below. There is very good agreement between these curves of the net storage function.

In Figure 1.23b, the required curves are given as a function of r and the pore radius r_R. The reference curve $\theta r(r)$ is calculated with the (here abbreviated) Kelvin equation (1.67) plus (1.68) from $\theta\varphi(\varphi(r))$.

$$\varphi(r) = exp\left(\frac{M_\gamma}{r}\right) \tag{1.67}$$

$$M_\gamma = -\frac{\overline{V}_L \cdot 2 \cdot \gamma_L \cdot cos(\theta)}{R \cdot T} \tag{1.68}$$

However, in order to determine realistic progressions of the inner surface on the basis of the net storage function, not a dependence on the mean pore radius r should be used, but the dependence on the "real" mean pore radius $r_R = r(\varphi) + t(\varphi) \cdot 10^{-9}$.

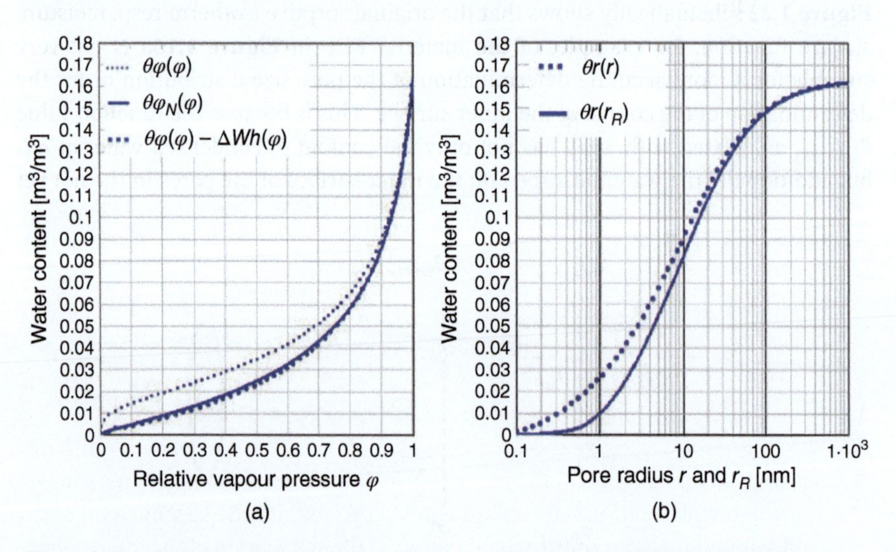

Figure 1.23 Measured sorption isotherm $\theta\varphi(\varphi)$ of the sample material REF and its accurate formulation $\theta\varphi(\varphi) - \Delta Wh(\varphi)$, furthermore the net storage functions $\theta r_{R\_N}(r_R)$ (= $\theta r(r_R)$ in figure (b)) calculated from Eq. (1.66). (a) Relative to vapor partial pressure. (b) Relative to the effective "real" pore radius r_R.

This net storage function, which depends on the effective radius, is plotted in **Figure 1.23b** as $\theta r_{R\_N}(r_R)$.

The net storage function calculated on the basis of the internal surface area via vapor sorption is obtained from Eq. (1.89), also shown in **Figures 1.23a and 1.32**.

1.4.4.2 Basic Equation for the Pore Size-Dependent Distribution of the Inner Surface

For a given radius r_R, the net storage function exhibits the volume increase $d\theta(r_R)/dr_R$. **The volume of associated pores** with largely unknown pore shapes can be generally formulated as $k_V \cdot r_R^3$, **the associated internal surface area of these pores** as $k_A \cdot r_R^2$.

The associated internal surface area increment is then equal to the theoretical number of pores of radius r_R in the pore area dr_R, to be calculated from the associated volume increment divided by the formulated pore volume $k_V \cdot r_R^3$.

The associated surface area increment is then obtained by multiplying the pore number by its assumed internal surface area $k_A \cdot r_R^2$.

The inner surface area of the pore system between two radii r_{R1} *and* r_{R2} is then obtained from the net storage function to

$$A(r_{R1}, r_{R2}) = \int_{r_{R1}}^{r_{R2}} \frac{d\theta r_{R\_N}(r_R)}{dr_R} \cdot \frac{k_A \cdot r_R^2}{k_V \cdot r_R^3} \cdot dr_R \tag{1.69}$$

Of importance now is the question of the pore shape factors to be applied, which can be combined into a common pore shape factor $k_p = k_A/k_V$.

For cylindrical pores, $A/V = 2/r$, i.e. $k_P = 2$, for conical from wide to narrow pores, A/V would be $\approx 1.01/r$, for spherical pores $A/V = 3/r$. Ridgway et al. [43], in modeling pore-size distributions, shows that by incorporating conical pore shapes, a better fit to "natural" pore progressions is obtained. Hereafter, $kp = 2$ is used as the initial value (the applicable value is determined in Section 1.4.4.4). **Equation** (1.69) **is then the basic equation for the internal surface area curve for the region between two pore radii** r_{R1} **and** r_{R2} assuming a largely matching pore structure in the volume region between these two pore radii.

$$A(r_{R1}, r_{R2}) = \int_{r_{R1}}^{r_{R2}} \frac{d\theta r_{R\_N}(r_R)}{dr_R} \cdot \frac{1}{r_R} \cdot kp \cdot dr_R \tag{1.70}$$

1.4.4.3 Application of the Basic Equations with Regard to the Material REF

Application of the basic equation for the calculation of the course of the internal surface specific calculations of the course of the internal surface of a material based on the net storage function with the integral (1.70) require the specification of the lower and upper limits for the area calculation. If the calculation of the total internal surface area is desired, the approximated water molecule diameter should be used as the lower bound for water as the initial radius, i.e. $r_R = 0.30 \cdot 10^{-9}$ [m]. As end radius or upper limit it is convenient to use the upper limit radius of the mercury porosimeter MIP, in the present case $1.50 \cdot 10^{-4}$ [m]. Since the area increment in the region of the upper radius of the pore system is relatively low for materials such as the REF used here as the sample material, the error is small if, for example, $1.50 \cdot 10^{-5}$ [m] is used instead. **The area calculation started from the lower boundary is then with variable upper boundary radius**:

$$A(r_{R1}) = \frac{1}{10^6} \int_{0.30 \cdot 10^{-9}}^{r_{R1}} \frac{d\theta r_{R\_N}(r_R)}{dr_R} \cdot \frac{1}{r_R} \cdot kp \cdot dr_R \tag{1.71}$$

For the upper limit $r_{R1} = 1.50 \cdot 10^{-5}$ [m], the pore shape factor $k_P = 2$ gives a total area for the pattern material REF of $A(1.50 \cdot 10^{-5}) = 109.7$ [m²/cm³]. The factor $1/10^6$ in front of the integral gives the conversion to [m²/cm³].

Including a bulk density of the REF material of 2.10 [g/cm³], this basic calculation yields a total internal surface area of $A_i = 52.2$ [m²/g].

An alternative calculation can be made by entering the boundaries as a function of relative air humidity φ. The integral for the course of the inner surface from any lower limit $r_R(\varphi 1)$ to the upper limit for example $\varphi \approx 1.0$ is then

$$A(\varphi 1) = \frac{1}{10^6} \cdot \int_{r_R(\varphi 1)}^{r_R(0.99999)} \frac{d\theta r_{R\_N}(r_R)}{dr_R} \cdot \frac{1}{r_R} \cdot kp \cdot dr_R \quad [\mathrm{m^2/cm^3}] \tag{1.72}$$

If $r_R(\varphi 1) = 0.30 \cdot 10^{-9}$ [m] is to be used as lower integration limit, as before in Eq. (1.71), the corresponding $\varphi 1$ value must be chosen for this start value with the help of Eq. (1.80) in such a way that $0.30 \cdot 10^{-9}$ [m] results from it. This value is $\varphi 1 = 0.0116$, as the result of the following equation (with $t(\varphi)$ from Eq. (1.50)) shows

$$r_{R(\varphi 1=0.0116)} = r(0.0116) + t(0.0116) \cdot 10^{-9} = 3.00 \cdot 10^{-10}$$

If one chooses instead approximately $\varphi 1(r = 0.30 \cdot 10^{-9}) = 0.031$ according to the standard Kelvin equation and from this the starting radius $r_R(\varphi 1 = 0.031) = 0.42 \cdot 10^{-9}$ [m], the Eq. (1.72) now yields the area 102.1 [m²/cm³] instead of the correct value 109.7 [m²/cm³].

1.4.4.4 Calibration of the Pore Shape Parameter and Calculation of the Internal Surface Distribution of the Material REF

In the previously commented basic integrals for the calculation of the course of the inner surface, the pore-shape factor $kp = 2$ of a cylindrical pore structure was used. However, the other theoretical pore shape factors given show that for realistic modeling, pore shape factor 2 is unlikely to be suitable for materials such as the REF material. It therefore suggests itself to look for a calibration possibility. A calibration based on MIP measurement results would be conceivable, but preferably based on the total internal surface area determined according to BET, as shown below.

The pore shape parameter kp **is adjusted to** k_N **such that the surface integration yields the total internal surface area** A_{BET} of the pore system as determined by the BET method :

$$k_N = kp \cdot \frac{A_{BET}}{A(r_{R1}, kp = 2)} = 2.0 \cdot \frac{72.4}{109.7} = 1.32 \tag{1.73}$$

With $k_N = 1.32$ and the lower limit $r_R(\varphi 1 = 0.0116)$, the Eq. (1.72) now gives a total inner area of 72.4 [m^2/cm^3]. With $k_N = 1.32$, also the Eq. (1.71) yields this **total internal area related to 1 [g] corresponding to** $A_i = 34.5$ [m^2/g].

It was also tested whether a calibration of the pore structure parameter was possible via the results of mercury intrusion porosimetry (MIP) measurements also leads to the target. The measuring range of the mercury porosimeter MIP used is $1.8 \cdot 10^{-9}$ to $1.51 \cdot 10^{-4}$. However, the internal surface area calculated by the porosimeter for this measurement range scatters between 8.5 and 12.5 [m^2/g] for the material REF, for example, in successive manufacturing series despite strict manufacturing quality control and little scattering of other properties.

The integration with Eq. (1.71) over the pore measurement range of of the MIP based on a mean MIP measurement result of 10.0 [m^2/g] leads after a k_{MIP}-determination analogously to Eq. (1.73) to the lower lines in **Figure 1.24a,b**, which do not satisfactorily represent the surface progression.

It should be noted that the pore shape parameter is used for the entire inner surface. However, the calculations for sorptive moisture storage subsequently based on this type of surface calculation yield satisfactory and comprehensible results. A further refinement of the modeling of the course of the inner surface could be achieved by adjusting the pore shape parameter section by section.

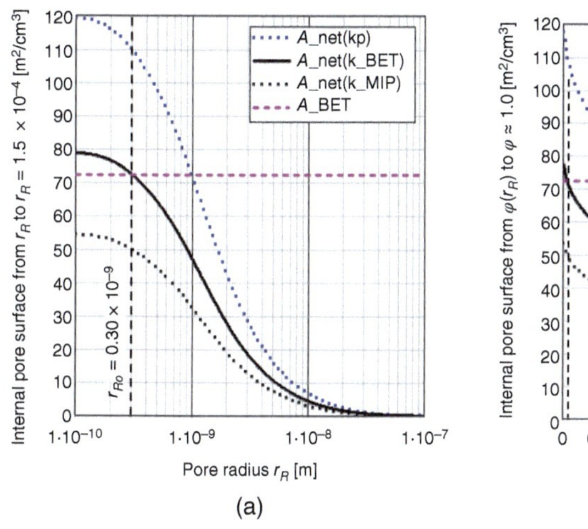

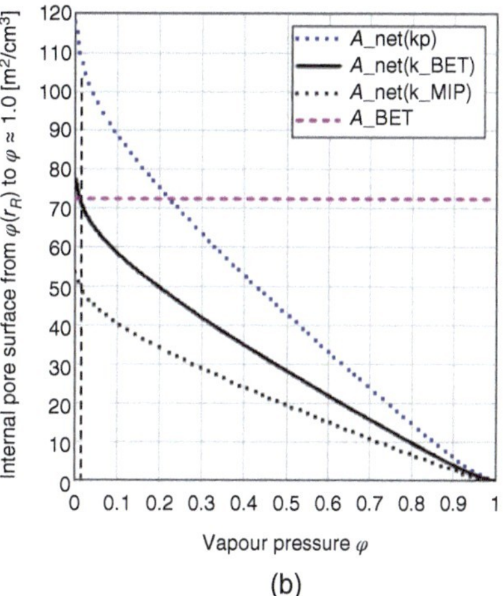

Figure 1.24 Course of the inner surface of the reference material REF above a given "real" pore radius r_R calculated via the net storage function with Eq. (1.71) as well as depending on the vapor partial pressure φ calculated with Eq. (1.72). The initial lines A_net(kp) shown were obtained with the area shape factors $kp = 2$, and the optimized lines A_net(k_BET) with the shape facto k_N and k_{BET}, respectively. Attempting calibration based on the MIP measurement results leads to the inapplicable lines A_net(k_MIP). (a) $A_{internal}$ between r_R and $r_R \geq 10^{-7}$. (b) $A_{internal}$ between φ and $\varphi \approx 1.0$.

For this purpose, an additional evaluation of adsorption isotherms of representative material compositions would be necessary, for example, for cement paste with a low W/C value with regard to the gel-pore structure and the possibly deviating pore-structure parameter.

1.4.4.5 Distribution of the Volume of the Internal Adsorbed Water Films

The results presented in Sections 1.4.2 to 1.4.4 on the sorption of water vapor or water molecules, especially on the inner surface of the pores of porous materials, as well as the knowledge of the course of their inner surface, now allow to obtain quantitative results on details of moisture storage, i.e. on the proportional extent of sorption and condensation.

Volume fractions of the moisture storage function from surface sorption can be determined as a function of relative air humidity φ or pore radius $r_R(\varphi)$ by multiplying inner surface area by the corresponding adsorption layer thickness. Note that **film thicknesses occupy a smaller volume** $\Delta V(\varphi r)$ than on the same planar surface $\Delta V(\varphi)$ for a given film thickness $t(\varphi)$, by a factor $k_h(\varphi)$, the value of which can range from 0.5 to 1.0, due to curvature on the inner surfaces. From this follows the adsorbed volume:

$$\Delta V(\varphi) = \Delta A(\varphi) \cdot t(\varphi) \cdot k_h(\varphi) \tag{1.74}$$

Depending on the mean net pore radius $r_R(\varphi)$ and the corresponding film thickness $t(\varphi)$ (film thickness in [m] $\leq 0.5 \cdot r_R(\varphi)$), $k_h(\varphi)$ can be estimated to

$$k_h(\varphi) = 1 - \frac{t(\varphi)}{2 \cdot r_R(\varphi)} \tag{1.75}$$

The previously determined pore shape parameter k_N **states that the mean pore shape deviates significantly from a cylindrical shape.** $k_h(\varphi)$ can be additionally adjusted by considering the pore shape parameter k_N.

$k_h(\varphi)$ is then more generally given by $t(\varphi)$ in [m]:

$$k1_h(\varphi) = \frac{1}{1 + (k_N - 1) \cdot \dfrac{t(\varphi)}{r_R(\varphi)}} \tag{1.76}$$

When the ratio of the film thickness to the respective pore radius is low or for **flat surfaces** $k1_h(\varphi)$ **takes the value 1,** when the film thickness reaches the pore radius, the value of $1/k_N$.

Based on the Eq. (1.72), which calculates a fraction of the internal surface area between a chosen $\varphi 1$ and $\varphi = 1.0$, the **following Eq. (1.77) is obtained to determine a stored moisture volume resulting from a condensate film of constant thickness** $t(\varphi 1)$ **in [m] on the internal surface area captured by the integral,** which is bounded by the two chosen pore radii $r_R(\varphi 1)$ and $r_R(\varphi 2)$ (as lower and upper integration limits).

$k1_h(\varphi)$ within the integral accounts for the curvature influence from Eq. (1.76) of the inner surface. The included radius r_R is variable within the integral and must not be made dependent on $\varphi 1$ there, while $t(\varphi 1)$ is constant here according to the value before the integral. $t(\varphi 1)$ (if necessary from Eq. (1.50) must be entered in [m].

k_N is the pore shape factor calibrated for the considered material via BET.

$$\Delta V(\varphi 1) = t(\varphi 1) \cdot \int_{r_R(\varphi 1)}^{r_R(\varphi 2)} \frac{d\theta r_{R\_N}(r_R)}{dr_R} \cdot \frac{1}{r_R} \cdot k_N \cdot k1_h(\varphi 1) \cdot dr_R \quad [\mathrm{m^3/m^3}] \tag{1.77}$$

1.4.5 Equations for Capillary Condensation Considering Adsorbed Liquid Films

In Section 1.3.2, the mechanism of capillary condensation was described, and it was shown how the famous Kelvin equations (1.37) and (1.38), respectively, can be determined. If one now determines the associated mean pore radius for a given external relative air humidity or, conversely, the associated air humidity from a given pore radius, one commits an error, at least formally, due to the fact that the meniscus radius is reduced by the liquid layer sorbed on the inner surface, and this results in a relative air humidity that is smaller than that belonging to the real pore radius, see **Figure 1.25a.**

Therefore, the Eqs. (1.37) and (1.38) are to be expanded with the local film thickness to the "real" pore radius r_R

$$r_R(\varphi) = r(\varphi) + t(\varphi) \tag{1.78}$$

with r = Kelvin radius and $t(\varphi)$ = film thickness in [m].

Alternatively, Eq. (1.78) can be expressed by the following equation:

$$r_R(\varphi) = \frac{M_\gamma}{ln(\varphi)} + t(\varphi) \quad \text{in [m]} \tag{1.79}$$

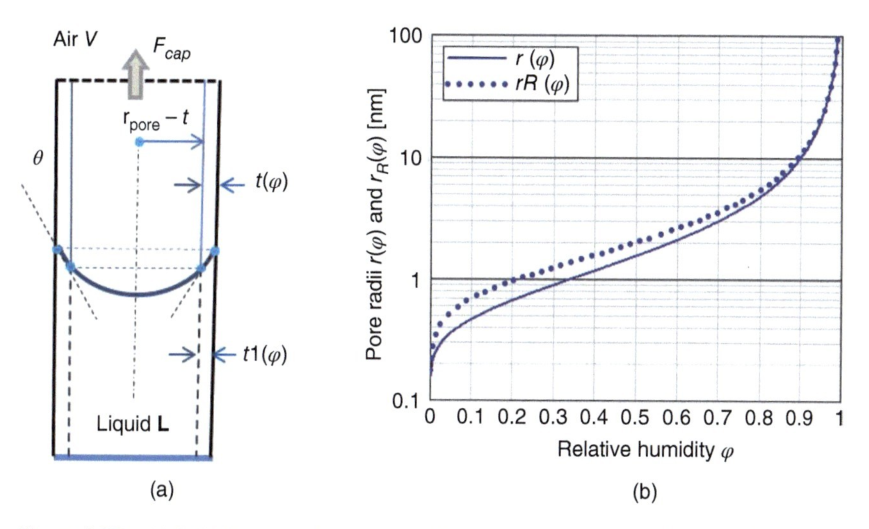

Figure 1.25 (a) Cylindrical capillary pore with meniscus and sorbed layer thicknesses at the pore walls $t(\varphi)$ in front of and $t1(\varphi)$ behind the meniscus. (b) Functions of vapor pressure-dependent pore sizes $r(\varphi)$ according to Kelvin equation (1.37) and real size $r_R(\varphi)$ according to Eq. (1.79) considering sorption layer thickness.

with the constant M_γ from (1.68) for constant temperature T.

If for a real radius r_R one searches the associated but initially unknown relative humidity $\varphi(r_R)$, one can proceed iteratively using the following equation:

$$\varphi(r_R) = e^{\frac{M_\gamma}{(r_R - t(\varphi))}} \tag{1.80}$$

For a radius r_R of 1.0 nm, $\varphi(r) = 35\%$ and $\varphi(r_R) = 23\%$, for $r_R = 2.5$ nm follows $\varphi(r) = 66\%$ and $\varphi(r_R) = 59\%$ RH, compare **Figure 1.25b**.

1.4.6 Modeling of Adsorption Film Thicknesses

In the following, it is first investigated to what extent the basic isotherm for non-porous hydrophilic surfaces worked out in Section 1.4.2 can also be derived or approximated on the basis of thermodynamic approaches.

Furthermore, it is investigated to what extent the film thicknesses formed depend on the curvature of the surfaces and which film thicknesses can be formed in the pores within porous materials.

1.4.6.1 Modeling of Vapor Adsorption on Flat Nonporous Surfaces

The basis function equation $t(\varphi)$ (Eq. (1.50)) indicates to what extent a sorption layer or film of water molecules can be deposited **on flat surfaces of mineral solids**. The surface tension at the surface of the film, which may be several molecule layers thick, decreases from the surface tension of the solid γ_{SV} to the surface tension of water γ_L as a function of the distance from the surface and of the relative humidity above the film. **Figure 1.26** of Wu/Zandavi/Ward [44] shows this for water vapor and for heptane when acting on silicon surfaces.

By Adolphs/Setzer [30] and Churaev et al. [45] and others, the energy change occurring during film formation is described by the already quoted Eq. (1.61) and

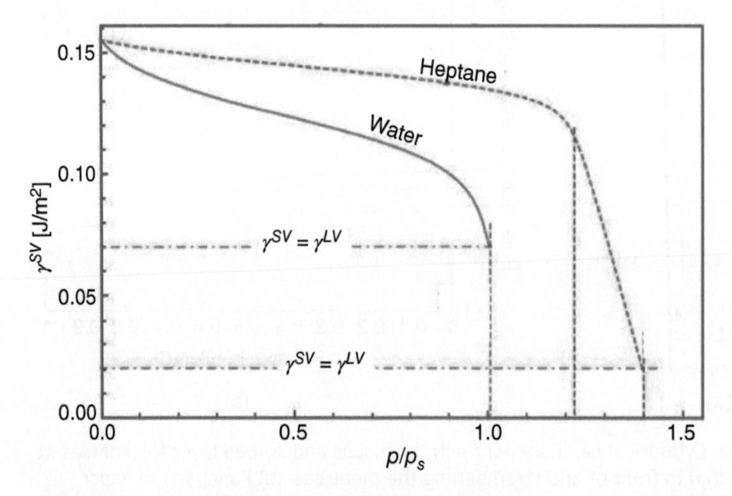

Figure 1.26 Course of surface tension or surface energy in sorption films upon exposure of a silicon surface to water vapor or heptane as a function of relative partial pressure. Source: Wu/Zandavi/Ward [44]/Royal Society of Chemistry.

the corresponding vapor pressure-dependent change in chemical potential (1.62), respectively, as follows:

$$\Delta\mu = RT \cdot ln\left(\frac{p}{p_S}\right) \quad [\text{J/mol}] \tag{1.81}$$

At the same time, it must be possible to express the chemical potential in terms of the layer thickness itself, compare Adolphs/Setzer [30] and Churaev et al. [45]:

$$\Delta\mu = \Delta\mu_0 \cdot exp\left(-\frac{n_{ads}}{n_{mono}}\right) = \Delta\mu_0 \cdot exp\left(-\frac{h}{h_m}\right) \quad [\text{J/mol}] \tag{1.82}$$

Where n_{ads} = adsorbed liquid volume of the layer, n_{mono} =mono layer volume, h and $h_m = n_{ads}/\overline{V}_L$ and $n_{mono}/\overline{V}_L$, respectively, $\overline{V}_L$ =molar volume of the liquid (water).

Since the change in chemical potential is also accompanied by a change in pressure (compare Eq. ((1.36)), Eq. (1.82) can also be rewritten **as an equation for the "separating pressure" resp. "disjoining pressure"** to (see Churaev et al. [45]) for flat surfaces:

$$\pi(h) = \pi_0 \cdot exp\left(\frac{h}{h_m}\right) \quad [\text{Pa}] \tag{1.83}$$

The chemical potential in the sorption film must be in equilibrium with the vapor pressure according to Eq. (1.81):

$$\Delta\mu_0 \cdot exp\left(-\frac{h}{h_m}\right) = RT \cdot ln(\varphi) \quad \text{with } \varphi = \frac{p}{p_S} \tag{1.84}$$

The initially unknown $\Delta\mu_0$ must be determined via additional boundary conditions. We now pretend that $\Delta\mu_0$ is to be determined in such a way that the isotherm Eq. (1.84) for **the relative humidity** $\varphi_{mono} = 0.219$ takes the value n_{mono}, which corresponds to the sample sorption isotherm $t(\varphi)$. Resolving Eq. (1.84) to h **yields the layer resp. film thicknesses on flat surfaces as** $h = h_{plane}$:

$$h_{plane}(\varphi) = h_m \cdot \left(ln\left(\frac{\Delta\mu_0}{RT}\right) - ln(ln(\varphi))\right) = h_m \cdot ln\left(\frac{\Delta\mu_0}{RT \cdot ln(\varphi)}\right) \quad [\text{m}] \tag{1.85}$$

Equation (1.85) is first used to determine $\Delta\mu_0$. With h_m = mono layer thickness = 0.30 [nm] and $h_{plane}(\varphi_{mono})/h_m = 1$ it follows:

$$\Delta\mu_0 = RT \cdot ln(\varphi_{mono}) \cdot exp(1) = -1.006 \cdot 10^4 \text{ at } T = 293.15\ [\text{K}] \quad \text{with}$$
$$exp(1) = \text{e} = 2.718.$$

Calculating now with this the resulting isotherm from Eq. (1.85), **we obtain the curve $h_{plane}(\varphi)$ shown in Figure 1.27 in comparison to the isotherm** $t(\varphi)$ from Eq. (1.50) according to [Franke]. There is surprisingly good agreement between the two curves.

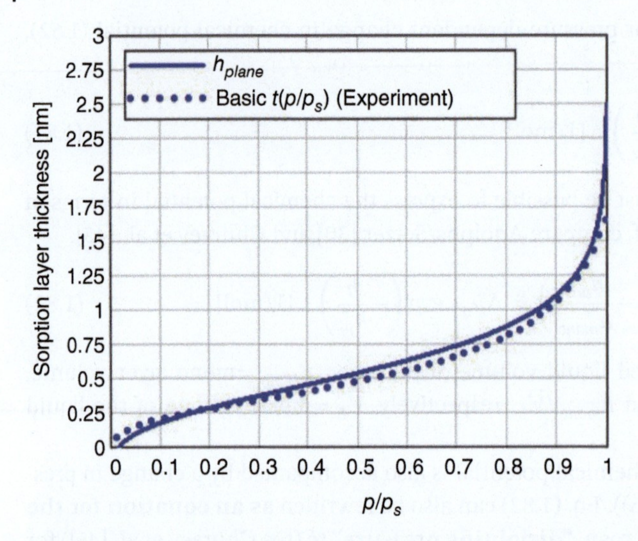

Figure 1.27 Comparison of sorption isotherms for plane surfaces of water vapor on mineral solids: h_{plane} according to Eq. (1.85) as well as the experiment-based basic isotherm according to Eq. (1.50).

1.4.6.2 Modeling of Sorption Film Thicknesses in Cylindrical Pores with Influence of Surface Curvature on the Thicknesses of Adsorbed Films

On concave surfaces of the inner wall of cylindrical pores influences not only the vapor pressure on the adsorbed film thicknesses but also an additional pressure from the surface tension of the curved sorbed liquid film, shown in **Figure 1.28a**. At equilibrium, the additional pressure component is Δp in [Pa] (**Figure 1.28b**).

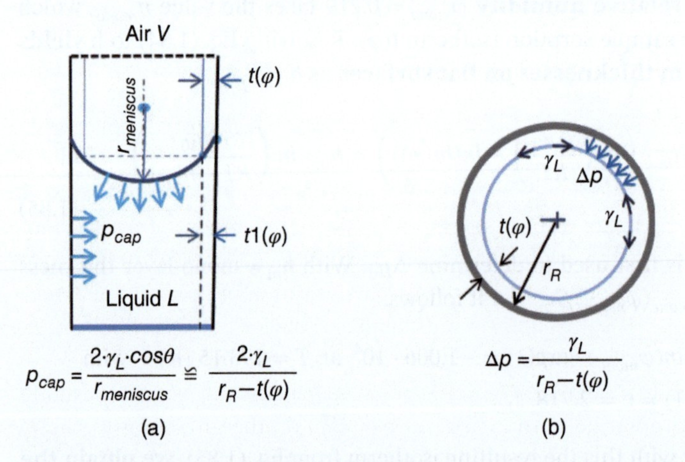

Figure 1.28 (a) Cylindrical capillary pore with meniscus and sorbed layer thicknesses at pore walls $t(\varphi)$ in front of and $t1(\varphi)$ behind the meniscus (b) functions of vapor pressure-dependent pore sizes $r(\varphi)$ according to Kelvin equation (1.37) and real size $r_R(\varphi)$ according to Eq. (1.79) considering sorption layer thickness.

The total pressure in the sorption layer can then be expressed as a change in chemical potential starting from the disjoining pressure for a concave curved surface.

$$\Delta\overline{G}_L = \overline{V}_L \cdot (\pi(h) - \Delta p) = -RT \cdot ln\left(\frac{p}{p_S}\right) \tag{1.86}$$

In contrast to the explanations of Churaev et al. [45] or Mattia et al. [46], here Δp has a negative sign, since the pressure reduction in the liquid film must lead to a reduction in the chemical potential.

Using the relations (1.83) and (1.84), it follows after transformation

$$\frac{\Delta\mu_0}{RT} \cdot exp\left(-\frac{h}{h_m}\right) = \frac{\gamma_L}{(r_R - h(\varphi))} \cdot \frac{\overline{V}_L}{RT} + ln(\varphi) \tag{1.87}$$

After logarithmizing and transforming this equation, we obtain the relation for the resulting film thickness:

$$h(\varphi, r_R) = h_m \cdot \left[ln\left(\frac{\Delta\mu_0}{RT}\right) - ln\left(+\frac{\gamma_L \cdot \overline{V}_L}{(r_R - h) \cdot RT} + ln(\varphi)\right)\right] \cdot 10^9 \quad [\text{nm}] \tag{1.88}$$

When calculating $h(\varphi, r_R)$, note that in the right part of the Eq. (1.87), h is also included. Therefore, in the 1st step, one calculates $h(\varphi, r_R)$ with an initial value $h = 0$ or $h = t(\varphi)$. One then repeats the calculation at least twice by inserting the result value $h(\varphi, r_R)$ of the equation into its right-hand part and then obtains a sufficiently high accuracy.

The disjoining-pressure part $\pi(h)$ in Eq. (1.86) is calculated in a number of publications using the **Hamaker constant** A_H. This constant describes the interaction between the involved molecules of solids and liquids and the influence of opposing material surfaces. The Hamaker constant A_H in the size of about 10^{-20} is in fact only approximately constant, compare literature Israelachvili [28]. The disjoining pressure fraction $\pi(h)$ in Eq. (1.86) formulated with the help of these constants can consist of up to three fractions depending on the physical influence taken into account with several physical parameters also known only approximately, compare calculations for example in literature Jing Li et al. [47] or the extensive explanations in Teletzke and Davis de L.E. Scriven [48].

This means that the calculated disjoining pressure values or resulting film thicknesses are also approximate values. Therefore, no further comments are made here with reference to the cited literature. However, the film thicknesses calculated in the manner described above obviously give a reliable indication of the existing dimension. Mattia et al. [46] also investigate the stability behavior of water films on the inside and outside of nano-capillaries made of quartz and carbon material. They use a Hamaker constant to model the disjoining pressure fraction and neglect the radius influence of the sorption layer thickness $h(\varphi)$ or $t(\varphi)$ on the growth of the films. Results on the behavior for convex cylinder surfaces are not reported.

On the convex outer surface of cylindrical material shapes, Eq. (1.86) also applies, but in this case, Δp must be given a positive sign, since now the surface tension generates a surface pressure. For the calculation of the corresponding

$h(\varphi, r_R)$, Eq. (1.87) also applies, whereby a minus sign must now be inserted before the term with γ_L. Proceed accordingly in Eq. (1.88). The calculation procedure then also corresponds to the procedure for concave surfaces.

The results of the film thickness calculations for concave and convex surfaces are shown in **Figures 1.29a,b and 1.31**. In **Figure 1.29a,b**, the film thickness profile is shown for cylindrical pores with constant radii of 2, 5, and 20 [nm] (and open ends) and corresponding external surfaces, respectively, as a function of relative humidity.

It can be seen that the film thickness at the inner surface of the pores increases faster with increasing relative humidity than for the reference curve plotted for flat

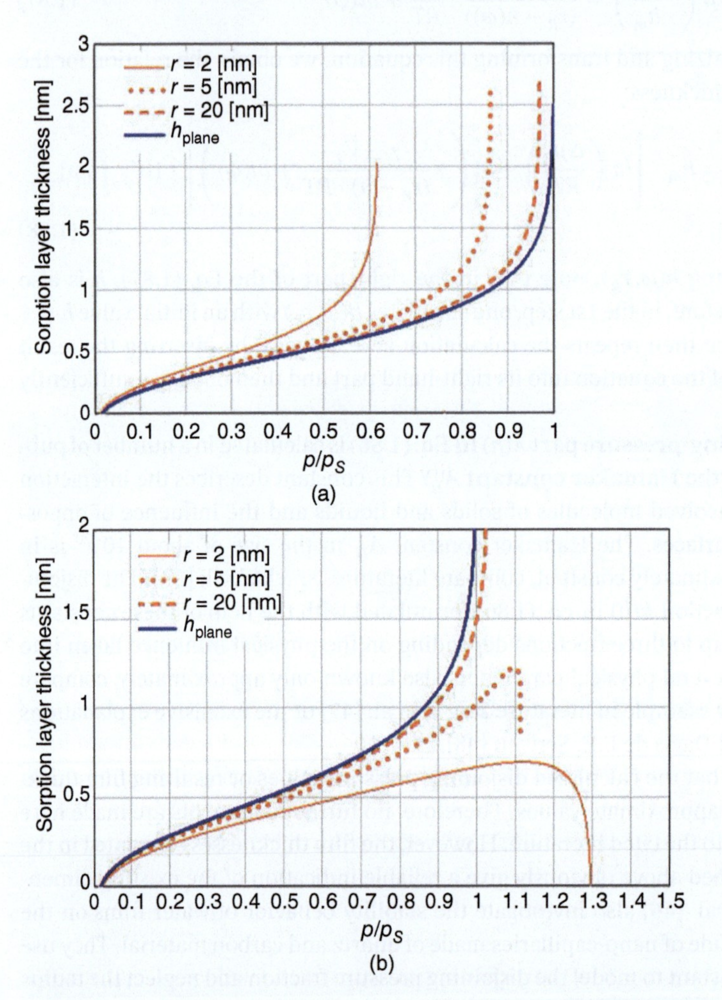

Figure 1.29 (a) **Influence of surface curvature resp. radial tensile stress on adsorbed film thickness** corresponding to **Figure 1.28b** in cylindrical pores at increasing RH compared to the film thickness function $h_{plane} \approx t(\varphi)$ for water vapor on flat surfaces (constant real pore radius $r_R = 2, 5$, and 20 [nm]). (b) Influence of increased external vapor pressure on convex external surface of cylindrical material shapes compared to film thickness function $h_{plane} \approx t(\varphi)$ for water vapor on flat surfaces.

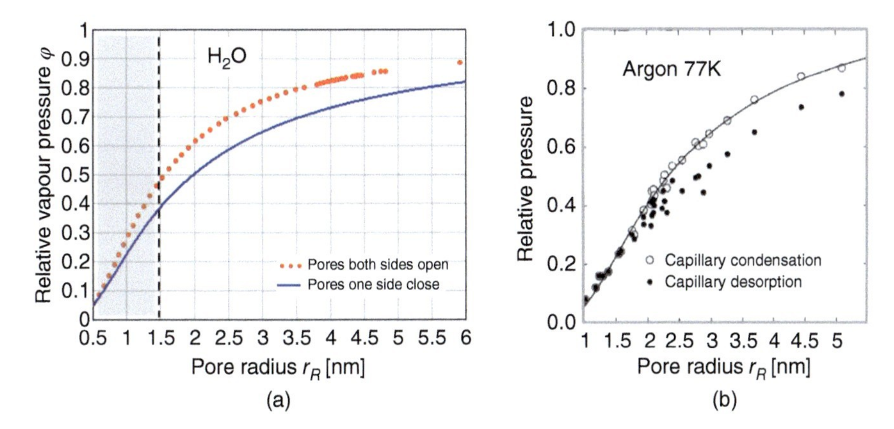

Figure 1.30 (a) Relative air humidities φ at which **filling of cylindrical pores** by condensed **water** occurs : **In pores open at both ends** up to radius of about 5 [nm] by collapse of inner surface film (calculated by Eq. ((1.88)), **in pores closed at one end** by classical capillary condensation from $r_R \approx 1.5$ [nm] for comparison. (b) **Experimental results** from [49] on the dependence of capillary condensation pressure and evaporation pressure on the pore radius in parallel cylindrical siliceous pores of MCM-41, for **argon** at 77 K.

surfaces for comparison $h_{plane} \approx t(\varphi)$. **Figure 1.29a** also shows clearly that, especially in the smaller pores, **film formation changes relatively spontaneously to pore filling in the sense of capillary condensation.**

As the calculation results according to Eq. (1.88) show, this occurs according to **Figure 1.30a** for cylindrical pores open on both sides up to a radius r_R of approx. 5 [nm] or an associated relative air humidity of up to approx. 85%. The comparison with cylindrical pores closed on one side, where a meniscus forms early starting from the closed end of the pore, using the extended Kelvin equation (1.80) in **Figure 1.30a** shows that capillary condensation starts earlier in such pores for a given pore radius r_R at lower vapor pressures or φ values as a function of the pore radius.

According to these results, for example, an initially empty longer cylindrical pore with a radius of 2 [nm] open at both ends fills completely with condensed water at about $\varphi = 62\%$, whereas the cylindrical pore closed at one end fills already at $\varphi \approx 50\%$. If we look at the desorption behavior of the pore open at both ends after filling, menisci now form at both ends, so that the onset of desorption corresponds to the blue curve in **Figure 1.30a**.

This results in a hysteretic adsorption–desorption behavior of such pores, which becomes more pronounced with increasing pore radius. The extent to which this mechanism is relevant to porous materials such as cement paste is discussed in Section 4.4.2.

If one considers the film thicknesses at the outer surface of cylindrical forms, the calculation yields the opposite result shown in **Figure 1.29b**. According to this, the film thicknesses are lower than for the reference curve $t(\varphi)$, especially at the smaller radii, as expected. There is no instability here up to relative vapor pressures of $\varphi = 1.0$. For this, vapor supersaturations $\varphi > 1.0$ are necessary, which can only exist under exceptional boundary conditions. For a surface radius of 2 [nm],

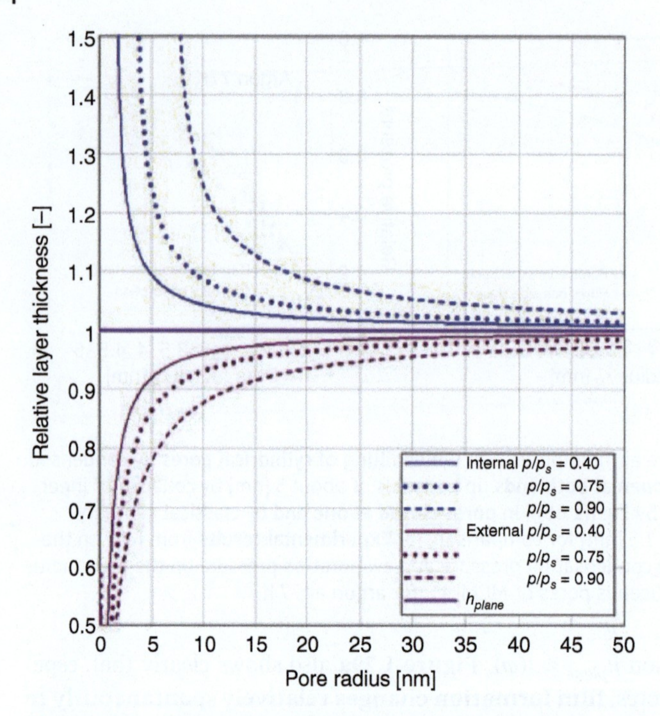

Figure 1.31 Comparison of relative film thicknesses on the inside of cylindrical pores and the outside of cylinders, taking into account capillary pressure (suction), as relative values related to the corresponding adhesion film thicknesses from water on a flat mineral surface (as a function of radius).

instability results at a φ value of 1.3, and for a radius of 5 [nm], instability results at φ about 1.1.

It was further investigated **the radius dependence of film thicknesses** in more detail for three selected relative vapor pressures, $p/p_S = 0.40,\ 0.75$, and 0.90. The calculation results are plotted relative to the layer thickness of the reference curve $t(\varphi)$ radius-dependent in **Figure 1.31**. One can see the strong tendency for increasing thickness differences at radii below about 10 [nm].

1.4.6.3 Volume of Adsorbed Water in the Not-yet-water-filled Pore Area

In the case of partial water saturation of the material, the Eq. (1.77) can be used to determine, for example, **the amount of water adsorbed in the not-yet-filled pore region** at different degrees of filling at the inner surface as a function of the relative air humidity $\varphi 1$.

For a given air humidity $\varphi 1$, Eq. (1.77) with lower limit $r_R(\varphi 1)$ and upper limit $r_R(\varphi 2) = r_R(0.9999)$ gives the adsorbed water volume stored in the unfilled pore region, denoted here by $\Delta Wh(\varphi 1)$.

The r_R-values of the integration limits for chosen φ are obtained from Eq. (1.79).

The dotted curve (color pink) in Figure 1.32 shows the calculated variation of sorbed moisture at increasing filling level or lower limit $\varphi 1$. At $\varphi = 1$, the sorbed volume is equal to 0, as expected, and exhibits the maximum at $\varphi \approx 20\%$ RH.

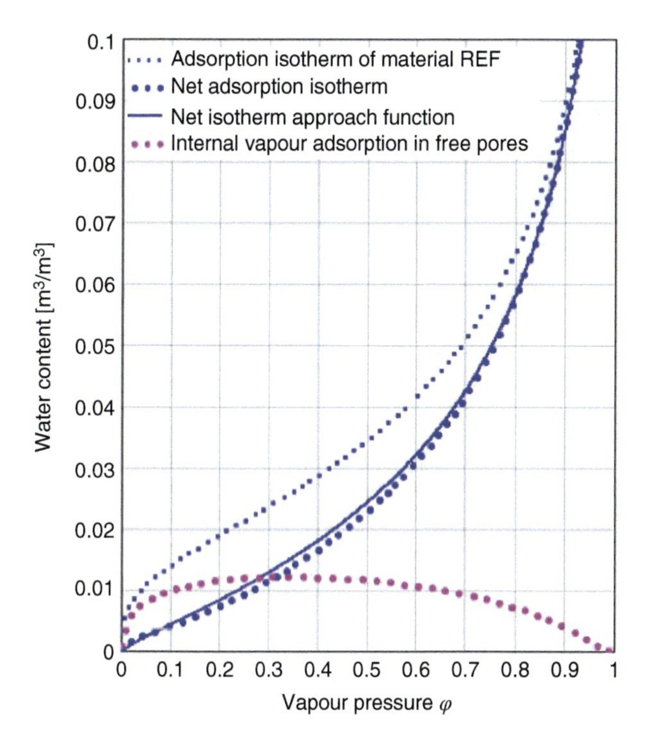

Figure 1.32 The adsorption isotherm of the material REF, the derived net storage function, and the curve of adsorbed vapor molecules from $\varphi 1$ to $\varphi 2 \approx 1.0$ (pink dotted) **in the non-water-filled pore region.**

The **net storage function**, which is important for all calculations of the internal surface area and moisture storage fractions, is calculated from the original storage function or adsorption isotherms by subtracting the fraction of the sorbed volume in the not yet filled pore region according to the following relation corresponding to **Figure 1.23a**:

$$\theta_N(\varphi) = \theta\varphi(\varphi) - \Delta Wh(\varphi) \tag{1.89}$$

In Figure 1.32, this yields the blue dotted curve as the net storage function for the material REF considered as an example, **calculated with the inner surface sorption**. The **dotted blue curve also included in the figure corresponds to the approximate curve used in Eq. (1.66)** in Section 1.4.4.1.

The curve $\theta r_{R\_N}(r_R)$ in **Figure 1.23b** is the net storage function Eq. (1.89) after conversion to r_R.

1.4.6.4 Film Thicknesses Due to Adsorption in the Water-filled Pore Region Behind the Meniscus, Estimation Using a Surface Energy Approach

In the large number of publications dealing with the question of film thicknesses on the inner surface of porous materials, the film thicknesses in the not yet water-filled pore region of the pores are always considered. The question of film thicknesses in the liquid-filled pore region behind a meniscus is not addressed, or

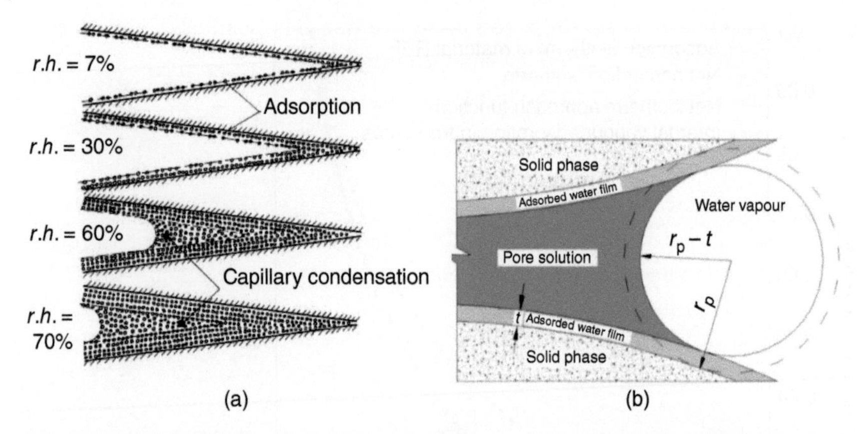

Figure 1.33 Schematic representation of capillary condensation in two different pore shapes with the assumption that the sorption film thickness behind the meniscus in the already water-filled pore region corresponds to the film thickness in front of the meniscus. (a) From Ahlgren [9]. (b) From Zhang et al. [50].

it is assumed that molecular layers sorbed there correspond to the total thickness in the nonliquid-filled pore region in front of the meniscus, compare the examples in **Figure 1.33a,b** from [9] and [50].

The international investigation results summarized in Section 1.4.7 largely agree in the statement that due to sorption of water molecules and ions in aqueous solutions film thicknesses of high toughness at least about 2 monolayers thick are formed. A consideration of this situation in the water-filled pore region behind the meniscus is therefore necessary, since according to these results, a probable **non-negligible influence in the modeling of moisture transport** arises.

To estimate the film thicknesses addressed, a theoretical consideration on film thickness evolution is given below, which starts from the following Gibbs–Duhem equation (1.90), derived from the well-known Gibbs energy equation with surface energy influence and describing the enthalpy at equilibrium for constant temperature and constant external pressure.

$$\sum n_i \cdot dn_i + A \cdot d\gamma = 0 \qquad A = surface \tag{1.90}$$

This equation makes clear the direct relationship between chemical potential and surface energy, which will be used below.

Dash and Peierls [51] gives the related equation $d\gamma = -n \cdot d\mu$ with μ as chemical potential and n as sorption layer thickness to characterize sorption film thicknesses. However, concrete layer thicknesses are not derived from Dash.

It is assumed that the chemical potential starting from the solid surface decreases as follows according to the well-known approach, for example, Churaev/Setzer [52] or the course of the disjoining pressure approach Eq. (1.83):

$$\Delta\mu(h) = \Delta\mu_0 \cdot exp\left(-\frac{h}{h_m}\right) \tag{1.91}$$

For this, it is necessary that the solid surface has a larger surface energy than the liquid or water, according to the condition for spreading given by Eq. (1.10). As shown by Wu/Zandavi/Ward [44], **compared to Figure 1.26**, it is assumed that the surface energy within the surface film must change from γ_{SV} to γ_{LV}.

The number of moles per m^2 area per monolayer is

$$z_m = \frac{h_m \cdot A}{\overline{V_L}}$$

With the surface energy γ_{SV} [Nm/m^2 = J/m^2] of the solid surface A, the associated change in the chemical potential of the 1st molecular layer related to the molar volume $\overline{V_L}$ is now formulated as follows:

$$\Delta\mu_0 = \frac{\gamma_{SV} \cdot A}{z_m} = -\frac{\gamma_{SV} \cdot \overline{V_L}}{h_m} \tag{1.92}$$

In the present approach, the chemical potential changes from the film thickness h_{max} to the value of bulk water γ_L, i.e. the chemical potential is reduced by the value $\Delta\mu_{h_{max}}$:

$$\Delta\mu_{h_{max}} = -\frac{\gamma_L \cdot \overline{V_L}}{h_{max}} \tag{1.93}$$

Substituting the Eqs. (1.92) and (1.93) into (1.91), we obtain the equation **for the total film thickness h_{max} sought for a plane solid surface covered with bulk water**:

$$exp\left(-\frac{h_{max}}{h_m}\right) = \frac{\gamma_L}{\gamma_{SV}} \cdot \frac{h_m}{h_{max}} \tag{1.94}$$

The results resulting from Eq. (1.94) are explained in the following:

First, the **influence of the capillary pressure from the meniscus, which is to separate the liquid filling in the considered pore region from the not yet filled pore region**, shall still be considered.

Let the capillary suction on the inner pore surface, which depends on the vapor partial pressure at the concave surface of the meniscus, be denoted by $\Delta p_{cap}(\phi)$ with $\phi = p/p_S$.

It is assumed that this results in a change of the chemical potential of the liquid and also in the influence of thickness h_{max} accordingly:

$$\Delta\mu 1_{h_{max}} = -\left(\frac{\gamma_L}{h_m} + \Delta p_{cap}(\phi)\right) \cdot \overline{V_L} \tag{1.95}$$

In it, with Eq. (1.34) and considering the Kelvin equation

$$\Delta p_{cap}(\phi) = \frac{2 \cdot \gamma_L}{r_{meniscus}(\phi)} = -\frac{RT \cdot ln(\phi)}{\overline{V_L}} \tag{1.96}$$

Substituting Eq. (1.96) into Eq. (1.95) and then combining it with Eqs. (1.91) and (1.92) gives the equation for **the film thickness h_{max} for the inner surface of water-filled cylindrical pores under capillary suction influence** at given vapor partial pressure $p/p_S = \phi$:

$$exp\left(-\frac{h_{max}}{h_m}\right) = \frac{\left(\frac{\gamma_L}{h_{max}} + \Delta p_{cap}(\phi)\right)}{\frac{\gamma_{SV}}{h_m}} \tag{1.97}$$

From the Eq. (1.97), the film thickness can now be estimated as a function of the capillary pressure in filled pores or the relative vapor pressure ϕ. To solve the equation, $h_{max} = h_m \cdot x$ is set. The results can be obtained from **Figure 1.34**.

The value $x = 2.5$ molecule layers would be assigned to narrow pores, 4.4 molecule layers or approx. 1.3 [nm] to planar, water-covered silicate surfaces.

The following values were used for calculation: $\gamma_{SV} = 1.4\,[\text{J/m}^2]$ for silicate hydrates and HCP, respectively, $\gamma_L = 0.073\,[\text{J/m}^2]$, $h_m = 0.30\,[\text{nm}]$ or $0.30 \cdot 10^{-9}\,[\text{m}]$ as monolayer thickness for water, $\overline{V}_L$ = molar volume of the liquid (water).

The measured value given by Zhao et al. in literature [53] for film thicknesses in bulk-water on muscovite mica is 0.9 [nm], compare **Figure 1.39b**. The **film thicknesses for water-filled pores** estimated here by calculation as a function of capillary pressure are 2.5–4.4 molecular layers and 0.75–1.3 [nm] for water on silicates, respectively. From these results, it can be concluded that a simple "extension" of **the film thicknesses determined in the non-water-filled pore region in front of a meniscus into the water-filled region is not readily possible and justified**. Therefore, the behavior of the film thickness trajectories assumed in **Figure 1.33a,b** cannot be adopted in this way, which is supported by the investigation results in Section 1.4.6.5.

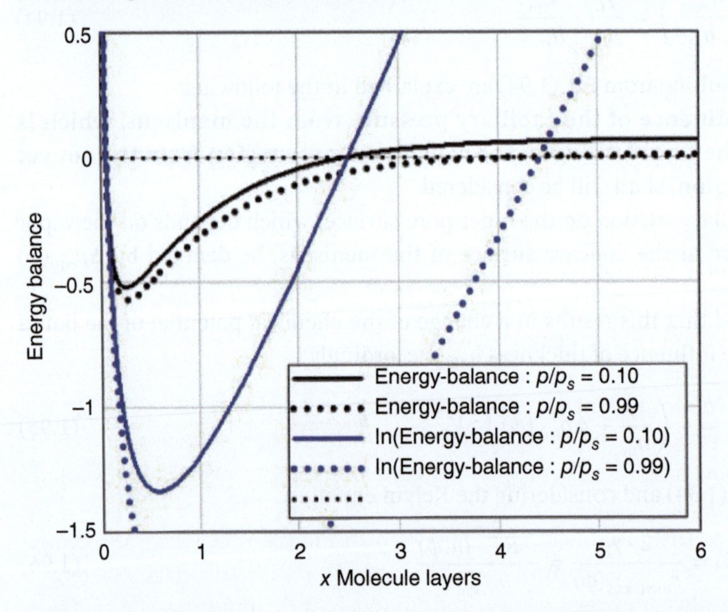

Figure 1.34 Calculated film thicknesses on the inside of water-filled cylindrical pores via a surface energy approach. The intersections of the curves with the zero line give the number x of molecular layers of the sought film thicknesses $h_{max} = h_m \cdot x$. For $\phi = 0.99$ RH: $x = 4.4$ layers and for $\phi = 0.10$ RH: $x = 2.5$ layers.

1.4.6.5 Volume of Aadsorbed Water Molecules on the Inner Surface in the Water-filled Pore Area

In Section 1.4.7, results of experimental investigations and molecular-dynamic (MD) studies are reproduced on the question of adsorbed water-molecule films at the inner surface in the water-filled pore region.

Reported are, among others, the results of capillary water transport experiments on nanopores, according to which stable surface films of 0.4–0.6 [nm] thickness (Xu et al. [54]) or 0.60 [nm] thickness (Gruener et al. [55]) are formed on the inner surfaces of the pores. These adsorbed films constrict the pore cross sections and do not themselves participate in fluid transport.

K. Wu et al. [56] evaluates 50 corresponding studies and finds a nonparticipating film thickness of about 0.7 [nm] and a "dramatic" increase in viscosity of the adsorbed films at pore diameters ≤ 1.4 [nm].

Antognozzi et al. [57], Zhao et al. [53], and Arai et al. [58] determine using AFM that tough molecular layers of at least 0.70 [nm] form on surfaces with behavior different from bulk water.

The theoretical consideration undertaken in Section 1.4.6.4 based on the surface energy of adsorbed films yields a sorption thickness between 2.5 and 4.4 molecular layers, on average order of magnitude 0.90 [nm].

Knowing the sorption film thicknesses in the water-filled pore regions is of high importance for understanding and modeling moisture transport in porous materials. Therefore, for the sample material REF, it was computationally determined which water content fraction results in partial or complete water saturation of the material, if constant adsorption film thicknesses are taken as a basis in the order of magnitude of the previously mentioned, experimentally and theoretically determined values.

Film thicknesses t_{film}= 0.30, 0.45, and $0.60 \cdot 10^{-9}$= const. [m] were used in the calculation.

The following relation was used for this purpose:

$$\Delta V_{film}(\varphi_1) = \theta_N(\varphi_0) + t_{film} \cdot \Delta A(\varphi_0, \varphi_1) \tag{1.98}$$

Then, to an initial value $\theta_N(\varphi_0)$ is added the water volume, which is the product of the film thickness t_{film} and the inner surface area $\Delta A(\varphi_0, \varphi_1)$ between the initial φ_0 and the running upper φ_1 boundaries. $\theta_N(\varphi_0)$ is the water content of the net adsorption curve at a pore radius of a water molecule $r_R(\varphi_0) = 0.30 \cdot 10^{-9}$ with $\varphi_0 = 0.0116$. $\Delta A(\varphi_0, \varphi_1)$ corresponds to the area integral (1.77).

For smaller φ values, the volume from film thickness and surface area partially exceeds the existing pore volume. It is then set equal to the corresponding volume of the net storage function.

The result of the calculations can be seen in Figure 1.35. It can be seen that for a film thickness of 0.30 [nm], the pores are filled with adsorbed water up to $\varphi \approx 20\%$, for a film thickness of 0.45 [nm] up to $\varphi \approx 40\%$, and for a film thickness of 0.60 [nm] at relative air humidity up to $\varphi \approx 55\%$. These conditions have an appreciable effect on moisture storage and transport in a pore system, which must be taken into account when modeling moisture transport.

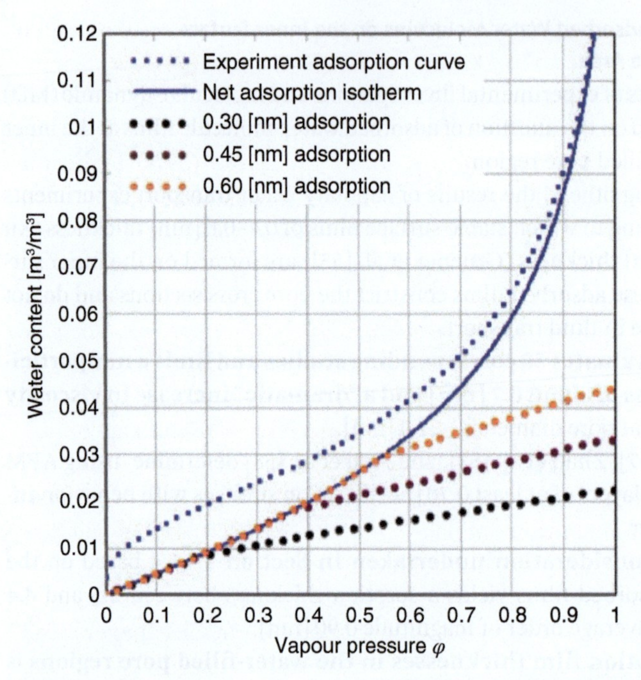

Figure 1.35 Adsorption isotherm of the material REF and net storage function and volume of adsorbed water molecules at the inner surface **in the water-filled pore region** with film thicknesses of 0.30, 0.45, and 0.60 [nm]. **The net adsorption isotherm** shows the water content of the pore system up to the considered value φ, without the moisture content adsorbed on the inner surface in the non-filled pore area above.

1.4.7 Molecular Simulations and Experimental Investigations on the Dimension of Adsorbed Film Thicknesses (International Research Results)

With the help of computer programs, that simulate via modeling, the interactions between molecules and atoms to their motions, mergers and attachments, it is possible to estimate the sorption of liquids on solids or resulting film thicknesses. For years, programs have been used for this purpose, known as **MD or MDS** as molecular dynamics simulations, which are further developed for simultaneous description of chemical reactions under the name molecular mechanics, in the last stage as **reactive molecular dynamics (RMD)**, which obviously require very large computational effort, for example, require the use of parallel computers, compare H.M. Aktulga et al. [59]. An overview of the state of development and available programs is given by K. Farah et al. [60].

Castrillon et al. [61] reports in his paper the results of **MD simulations** on sorption of water films on quartz surfaces. Among other things, the distribution of density and local potential energy near the surface is also reported. Castrillon, extending his calculations to water molecule thicknesses corresponding to about 15 monolayers, finds an influence on the water-molecule orientation up to a distance of about 1.4 [nm] from the surface and **a firmer bond or stiffness (H-bonding) of the water-molecule layers up to a thickness of 0.4–0.6 [nm].**

However, the results of calculations with current computer programs always require confirmation by the results of experimental investigations:

The development of AFM or surface force apparatus (SFA) technique now allows additional insight into the formation of film thicknesses at interfaces. The paper by Israelachvili et al. [62] provides an overview of the status of this development up to 2010. In these devices, an upper tip (guided by special piezoelectric crystals) is gradually moved in small air-conditioned chambers to the material surface to be tested, which may be covered with a liquid film thickness. The exact distances are measured using light wavelengths or laser technology.

Figure 1.36 shows schematically the structure of an AFM. The caption provides more information about the function. A recent overview of the state of AFM technique can be found from Peng et al. [63]. Problems concerning the interpretation of the results of conventional AFM technique with tip insertion are discussed in [64]. An overview of other investigation techniques used, such as NMR and X-ray techniques, can be found in Monroe et al. [65].

In the work of Xu et al. [54], permeability experiments and subsequent MD simulations are reported. In particular, transport experiments are performed on fine porous material of Vycor glass with mean pore diameters of 3.4 and 7.2 [nm] and on porous quartz material Xerogel with mean pore diameters of 3.4 [nm]. Aqueous solutions with 1M NaCl and 1M $CaCl_2$ are also used for comparison.

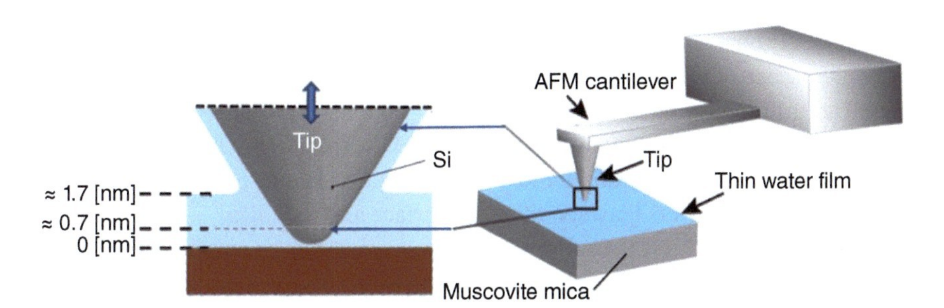

Figure 1.36 Schematic representation of the principle of an AFM (atomic force microscope) inside a temperature and humidity adjustable measuring chamber. A very fine tip attached to a cantilever is moved over the area of the interface to be measured, horizontally in 2D or/and vertically in 3D scanning. The tip or curved surface can be guided into the liquid film in Ångström steps. The path of the tip is measured by a laser apparatus or a light wave interference system, and the ultra-small forces occurring at the tip are measured via the spring stiffness of the cantilever. The cantilever can also operate above the interface with the solid submerged in the liquid. The cantilever can be operated in static mode or in dynamic mode (near its resonance frequency), in the latter case frequency-modulated as FM-AFM or amplitude-modulated as AM-AFM. Source: The pictorial elements shown here are taken from Arai et al. [58]/Springer Nature/CC BY 4.0. The measurements of Arai et al. were made in saturated water vapor atmosphere on interface to muscovite mica. The nanometer dimensions given in the figure are measurement results: up to about 0.7 [nm] a tough molecular layer was measured, covered up to about 1.7 [nm], with bulk water.

In both the experiments and the MD simulations, respectively, it is found that the material pores have a 2-layer water-molecule film of about 0.6 [nm], which leads to a corresponding narrowing of the pores during moisture transport. There were no significant changes when sodium chloride and calcium chloride solutions were used.

S. Gruener et al. [55] perform capillary water uptake experiments on Vycor glass with mean pore diameters of 3.4 and 4.9 [nm]. The analyses revealed surface films not participating in moisture transport on the order of 2 molecule layers.

Asay and Kim [66] performs infrared spectroscopic ATR (attenuated total reflection) studies on flat silicate surfaces at 20 °C and increases the relative humidity of air and finds that an approximately 3 monolayer thick water film of an "icelike" nature forms on the silicate surface.

K. Wu et al. [56] evaluates the results of more than 50 different studies on the question of the formation of water-film thickness using a model based on Hagen-Poiseuille's theory to describe water transport in cylindrical pores in the presence of film thicknesses on the inner pore walls. The theory is applicable to pore diameters >1.4 [nm], since a dramatic increase in viscosity occurs for smaller diameters. Wu's comparative calculations lead to the conclusion that, especially for pore diameters <10 [nm], there is an increase in viscosity for hydrophilic materials compared to bulk water, while a large decrease in transport resistance occurs for transport processes in hydrophobic materials. In hydrophilic materials, according to Wu, an approximately 0.7 [nm]-thick stable film thickness is formed on the inner surfaces.

Antognozzi et al. show in [57] results of measurements on thickness-dependent shear viscosity on a glass/mica combination, determined with a transverse dynamic force microscope **(TDFM)**. A section of an optical glass fiber with plane-cut surfaces is used as a tip and guided in dynamic shear motions over water films of different thicknesses on a muscovite mica surface. The experiments show, that below about 0.9 [nm], there is a dramatic increase in shear viscosity or shear stiffness of the water-molecule layers between the glass surface and the mica.

This increase in stiffness therefore occurs below a film thickness of 0.45 [nm] with respect to one of the two material surfaces.

Figure 1.37 shows AFM results and MD calculations from Kimura et al. [67]. These analyses were performed at the interface between muscovite mica and an upstanding 1M KCl aqueous solution. According to the results of Martin-Jimenez et al. [68], ion concentrations ≤ 1M have only a minor influence on the structure of the water-molecule layers, compare the following remarks. From the results shown in **Figure 1.37**, it can be concluded that a film thickness of 0.45 [nm] was also present, according to the results of Antognozzi et al. From these results, it can be concluded that, at least for low-concentrated electrolyte solutions (or pure water), a primary sorption layer thickness of about 0.45 [nm], already determined by Kimura et al. and Antognozzi et al., can be assumed. Furthermore, it follows from the present work that primary layers of greater thickness result at higher electrolyte concentrations.

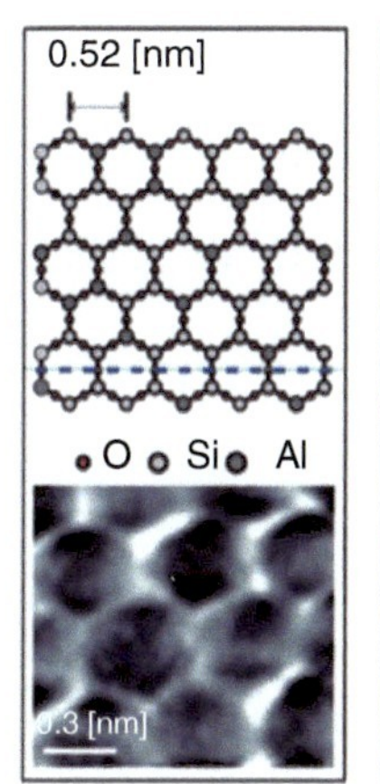

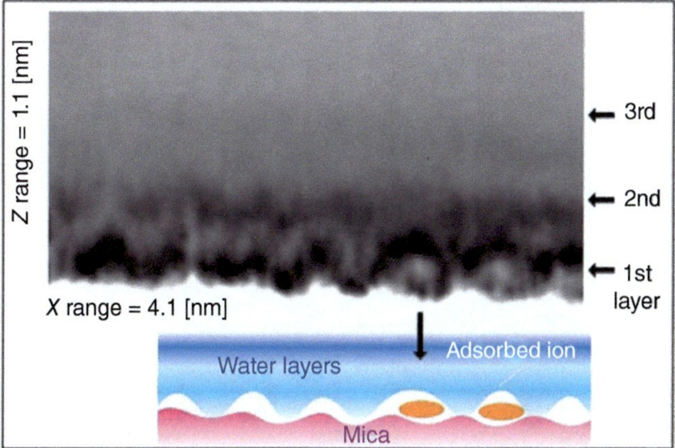

Figure 1.37 Results of 2D force scanning of the muscovite mica material interface perpendicularly into the overlying 1M KCl aqueous solution using FM-AFM. **The left part** schematically shows the crystalline structure of the mica. Below is the microscopic AFM image of the corresponding surface when scanning the surface. This cleaved mica surface corresponds to a structural section corresponding to the blue-dashed line in the schematic structure image above. **The right part** shows the registered structure in the aqueous solution above the mica surface. At the muscovite mica/water interface, the water molecule layer penetrates into the center of the hexagonal structure schematically illustrated on the left. In the upper layers, fluid water molecule layers are observed, and they show a smaller correlation with the surface structure. The layers registered by the authors are indicated at the right edge of the image. The total image height is 1.1 [nm]. The yellow particles in the sub-image are likely to be K^+ ions or hydrated K^+ ion complexes adsorbed on the surface. Source: Kimura et al. [67]/AIP Publishing.

By Martin-Jimenez et al. [68] present further insightful investigation results on 3D structure at interfaces. **Figure 1.38** contains some results of these investigations and related details. Further results of this work can be reproduced as follows:

The interaction with the mica is found to have a nearly complete monolayer of cations at low salt concentrations (0.2M), as indicated by the AFM images. The K^+ ions from the solution occupy the positions in the cleaved plane previously occupied by K^+ ions in the bulk mica crystal. The water molecules from the first hydration layer are tightly bound to the solid surface, occupying the space between the cations. The second hydration layer is placed 0.25 [nm] above the adsorbed K^+ ions and it follows its atomic corrugation. The water molecules in the 2nd layer are centered on top of the cations of the first layer, but with a larger lateral spreading. The accompanying DFC simulation predictions show a significant reduction of the cation coverage of the mica at low salt concentration.

At higher salt concentrations (3–5M), the interfacial layer has a different structure and composition. It is thicker and shows a crystal-like structure.

The observed phenomenology is not restricted to KCl solutions. It also applies to other alkali halide electrolyte solutions.

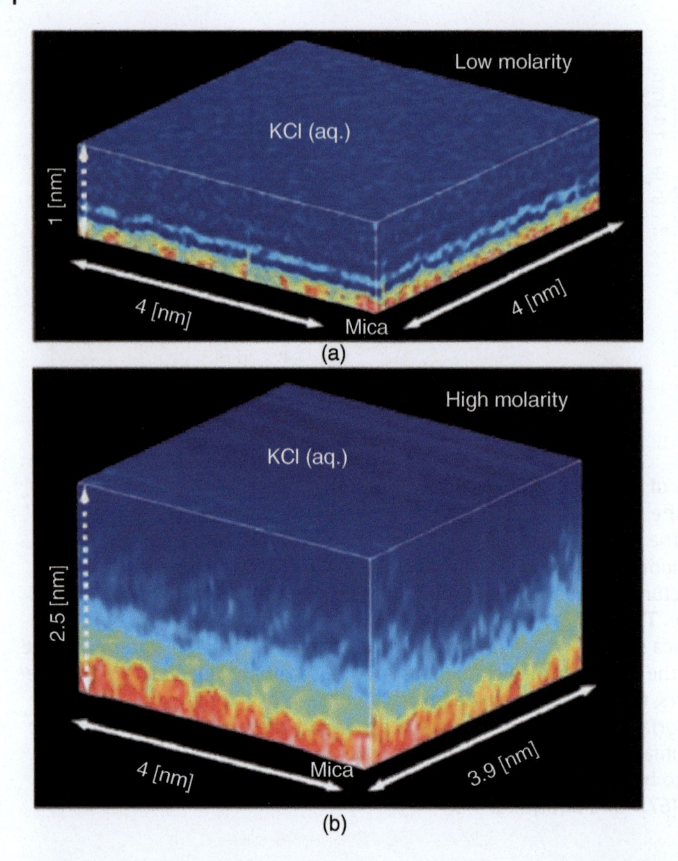

Figure 1.38 Three-dimensional AFM-images of muscovite mica/electrolyte solution Interfaces. (a) 3D AFM image of a KCl (aq.) solution (0.2M KCl). The image shows a monolayer resp. interface of partial adsorbed K^+ ions (light red) topped by two hydration layers (lighter stripes), following the atomic corrugation of the mica surface. (b) 3D AFM image of a mica-KCl (aq.) interface (4M KCL). The interface is divided into two main regions: an ordered liquid layer extending up to 2 [nm] from the mica surface into the bulk solution above it. These 3D maps show the variations of the phase shift of the tip's oscillation. The 3D AFM experiments were performed at 300 K. Source: Martin-Jimenez et al. [68]/Springer Nature.

The results reported by G. Zhao et al. in [53] are particularly remarkable as well. **Figure 1.39a,b** shows selected results of the investigations with a SFA 2000 apparatus.

These AFM investigations have the special feature that the conventional tip configuration was not used, but instead 2 cylindrical surfaces (tube sections with a radius of 2 [cm]) were arranged in a crossed configuration opposite each other. Both crossed cylindrical surfaces were covered with a cleaved layer of muscovite mica. The mutual distance of both surfaces was measured in the AFM in steps of 0.1 [nm] using the MBI lightwave interference technique FECO, using a video camera to allow direct monitoring of the surface separation and phase change of the adsorbed water between two mica surfaces. More Details of the measurements

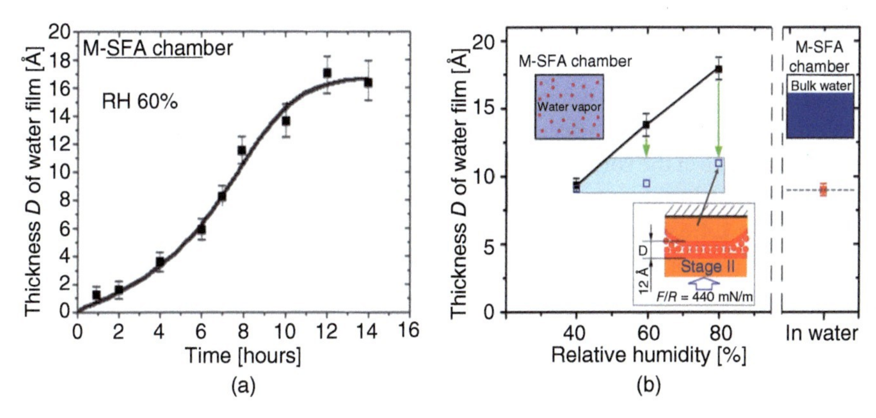

Figure 1.39 Measurement results of G. Zhao et al. from [53] **on water vapor sorption on muscovite** mica (measured with SFA 2000 measuring apparatus). (a) Time-dependent evolution of sorption film thickness from water on muscovite mica under constant vapor pressure. (b): The black squares give the measured values of the formed film thicknesses as a function of vapor pressure at sufficient distance of the opposing surfaces. When the surfaces are automatically moved together, part of the original water film is reduced. This process comes to a halt due to corresponding counterforce at a mutual distance of the muscovite surfaces in the order of $D = 0.90$ [nm] after 10 hours, which leads to a layer thickness of $D/2 = 0.45$ [nm] related to each of the two surfaces, if the feed force is not significantly increased. The solid square symbols indicate the initial thickness of the measured adsorbed water film at the moment of two mica surfaces contacting together. The hollow square symbols indicate the film thickness reduced to the light blue area under the action of the external load schematically shown by the sub-picture below. **In the right part of the figure (b)** the corresponding measured value of approx. 0.9 [nm] after approx.10 hours pressure at 80% RH results, when the experiment is carried out **in water**. This result also corresponds to an individual layer thickness of 0.45 [nm] per surface. Source: Zhao et al. [53]/AIP Publishing.

can be taken from **Figure 1.39b** or the corresponding figure caption. Some relevant details will be presented here in addition:

After vapor adsorption process, over hours at sufficient distance between the mica surfaces, the opposing mica surfaces are pressed together with the pressure value mentioned in the sub-figure of **Figure 1.39b**. A film of water molecules of thickness of about 0.45 [nm] is formed under this pressure. The authors describe the state of this film as similar to solid that cannot flow easily.

In following experiments, the stability of this film under sustained compression was tested. It is found that creeping occurs at which after hours, the layer thickness was reduced approximately to half. They conclude that the primary layer (of 0.45 [nm]) of the adsorbed water is unstable under the action of an external load and its property is similar to an ice-crystal structure that will creep under the action of an external load even with a smaller pressure.

Separation tests after compression were also performed. In this case, the separation between two rigid solid surfaces happens without any capillary condensation. It is found that the adhesion force increases with the relaxation time when the mica surfaces are bridged by the first layer of the adsorbed water film.

Figure 1.39a shows the results of time-dependent adsorption experiments: after 10 hours at 60% RH/25 °C an adsorption layer thickness of $D/2 \approx 0.70$ [nm] was formed on the mica surface due to the vapor pressure in the test chamber. These results on 12 independent measurements, each with new muscovite surfaces, show a time dependence of the sorption process. Information on the related air/vapor velocity in the measuring chamber was not provided.

In Section 1.4.6.4, the question was asked at the beginning: Can one really assume that the sorption film thicknesses at the pore walls to be expected in the water-filled pore region correspond to the vapor pressure-dependent adsorption film thicknesses that are present in the open pore fraction not yet filled with water?

From the present book section and the preceding Sections 1.4.6.4 and 1.4.6.5, **it is now clear that there is no correlation between the sorptive film thicknesses in these two regions.** Films from vapor sorption with thicknesses of up to about 1.7 [nm] can be expected at high RH, while in the water-filled region films formed on the pore walls have a structure deviating from bulk water up to thicknesses of at most about 0.70 [nm], and which exhibit ice-crystalline properties at thicknesses ≤ about 0.45 [nm]. According to previous findings, this is true for hydrophilic interfaces to pure water or aqueous electrolyte solutions with concentrations ≤ approx. 1M.

For higher concentrated aqueous electrolyte solutions, significantly higher film thicknesses must be assumed, compare [68].

The analyses of ion concentrations in the pore solution of HCP from Portland cements according to EN-197 yield concentrations of at most ≈ 0.30M, compare Lothenbach et al. [69] and Kenny/Katz [70]. The vast majority of ions in the hydrated state of the cements consist of potassium and sodium ions. The ion concentration in hydrated hardened cement paste decreases drastically when increasing amounts of fly ash are used [70].

From the experimental studies, especially by G. Zhao et al. [53], it appears that mutual compression from opposing surfaces makes the adsorption films thinner, but they apparently adopt an even more stable consistency.

Therefore, the question is to be asked whether, for example, the **decrease in permeability during a pressure test** with high pressures when determining the hydraulic conductivity can also be attributed to a parallel solidification of formed adsorption films, compare Section 3.1.8.1.

1.4.8 Influence of Adsorption Films on Meniscus Formation and on Capillary Pressure in Capillary Pores

In Section 1.3, the basic relationships of fluid uptake in porous materials as a result of surface energetic conditions were discussed. A distinction was made between fluid uptake due to capillary suction and fluid storage due to condensation from the penetrating vapor phase. The investigations in the previous sections showed that **different adsorption layer thicknesses can form behind a meniscus in the water-filled pore region and in front of the meniscus**, which meet each other in the region of the meniscus.

At $\varphi = 50\%$, according to Section 1.4.6.5, an adsorption film thickness of about 0.7 [nm] is present behind the meniscus, while in front of the meniscus, the adsorption film thickness from the vapor phase is about 0.50 [nm] (**Figure 1.28a**). This raises the question of the **consequences for the formation of the meniscus and its edge angle to the pore surface**, and thus for an influence on the capillary pressure. Related to this is the question of fluid storage potential by suction or capillary condensation.

Numerous publications have investigated **what shape the fluid surface of a meniscus takes at the transition to the pore wall**. Cited here are only the papers by Churaev et al. [52] and Kuchin and Starov [72], from which **Figure 1.40a,b** were taken.

In Figure 1.40a, the shape of the menisci at complete wetting (2) and partial wetting (1) with associated edge angle is given in the figure scale. Furthermore, an adsorption film h_e at the pore inner surface of thickness 1.85 [nm] is assumed there. This is a film thickness present at the basic isotherms for flat surfaces used here at a relative vapor pressure of $\varphi \approx 99\%$ (compare **Figure 1.27**). The corresponding calculated shapes of the liquid surface at the transition of the meniscus to the pore wall are given in detail for the partial wetting case in **Figure 1.40b**.

Particularly in the presence of larger film thicknesses of, for example, $t = 1.5$ [nm] (in front of the meniscus), the question remains, despite the theoretical modeling results of Kuchin (**Figure 1.40b**), for example, to what extent a force transfer from a meniscus to the pore wall in the longitudinal direction of the pores into the membrane surface of the film may be assumed.

This mainly concerns situations where liquids are transported by capillary suction in a pore system whose pore walls are covered by said liquid film. It must be taken into account that the liquid molecules below the film surface are largely in the bulk state in the case of thicker films. Notable forces in the form of shear stresses cannot

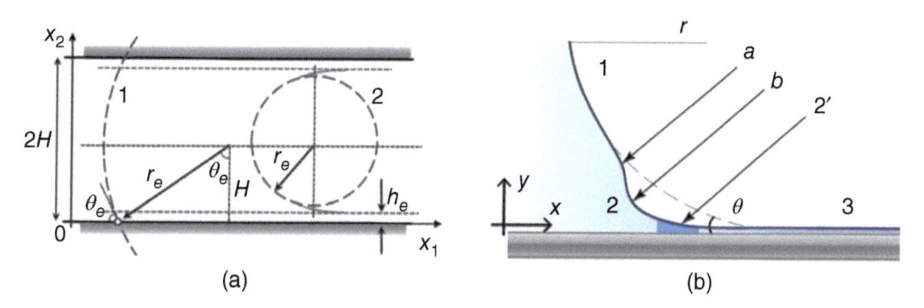

Figure 1.40 Figures and results of modeling from Kuchin et al. [71] and Kuchin and Starov [72]. (a) Representation of the shape of the menisci in total wetting (2) and in partial wetting with the corresponding contact angle θ_e. (1) (b) Liquid profile **in a capillary in the case of partial wetting** in the state of local equilibrium at excess pressure $P \neq Pe$; r and θ are the radius of the spherical meniscus in the central part of the capillary and the new local equilibrium contact angle ($\theta \neq \theta e$) (1) Spherical meniscus of a new radius r, where $r \neq re$; (2) profile of part of the transition zone at local equilibrium with the meniscus; (3) flat equilibrium liquid film of thickness he, with old equilibrium excess pressure Pe; (2′) a flow zone inside the transition region. Source: Kuchin et al. [71] and Kuchin and Starov [72]/American Chemical Society.

be transmitted to the pore wall in this region, if the results of research work are interpreted accordingly.

According to the results, for example of [73], [74], and [75], when shear stresses are applied, **slipping takes place between the film surface and the firmer approximately 2 molecule-layer-thick layer** on the material surface. Only when approaching the pore wall does the surface free energy in the adsorbed film increase according to the Sections 1.4.7, 1.4.6.1, or 1.4.6.4, and thus the ability to absorb the edge stress from the meniscus.

This is also clear in the approaches to disjoining pressure in adsorption films by, for example, Churaev et al. [52] and Kuchin et al. [71]. Also, Schimmele et al. [76] assumes that, especially at contact angles $\geq 0^o$, the meniscus edge force is introduced into the "load-bearing" pore wall.

From this, the assumption could be derived that the force transfer from a meniscus in the mentioned transport situation (with the occurring molecular slipping) is always transferred into the "solidly" adsorbed surface layer about 2 molecule layers thick, according to **Figure 1.41a,b**.

Systematic investigations of this question on capillaries with different inner film thicknesses, generated from different relative humidities of air, have probably not been realized so far because of experimental difficulties.

Behavior as assumed in **Figure 1.41a,b**, on the other hand, could be **unjustified in the case of capillary condensation in pore systems**, where increasing pore filling occurs not by suction but by condensation from the vapor phase. In this case, a stress equilibrium between meniscus edge force and pore-film surface is conceivable in the region of the meniscus transition, depending on local conditions, without any appreciable effect of a slipping influence.

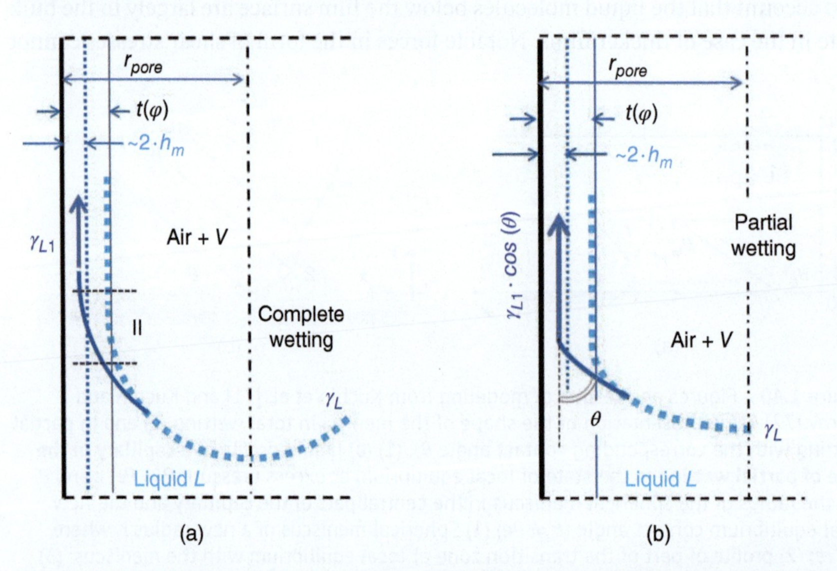

Figure 1.41 Modeling of the force transmission from the meniscus edge into the area of the contact zone. (a) Complete wetting with contact angle $\theta \approx 0^o$. (b) Partial wetting with contact angle $\theta \geq 0^o$.

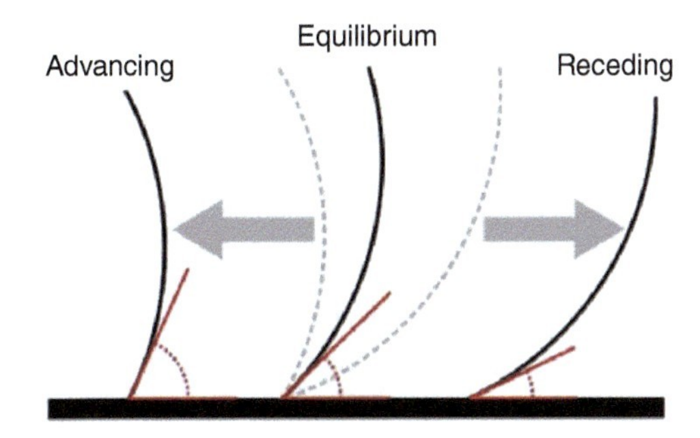

Figure 1.42 Schematic illustrated behavior of the shape of the menisci and the contact angles in a pore system during fluid adsorption (advancing meniscus) or fluid desorption (receding meniscus). Source: Andrew et al. [78]/with permission of Elsevier.

However, compared to equilibrium, a flattening of the meniscus takes place in the case of water absorption by vapor sorption and an increase of the curvature in the case of drying processes. Many publications attempt to model this process, see Starov and Velarde [77] and Kuchin and Starov [72]. Measured values on the phenomenon are reproduced, for example, in [78] and [79].

With capillary fluid uptake, a flatter meniscus and larger edge angle usually establishes itself because of the lower initial transport resistance, which transitions to the equilibrium position when transport stops. The degree of change of the edge angle depends on the particular transport resistance, see Popescu et al. [80] and **Figure 1.42**, for instance. Many publications are concerned with this question resp. the **shape of the meniscus and the contact angle as a function of liquid uptake or liquid release** in the pore system.

Different capillary pressures are associated with each of these meniscus shapes, which can differ significantly in the adsorption and desorption phases. Nevertheless, no influence on the hysteretic behavior of the moisture storage can be derived from these transient phases, since the respective equilibrium states are assigned to this behavior with the associated stationary menisci and contact angles. The condensate volume at the inner pore surfaces of the pore system forms according to Section 1.4, and because of the small influence of the discussed relation on the pore radii, **the dependencies between pore radius and capillary pressure or of Section 1.3 should be applied unchanged**.

References

1 Thommes, M., Kaneko, K., Neimark, A.V. et al. (2015). Physisorption of gases, with special reference to the evaluation of surface area and pore size distribution (IUPAC Technical Report). *Pure and Applied Chemistry* 87: 160.

2 Fagerlund, G. (2004). A Service Life Model for Internal Frost Damage in Concrete. Division of Building Materials. *Report TVBM*; Vol. 3119. Lund University.

3 Eriksson, D., Gasch, T., and Ansell, A. (2018). A hygro-thermo-mechanical multiphase model for long-term water absorption into air-entrained concrete. *Transport in Porous Media* 20: 573.

4 Carmeliet, J. and Roels, S. (2002). Determination of the moisture capacity of porous building materials. *Journal of Building Physics* 25: 209–237.

5 Krus, M. (1996). Moisture transport and storage coefficients of porous mineral building materials: theoretical principles and new test methods. *Reprint Dissertation (Stuttgart University) by Fraunhofer IRB Verlag.*
6 Scheffler, G.A. (2008). Validierung hygrothermischer Materialmodellierung unter Berücksichtigung der Hysterese der Feuchtespeicherung. Dissertation.Technische Universitt Dresden.
7 Espinosa, R.M. and Franke, L. (2006). Inkbottle Pore-Method: prediction of hygroscopic water content in hardened cement paste at variable climatic conditions. *Cement and Concrete Research* 36: 1954–1968.
8 Feldman, R.F. and Sereda, P.J. (1968). A model for hydrated Portland cement paste as deduced from sorption-length change and mechanical properties. *Materiaux et Constructions* 1: 509–520.
9 Ahlgren, L. (1972). Moisture Fixation in Porous Building Materials. *Report 36.* The Lund Inst. of Technology.
10 Welcome to DataPhysics-Instruments (2019). *Collection of Surface-Tension values of Liquids and Solids.* www.dataphysics-instruments.com (accessed 27 December 2023).
11 Pérez-Díaz, J.L., Álvarez-Valenzuelab, M.A., and García-Pradab, J.C. (2012). The effect of the partial pressure of water vapor on the surface tension of the liquid water–air interface liquid water–air interface. *Journal of Colloid and Interface Science* 381 (1): 180–182.
12 Portuguez, E., Alzina, A., Michaud, P. et al. (2017). Evolution of a water pendant droplet: effect of temperature and relative humidity. *Natural Science* 9: 1–20.
13 Chen, H., Li, Z., Wang, F. et al. (2017). Extrapolation of surface tensions of electrolyte and associating mixtures solutions. *Chemical Engineering Science* 162: 10–20.
14 Kinloch, A.J. (1980). Review the science of adhesion Part 1 : Surface and interfacial/aspects. *Journal of Materials Science* 15: 2141–2166.
15 van Honschoten, J.W., Brunets, N., and Tas, N.R. (2010). Capillarity at the nanoscale. *Chemical Society Reviews* 39: 1096–1114.
16 Barnat-Hunek, D. and Smarzewski, P. (2015). Surface free energie of hydrophobic coatings of hybrid-fiber-reinforced high-performance concrete. *Materiali in Tehnologije / Materials and Technology* 49: 895–902.
17 Kumikov, V.K. and Khokonov, K.B. (1983). On the measurement of surface free energy and surface tension of solid metals. *Journal of Applied Physics* 54: 1346–1350.
18 Yuan, Y. and Lee, T.R. (2013). Contact angle and wetting properties. In: *Surface Science Techniques, Springer Series in Surface Sciences 51*. Springer.
19 Lauth, G.-J. and Kowalczyk, J. (2016). *Einführung in die Physik und Chemie der Grenzflächen und Kolloide. Experimentelle Methoden: Messung der Grenzflächenspannung*. Springer-Spektrum.
20 Cwikel, D., Zhao, Q., Liu, C. et al. (2010). Comparing contact angle measurements and surface tension assessments of solid surfaces. *Langmuir* 26 (19): 15289–15294.

21 Extrand, C.W. and Moon, S.I. (2012). Which controls wetting? Contact line versus interfacial area: simple experiments on capillary rise. *Langmuir-Journal of the American Chemical Society* 28: 15629–15633.

22 Matsuoka, H., Fukui, S., and Kato, T. (2002). Nanomeniscus forces in undersaturated vapors: observable limit of macroscopic characteristics. *Langmuir* 18: 6796–6801.

23 Kim, S., Kim, D., Kim, J. et al. (2018). Direct evidence for curvature-dependent surface tension in capillary condensation: Kelvin equation at molecular scale. *Physical Review X* 8: 041046.

24 Blokhuis, E.M. and Kuipers, J. (2006). Thermodynamic expressions for the Tolman length. *Journal of Chemical Physics* 124: 74701.

25 Eslami, F. and Elliott, J.A.W. (2019). Gibbsian thermodynamic study of capillary meniscus depth. *Scientific Reports* 9: 657.

26 Maris, H. and Balibar, S. (2000). Negative pressures and cavitation in liquid helium. *Physics Today, American Institute of Physics* 53 (2): 29–34.

27 Caupin, F., Arvengas, A., Davitt, K. et al. (2012). Exploring water and other liquids at negative pressure. *Journal of Physics: Condensed Matter* 24: 284110

28 Israelachvili, J.N. (2011). *Intermolecular and Surface Forces*, 3e. Elsevier.

29 Rouquerol, F., Rouquerol, J., and Sing, K.S. (2014). *Adsorption by Powders and Porous Solids, Principles, Methodology and Applications*. Elsevier.

30 Adolphs, J. and Setzer, M.J. (1996). A model to describe adsorption isotherms. *Journal of Colloid and Interface Science* 180: 70–76.

31 Brunauer, S., Emmett, P.H., and Teller, E.J. (1938). Adsorption of gases in multimolecular layers. *Journal of the American Chemical Society* 60: 309.

32 Badmann, R., Stockhausen, N., and Setzer, M.J. (1981). The statistical thickness and the chemical potential of adsorbed water films. *Journal of Colloid and Interface Science* 82: 534–542.

33 Zandavi, S.H. and Ward, C.A. (2014). Clusters in the adsorbates of vapours and gases: Zeta isotherm approach. *Physical Chemistry Chemical Physics* 16: 10979–10989.

34 Zandavi, S.H. (2015). Vapours adsorption on non-porous and porous solids: Zeta Adsorption Isotherm Approach. Dissertation.University of Toronto.

35 Funk, M. (2012). Hysteresis der Feuchtespeicherung in porösen Materialien. Dissertation. Technische Universitt Dresden.

36 deBoer, J.H., Linsen, B.G., and Osinga, T.J. (1965). Studies on pore systems in catalysis. VI. The universal t-curve. *Journal of Catalysis* 4(6): 643–648.

37 Snoeck, D., Velasco, L.F., Mignon, A. et al. (2014). The influence of different drying techniques on the water sorption properties of cement-based materials. *Cement and Concrete Research* 64: 54–62.

38 Kumar, A., Ketel, S., Vance, K. et al. (2014). Water vapor sorption in cementitious materials—measurement, modeling and interpretation. *Transport in Porous Media* 103: 69–98.

39 Brunauer, S., Skalny, J., and Bodor, E.E. (1969). Adsorption on nonporous solids. *Journal of Colloid and Interface Science* 30: 546–552.

40 Pavlik, Z., žumár, J., Medved, I., and Černý, R. (2012). Water vapor adsorption in porous building materials: experimental measurement and theoretical analysis. *Transport in Porous Media* 91: 939–954.

41 Adolphs, J. and Setzer, M.J. (1996). Energetic classification of adsorption isotherms. *Journal of Colloid and Interface Science* 184: 443–448.

42 Adolphs, J. and Setzer, M.J. (1998). Description of gas adsorption isotherms on porous and dispersed systems with the excess surface work model. *Journal of Colloid and Interface Science* 207: 349–354.

43 Ridgway, C.J., Schoelkopf, J., Matthews, G.P. et al. (2001). The effects of void geometry and contact angle on the absorption of liquids into porous calcium carbonate structures. *Journal of Colloid and Interface Science* 239: 417–431.

44 Wu, C., Zandavi, S.H., and Ward, C.A. (2014). Prediction of the wetting condition from the Zeta adsorption isotherm. *Physical Chemistry Chemical Physics* 16: 25564–25572.

45 Churaev, N., Starke, G., and Adolphs, J. (2000). Isotherms of capillary condensation influenced by formation of adsorption films. *Journal of Colloid and Interface Science* 221: 246–253.

46 Mattia, D., Starov, V., and Semenov, S. (2012). Thickness, stability and contact angle of liquid films on and inside nanofibres, nanotubes and nanochannels. *Journal of Colloid and Interface Science* 384 (1): 149–156.

47 Li, J., Li, X., Wu, K. et al. (2016). Water sorption and distribution characteristics in clay and shale: effect of surface force. *Energy Fuels* 30: 8863–8874.

48 Teletzke, G.F. and Davis de L.E. Scriven, H.T. (1988). Wetting hydrodynamics. *Physical Review Applied* 23: 989–1007.

49 Kruk, M. and Jaroniec, M. (2002). Determination of mesopore size distributions from argon adsorption data at 77 K. *Journal of Physical Chemistry B* 106: 4732–4739.

50 Zhang, Y., Ouyang, X., and Yang, Z. (2019). Microstructure-based relative humidity in cementitious system due to self-desiccation. *Materials* 12: 1214.

51 Dash, J.G. and Peierls, R. (1982). Characteristics of adsorbed films. *Physical Review B* 25: 8.

52 Churaev, N.V., Setzer, M.J., and Adolphs, J. (1998). Influence of surface wettability on adsorption isotherms of water vapor. *Journal of Colloid and Interface Science* 197: 327–333.

53 Zhao, G., Tan, Q., Xiang, L. et al. (2015). Structure and properties of water film adsorbed on mica surfaces. *Journal of Chemical Physics* 143: 104705.

54 Xu, S., Simmons, G.C., Mahadevan, T.S. et al. (2009). Transport of water in small pores. *Langmuir* 25: 5084–5090.

55 Gruener, S., Hofmann, T., Wallacher, D. et al. (2009). Capillary rise of water in hydrophilic nanopores. *Physical Review E-APS* 79: 1–5.

56 Wu, K., Chen, Z., Li, J. et al. (2017). Wettability effect on nanoconfined water flow. *Proceedings of the National Academy of Sciences of the United States of America* 114 (13): 3358–3363.

57 Antognozzi, M., Humphris, A.D.L., and Miles, M.J. (2001). Observation of molecular layering in a confined water film and study of the layers viscoelastic properties. *Applied Physics Letters* 78 (3): 300–302.

58 Arai, T., Sato, K., Iida, A., and Tomitori, M. (2017). Quasi-stabilized hydration layers on muscovite mica under a thin water film grown from humid air. *Scientific Reports* 7: 4054.

59 Aktulga, H.M., Fogarty, J.C., Pandit, S.A., and Grama, A.Y. (2012). Parallel reactive molecular dynamics: numerical methods and algorithmic techniques. *Parallel Computing* 38: 245–259.

60 Farah, K., Müller-Plathe, F., and Böhm, M.C. (2012). Classical reactive molecular dynamics implementations: state of the art. *ChemPhysChem* 13: 1127–1151.

61 Castrillon, S.R.-V., Giovambattista, N., Aksay, I.A., and Debenedetti, P.G. (2011). Structure and energetics of thin film water. *Journal of Physical Chemistry C* 115: 4624–4635.

62 Israelachvili, J., Min, Y., Akbulut, M. et al. (2010). Recent advances in the surface forces apparatus (SFA) technique. *Reports on Progress in Physics* 73: 036601.

63 Peng, J., Guo, J., Ma, R., and Jiang, Y. (2022). Water-solid interfaces probed by high-resolution atomic force microscopy. *Surface Science Reports* 71: 100549.

64 Fukuma, T. (2015). Mechanism of atomic force microscopy imaging of three-dimensional hydration structures at a solid-liquid interface. *Physical Review B* 92: 155412.

65 Monroe, J., Barry, M., DeStefano, A. et al. (2020). Water structure and properties at hydrophilic and hydrophobic surfaces. *Annual Review of Chemical and Biomolecular Engineering* 11: 523–557.

66 Asay, D.B. and Kim, S.H. (2005). Evolution of the adsorbed water layer structure on silicon oxide at room temperature. *Journal of Physical Chemistry B* 109: 16760–16763.

67 Kimura, K., Ido, S., Oyabu, N. et al. (2010). Visualizing water molecule distribution by atomic force microscopy. *Journal of Chemical Physics* 132: 194705.

68 Martin-Jimenez, D., Chacon, E., Tarazona, P., and Garcia, R. (2016). Atomically resolved three-dimensional structures of electrolyte aqueous solutions near a solid surface. *Nature Communications* 7: 12164.

69 Lothenbach, B., Winnefeld, F., Alder, C. et al. (2007). Effect of temperature on the pore solution, microstructure and hydration products of Portland cement pastes. *Cement and Concrete Research* 37: 483–491.

70 Kenny, A. and Katz, A. (2020). Cement composition's effect on pore solution composition and on electrochemical behavior of reinforcing steel. *Israel Institute for Technology, Haifa, Israel, Reprint.*

71 Kuchin, I.V., Matar, O.K., Crasterc, R.V., and Starov, V.M. (2014). Modeling the effect of surface forces on the equilibrium liquid profile of a capillary meniscus. *Soft Matter* 10: 6024–6037.

72 Kuchin, I.V. and Starov, M. (2016). Hysteresis of the contact angle of a meniscus inside a capillary with smooth, homogeneous solid walls. *Langmuir* 32: 5333–5340.

73 Zhang, T., Li, X., Sun, Z. et al. (2017). An analytical model for relative permeability in water-wet nanoporous media. *Chemical Engineering Science* 174: 1–12.

74 Wang, F.-C. and Zhao, Y.-P. (2011). Slip boundary conditions based on molecular kinetic theory: the critical shear stress and the energy dissipation at the liquid–solid interface. *Soft Matter* 7: 8628–8634.

75 Feng, D., Li, X., Wang, X. et al. (2018). Capillary filling of confined water in nanopores: coupling the increased viscosity and slippage. *Chemical Engineering Science* 186: 228–239.

76 Schimmele, L., Napiórkowski, M., Dietrich, S. et al. (2007). Conceptual aspects of line tensions. *Journal of Chemical Physics* 127: 164715.

77 Starov, V.M. and Velarde, M.G. (2009). Surface forces and wetting phenomena. *Journal of Physics: Condensed Matter* 21: 464121.

78 Andrew, M., Bijeljic, B., and Blunt, M.J. (2014). Pore-scale contact angle measurements at reservoir conditions using X-ray microtomography. *Advances in Water Resources* 68: 24–31.

79 Siebold, A., Walliser, A., Nardin, M. et al. (1997). Capillary rise for thermodynamic characterization of solid particle surface. *Journal of Colloid and Interface Science* 186: 60–70.

80 Popescu, M.N., Ralston, J., and Sedev, R. (2008). Capillary rise with velocity-dependent dynamic contact angle. *Langmuir* 24: 12710–12716.

2

Real Pore Structure and Calculation Methods for Composition Parameters

2.1 Illustration of the Pore Structure of Selected Materials

The majority of porous materials, whose fluid storage or fluid transport are considered in this book, exhibit within a material section at most partially the cylindrical pore or slit pore shapes previously assumed theoretically. Therefore, it must always be checked to what extent the modeling of a material behavior, which is based on the described idealized pore properties, is actually applicable to the real pore system of the material or where the limits of the statement are. Therefore, it is important to know the real microstructure or the real pore structure as far as possible or necessary.

2.1.1 Pore Structure of the Synthetic Material MCM-41 and Selected Ceramic Bricks

First of all, the material MCM-41 should be mentioned, which consists of a densely packed, regular arrangement of unidirectional capillaries with a circular cross-section. This synthetic material MCM-41 appears to be predestined for the study of effects from physisorption.

Figure 2.1a shows a TEM image of a relevant material cross-section, as shown in **Figure 2.1b** of Reichinger [1] measured adsorption isotherms for nitrogen are reproduced.

Here, for the material with the cylindrical pores open on both sides with a radius of 2.2 [nm], a strong increase in adsorption is seen at a partial pressure of about 40%, followed by a nearly horizontal branch of the sorption line. This behavior corresponds to that described in Section 1.4.6.2 (compare **Figure 1.30a**, there for water vapor).

This is the spontaneous pore filling from condensation in the case of small cylindrical pores open on both sides. The rapid curve rise at 40% partial pressure corresponds to the partial collapse of the sorption layers from the concave inner

Moisture Storage and Transport in Concrete: Experimental Investigations and Computational Modeling, First Edition. Lutz H. Franke.

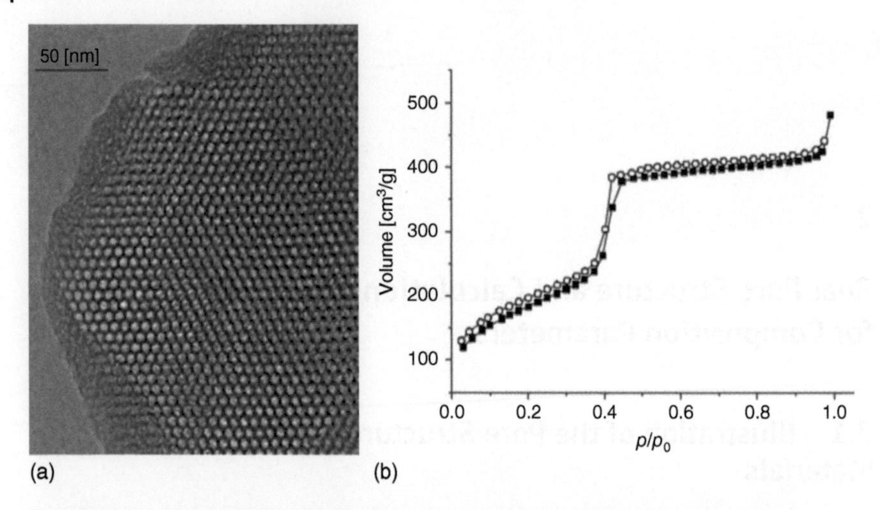

Figure 2.1 (a) TEM image of Reichinger of synthetic silicate MCM-41 with parallel cylindrical pores of constant radius of about 2.2 [nm] in uniform hexagonal material configuration. Source: Reichinger [1]/Ruhr University. (b) N_2-sorption isotherm of MCM-41, Rectangles = adsorption, circles = desorption. Source: Reichinger [1]/Markus Reichinger.

surface, which in the present case at about 50% partial pressure leads to extensive pore filling along the entire length.

The material group of ceramic bricks and brick products is known to have widespread use in construction, so for comparison, the microstructure and especially the porosity of these products will be considered here.

An impression of the microstructure of the surface produced by crushing of two ceramic bricks sintered to different degrees is shown in **Figure 2.2a,b** of Gomes-Sousa [2]. In particular, in the material of **Figure 2.2b**, one can see low porosity and concave inner surfaces of the visible pores. This is the typical microstructure of a clinker. After drying, the clinkers are fired at approx.

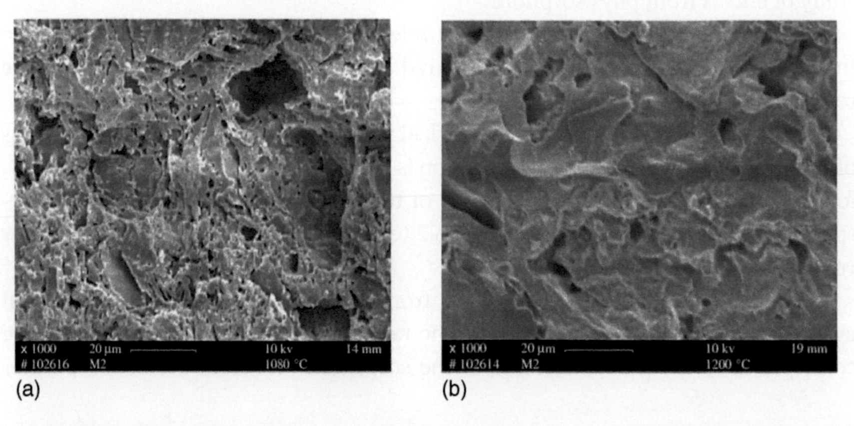

Figure 2.2 Microscopic pictures from fractured surfaces of partially sintered bricks (a) Specimens sintered at 1080 °C. (b) Specimens sintered at 1200 °C. Source: (a,b) Sousa and de Holanda [2].

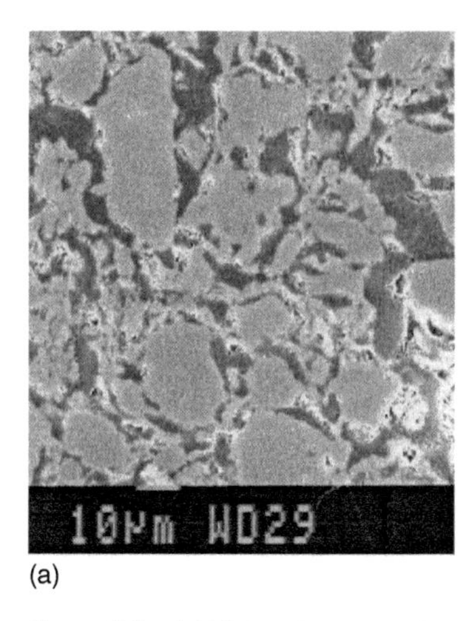

(a) (b)

Figure 2.3 (a) Light microscope image of the polished section of a ceramic brick, pores made visible by resin impregnation /TUHH. (b) Porosity of a standard brick from a historical facade, not polished cross section, large single pore $d \approx 150$ [µm] with grown gypsum crystals.

1100–1300 °C. For normal bricks, the firing temperature is usually only 800–1000 °C. The higher firing temperatures mentioned produce the sintered state with glassy, smoother, largely concave pore surfaces.

Figure 2.3a shows the surface impregnated with EP resin and subsequently polished of a brick material fired at about 950 °C with visibly higher pore volume. In this material, the pore surfaces are classified as slit-pore-like, with a significant proportion of concave pores.

Figure 2.4a,b shows two groups of pore size-dependent cumulative pore volume curves as a function of frost resistance, measured with MIP by Bentrup/TUHH, respectively. All brick samples were taken from buildings of different ages. The aim of the project was to compare the frost resistance of the brick material in practical

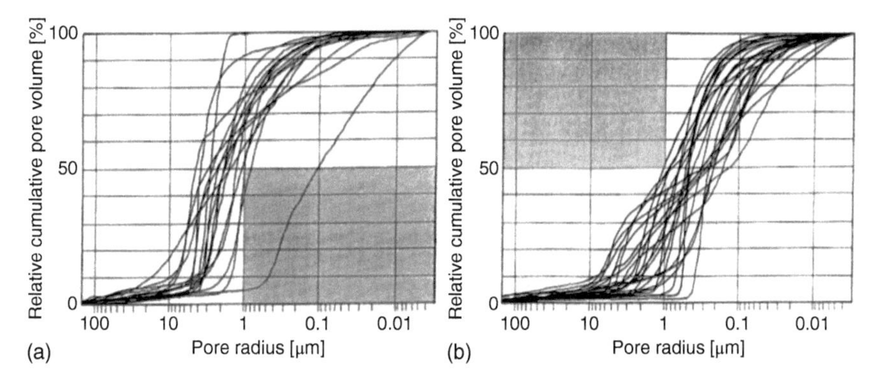

Figure 2.4 Two groups of MIP cumulative pore volume curves, relative to their total volume in %, results from Franke and Bentrup/TUHH [3],[4] (a) MIP curves of frost-resistant ceramic brick groups (b) MIP curves of frost-sensitive brick groups.

use with the MIP measurement results and to be able to estimate the frost resistance in a relatively simple way on the basis of MIP measurement results even before the material is used.

2.1.2 Results on the Pore-microstructure of Cement-Bound Materials

In the following, selected results of investigations on the question of pore formation and hydrate-phase structure of cement-bound materials are presented in extracts. These results are of importance, for example, for the assumption of pore-size boundaries between gel pores, capillary pores, and air pores and the interpretation of MIP measurement results, for the applicability of the Powers–Hansen method (Section 2.2), as well as for the applicability of the theoretical derivations or formulas according to Section 1.2 or for the modeling of moisture storage and transport (**Figure 2.5**).

In the following, Portland cement will be considered, which, together with the known additives, is the basis for the large number of cements in practice. The fineness of grind is important for the pore structure formed in the course of hydration. The starting material of Portland cement is cement clinker, which is obtained in pieces several centimeters in size from the burning process in the cement plant. The clinker consists mainly of the known mineral phases C_3S, C_2S, C_3A, and C_4AF. The phases C_3S (Alite), C_2S (Belite), and C_3A crystallize into granular structures as seen in the cement clinker in the ESEM image **Figure 2.6**.

After grinding, Portland cements CEM I 32.5 has grain sizes of ≤100 [μm] (about 35 [μm] on average), and CEM I 52.5 has grain sizes of ≤30 [μm]. It follows that the cement grains do not usually consist of a mixture of phases, but in large proportions of the individual phases alite or belite, which are contained in the clinker in proportions of about 60% and about 20%, respectively.

After the addition of water, the cement grains hydrate to form the hardened cement paste (HCP), the structure of which can be imagined approximately as

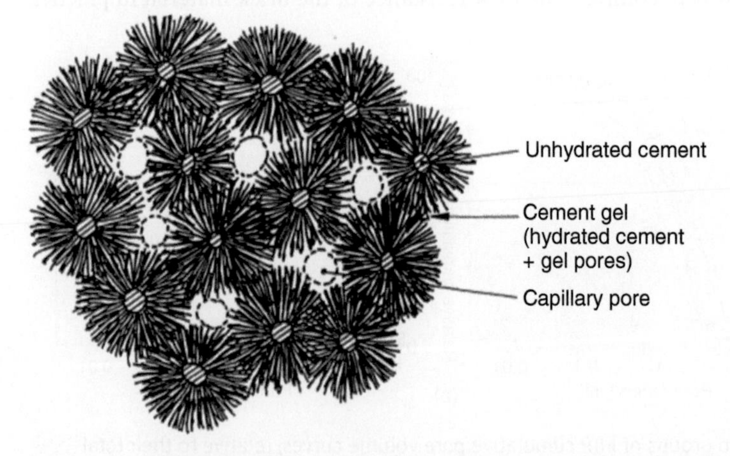

Figure 2.5 Microstructure of hardened cement paste HCP schematical. Source: Hedenblad [5]/Lund Institute of Technology/CC BY 4.0.

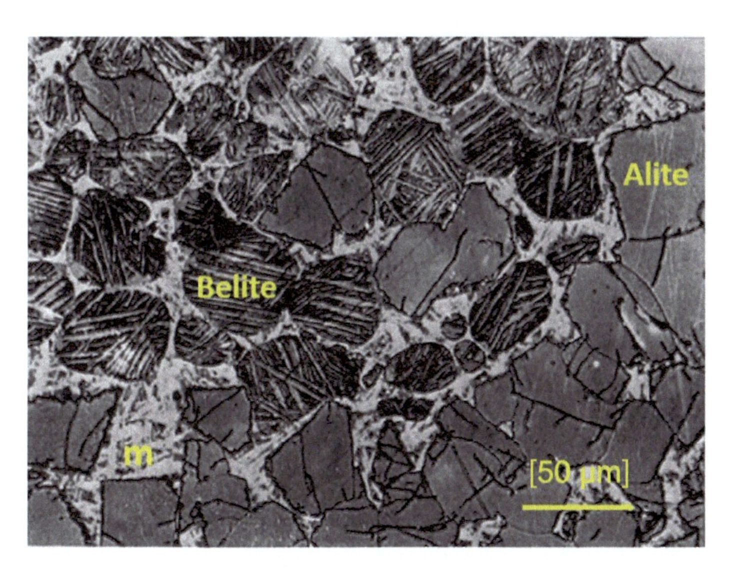

Figure 2.6 Microscopic picture of Portland cement clinker structure before grinding, showing the grains of crystallized C_3S (**Alite**) and C_2S (**Belite**) and the ground mass of solidified melt **(m)** of C_4AF (light gray) and C_3A (dark particles).

shown in **Figure 2.5**. In the meantime, numerous research results and publications are available to clarify the actual structure. Microscopic and spectroscopic investigations of the developing microstructure during the hydration of the cement to HCP are often carried out on selected cement grains, which are chosen via the microscope, predominantly alite grains because of their abundance, but also on synthetic C_3S.

Results of such microscopic studies are shown **in Figure 2.7a,b** [6]. Here, SEM is used to study the growth of an alite grain as a function of water–cement ratio, hydration time, and temperature. These two images show the hydration progress of the alite grain after one day and 90 days of hydration time.

These observations clearly show and confirm the view that a thin boundary zone forms at the initial surface of the cement grain, separating the structure of the CSH phases formed with increasing hydration time into a finer microcrystal structure within the cement grain and making the microcrystals more easily identifiable in ESEM or TEM growing out of the cement grain into the water-filled space.

Richardson and Groves, for example, in their 1993 [7] and 1999 [8] papers, emphasize this boundary zone with the product formed inside the zone Ip (Inner product) and outside Op (Outer product). **Figure 2.8** of Richardson 2004 [9] shows a TEM image from a C_3S cement paste hydrated over eight years with a W/C value of 0.40. Highlighted is the boundary between Ip and Op.

The fiber-like fine-porous fiber structure in the IP region and the band-like structure of the CSH phases growing into the water-filled pore space are clearly visible. The TEM image (**Figure 2.9**) shows the same band structure of the CSH phases in a HCP of β-C_2S hydrated over three months, with a W/C value of 0.40.

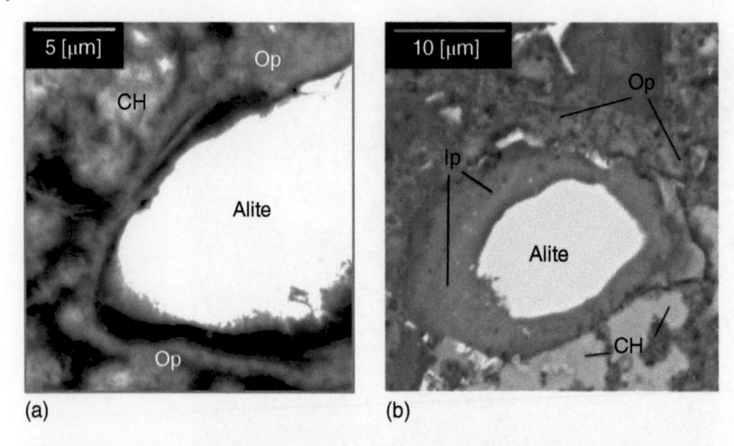

Figure 2.7 SEM pictures of alite grains in HCP of Portland cement with a W/C-ratio of 0.40. (a) Polished sample after one day curing; beginning of the Op- and Ip-zones around the grain. (b) Op- and Ip-zones after 90 days of hardening, areas of calcium hydrate CH. Source: (a,b) Rossen and Scrivener [6]/with permission from Elsevier.

Figure 2.8 Sample of hardened C_3S paste, $W/C = 0.40$, hydrated for eight years at 20 °C; the white arrows show the transition from the "inner product" Ip (top left) to the "outer product" Op (right), TEM. Source: Richardson [9]/with permission from Elsevier.

Richardson describes the C_3S-fiber structure in these images or in HCP from Portland cement as rather "fibrillar." He also notes that the morphology of CSH fibers is affected in the free pore space as they grow [9].

In slag-containing and pozzolan-containing systems, where the Op CSH phases have lower Ca/Si content, there is rather a "foil-like" morphology [9]. However, the TEM images also indicate that these are not foil surfaces in the true sense but rather disordered, intertwined ribbon structures.

Rossen et al. [10] show TEM images of 90-day hydrated cements of CEM I 32.5 with $W/C = 0.40$ (**Figure 2.10a**) and HCP of 60 C̃EM I, 20% Silica Fume, and 20% Quartz

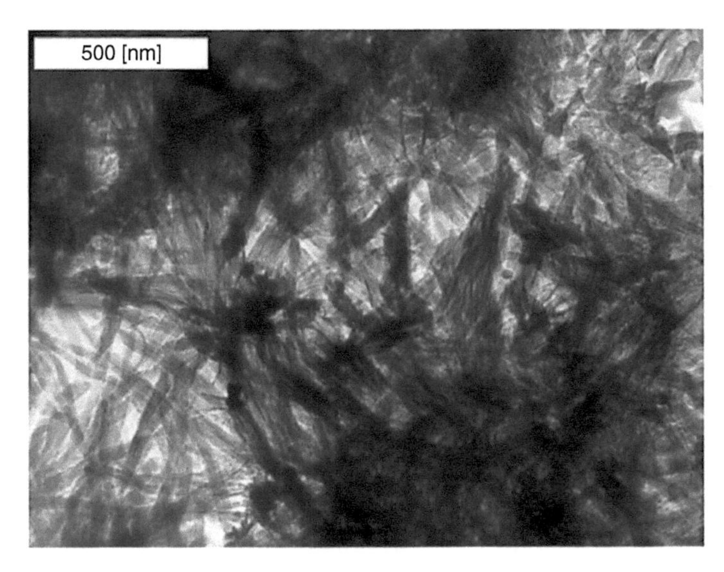

Figure 2.9 Op-CSH in a sample of hardened β-C_2S paste, $W/C = 0.40$, hydrated for three months at 20 °C; TEM. Source: Richardson [9]/with permission from Elsevier.

with $W/C = 0.50$ (**Figure 2.10b**). In particular, **Figure 2.10b** and other images included in [10] tend to suggest interwoven ribbon-like structures in the Op region labeled "foils." On the other hand, Taylor et al. [11] via TEM images on 20-year-old samples with very high blast furnace slag content found CSH structures that can be labeled as **foil-like**.

A significant question is whether the CSH phases described are amorphous or microcrystalline phases. X-ray diffractometer XRD studies show that the cement paste phases should be largely amorphous (X-ray amorphous), compare also Stark and Wicht [12]. From TEM studies, Richardson 1997 [13] deduces that CSH phases formed by water activation are nearly amorphous, while alkali activation (of fly ash, reactive silica, or blast furnace slag) results in semi-crystalline structures.

Rösler et al. [14] discover that CSH phases from C_3S hydration can also exhibit a crystalline structure at a water/binder ratio of 0.53 at 20 °C. A prerequisite for this finding was that, in order to prevent a structural change during the measurement, the radiation dose was reduced to a limiting minimum in the electron diffraction measurements on the TEM used for this purpose. They measure a tobermorite structure and assume that the CSH phases from C_3S-hydration have a (2-layer) layer silicate structure, although at present only models are used to describe the structure of CSH phases from hydrated alite or belite, and only a partial crystalline structure is expected.

In the hydration of Portland cements, i.e. not pure alite or belite, a predominantly amorphous gel structure of the CSH phases can still be assumed. This is underscored by the extensive studies of Fylak [15] on commercial Portland cements CEM I 32.5 to CEM I 52.5 and composite cements. **Figure 2.11a** shows the result of his cryoSEM observations on an HCP of CEM I 52.5, with further explanation.

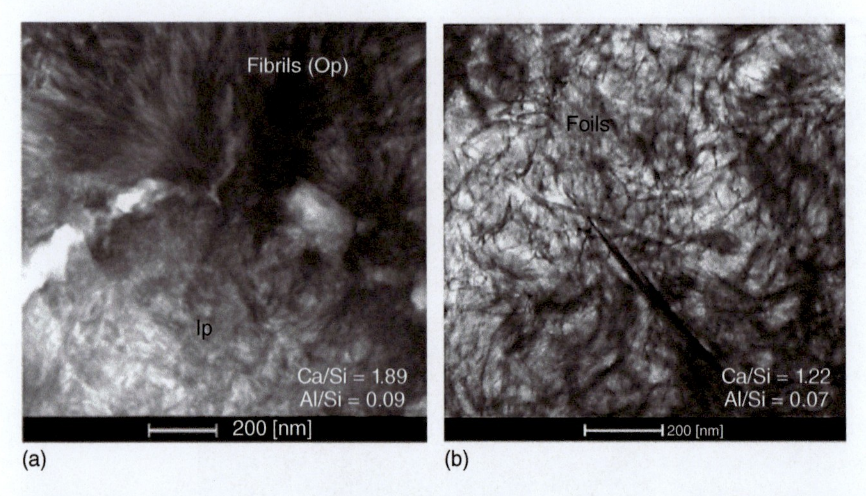

Figure 2.10 TEM pictures to characterize the different microstructure of CSH phases of the Op- and the Ip - zones. (a) HCP of CEM I 32.5, $W/C = 0.40$, 90 days of hydration at 20 °C. (b) Sample of HPC of 60% CEM I 32.5, 20% Silica Fume, and 20% Quartz, $W/B = 0.50$, 90 days of hydration at 20 °C; assumed "outer product" Op. Source: (a,b) Rossen et al. [10]/with permission from Elsevier.

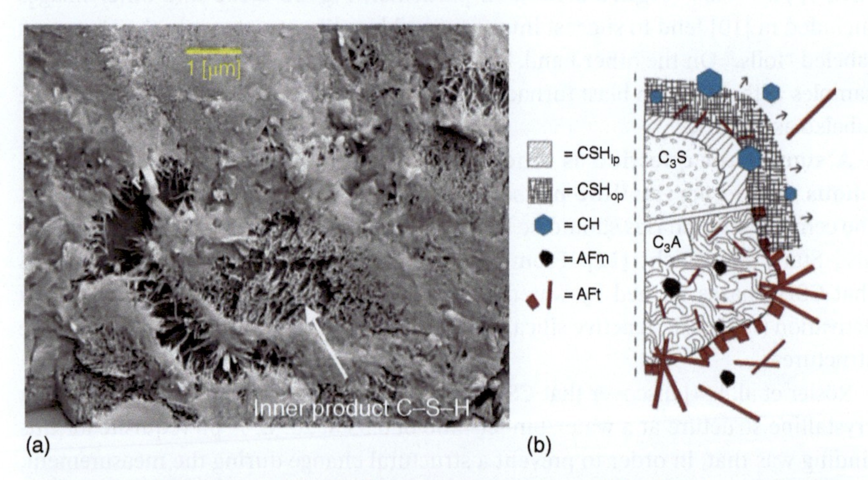

Figure 2.11 Hydration of HCP of **ordinary Portland cement CEM I 52.5**. Source: Results from Fylak [15]/University and State Library of Saxony-Anhalt/CC BY 4.0. (a) CryoSEM, $W/C = 0.50$, 1d hydration at 22 °C. (b) Schematic description of Op- and Ip formation: CSH of fiber-like morphology grows from clinker grain surfaces into the pore water space and forms the Op-CSH. At the same time, the corrosion and dilution processes of C_3S and C_2S particles forming open-pored Ip-CSH, grow into the grain interior. Depending on the cement, hydrate shells of different densities are formed around the clinker grains. These form diffusion barriers and slow down the reaction, followed by further formation of Op-CSH and Ip-CSH.

Figure 2.11b contains a schematic representation of the hydration of an alite grain with the hydrate phases formed, in particular also Op and Ip.

Results of transmission microscopic investigations on UHPC can be taken from Schaan et al./TUHH [16]. The W/C value of the investigated material is 0.24. More details on the composition can be obtained from **Figure 2.12**. The material samples investigated in STEM were prepared from test specimens (60 [mm] diameter) cylindrical in the center after at least 56 days of water storage. Separate drying of the material samples did not take place.

Despite the very high magnification, the structure can be identified down to a few nanometer, **Figure 2.13b**.

The STEM images of **Figures 2.13a,b and 2.14** are from the same UHPC material. This microstructure was selected from the region between two larger quartz-flour grains, where few micro-silica particles happened to be deposited when the material was mixed, whereas **Figure 2.14** was selected from a region with stronger concentration of micro-silica.

Regarding the hydration behavior of these UHPC specimens, it can be assumed that the investigated material specimens hardened without external water supply despite the water storage of the cylindrical test specimens from whose core they were taken.

From the given UHPC composition, the given W/C value = 0.24, and a degree of hydration of 0.40–0.50 (see **Figure 2.20**), a **shrinkage pore fraction from chemical shrinkage** of 16–21% can be estimated computationally for the material range of **Figure 2.13a,b**. These shrinkage pores fill with air but are not air pores from material fabrication, especially since these air pores have pore radii above about 2000 [nm] according to Section 4.5.

Therefore, there is much to suggest that the large pores of the order of 200–300 [nm] (visible in **Figure 2.13a,b**) are pores from chemical shrinkage.

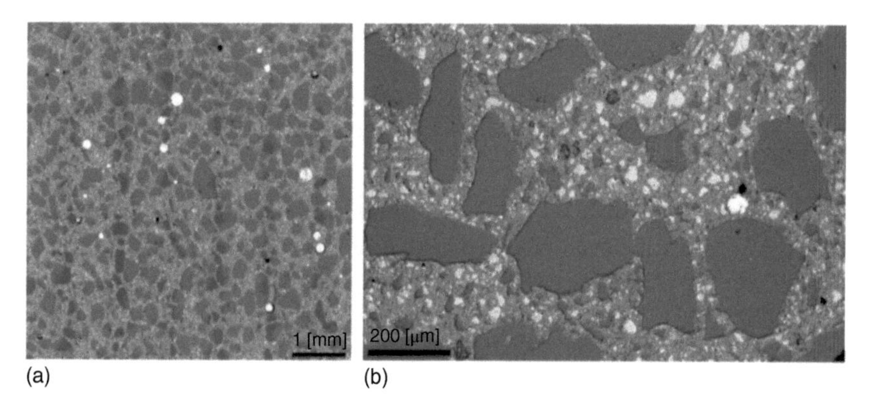

Figure 2.12 SEM overview images of the UHPC used; UHPC Composition: Portland cement CEM I 52.5 R, $W/C = 0.24$, aggregates quartz sand with grain size ≤300 [μm] and quartz powder ≤20 [μm] and silica fume with grain size ≤200 [nm], plasticizer carboxylatethter; samples cut from cylindrical specimens after 56d water prestorage; dark-gray irregularly shaped quartz grains and quartz powder, small silica fume particles not visible; gray ground mass of hardened cement, unhydrated light-gray cement areas. White points left side from gold sputtering. Source: Schaan et al. [16]/Technical University Hamburg.

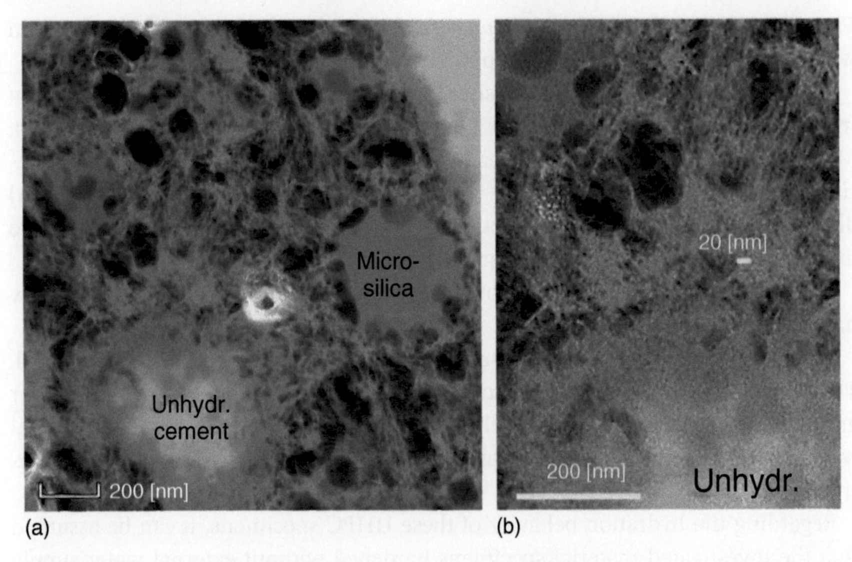

Figure 2.13 STEM high angle annular dark field (HAADF) images from a UHPC sample (composition and storage see **Figure 2.12** Discussion of the images in the text. (b) Detail of (a). Source: (a,b) Schaan et al. [16]/Technical University Hamburg.

Due to the cement used and the relatively long hydration time, a final state of hydrate phase formation can be largely assumed. Of particular note, in addition to the relatively high porosity of the material, is the round shape of the larger pores, which have a pore width (approx. pore diameter) between 40 and 100 [nm]. These pores are reached via finer pore entrances with pore widths of ≤ approximately 30 [nm]. In the Op region, the hydrate phases consist of fine fibers with thicknesses ≤ approximately 5 [nm]. The Ip region is visible but cannot be further structurally resolved. Porosity can be detected between the fibers in the op-region with pore widths of approximately 1–5 [nm].

2.1.3 Applicability of Theoretical Modeling to Cement-Bound Material Possessing the Morphology Shown

For the interpretation of the water uptake and water transport experiments and the acting mechanisms, the knowledge and, if necessary, definition of certain pore sizes of a capillary-porous system are helpful. This can be the size range of gel pores, an equivalent pore size for defining the beginning of capillary pores, the maximum capillary pore radius effective in a system, the effective boundary pore radius that can still be filled sorptively at a relative air humidity close to 100%, etc. Some of these pore sizes will be determined or defined below, preferably in context with the real pore structure presented in Section 2.1.2.

The fraction of closed pores in a pore system, which is formed in the course of hydration and cannot be reached by air or water transport, is also of importance. For this purpose, measurements were carried out, for example, Gluth [17] on different cements with different W/C values. Gluth confirms that porosity is strongly

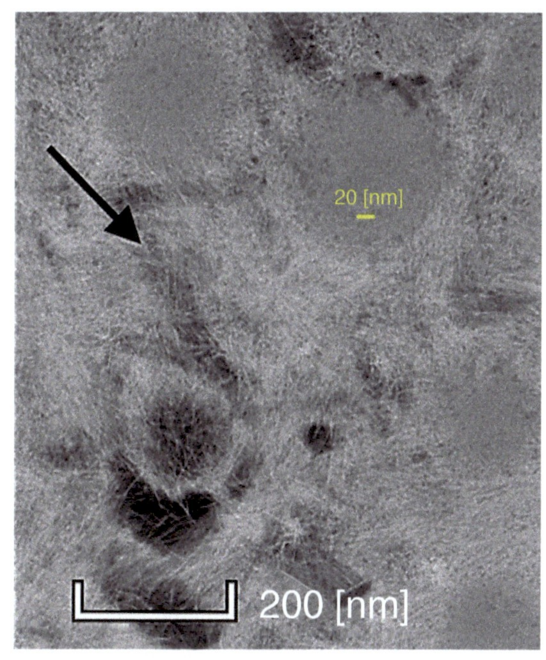

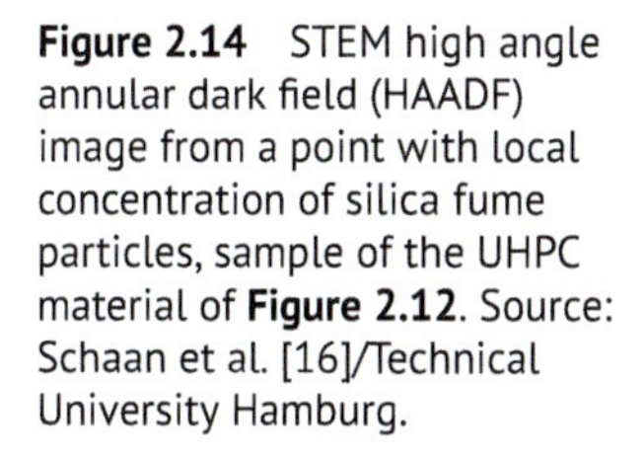

Figure 2.14 STEM high angle annular dark field (HAADF) image from a point with local concentration of silica fume particles, sample of the UHPC material of **Figure 2.12**. Source: Schaan et al. [16]/Technical University Hamburg.

dependent on the W/C value and the degree of hydration. The HCPs designated in **Figure 2.15** have a closed pore volume of about 5–10% at; however, different degrees of hydration. Significantly higher closed pore fractions were found for composite cements, especially at W/C values ≤ 0.30, as indicated, for example, in **Figure 2.14**.

The illustrations of the microstructure of the hydrated phases from C_3S (**Figure 2.8**), from C_2S (**Figure 2.9**), from HCP CEM1-52.5 (**Figure 2.11a**), and from UHPC (**Figure 2.13a,b**) show ribbon- and fiber-like shapes.

The question can therefore be **asked whether and to what extent the theoretical considerations carried out previously and equations derived for idealized pore shapes and structures are capable of accurately describing moisture uptake and transport** in material bodies from the microstructures shown.

Decisive for liquid uptake and storage is the effect of surface energies, shown on the spreading of droplets on solid material surfaces (Section 1.2.3), the transferability of this physical phenomenon to liquid uptake in pores, and the corresponding ability to overcome increasing potential energy resp. frictional resistance in material pores by adhesion work on the inner surface (Section 1.3.1). Model pores, as in Section 1.3.1, are used to characterize these processes in pores. Essential for the transferability of the theoretical descriptions obtained on model pores from Section 1.3 is the requirement that curved fluid surfaces ("menisci") are formed in the real pore systems, which are minimal surfaces and obey Laplace's law. As all experimental results show, this is also the case for the present real materials.

The "menisci" span in the irregular pores between the particles of the pore edges is canvas-like in the form of complex minimal surfaces with concave curvatures, so that capillary condensation takes place. Then, in principle, the Kelvin equation also applies, where r_{pore} can be considered as an equivalent "equivalent pore radius" according to Eq. (1.37). In this case, capillary pressures can also be specified for these assigned effective capillary pore radii, and capillary condensation can be calculated and measured according to Section 1.3.2.

In Section 1.4.6, approaches to describe the absorption film thicknesses arising in the pore system are presented, which are partly pore radius and curvature dependent, and need to be questioned with respect to the real structure of the pores:

Section 1.4.6.2 considers the influence of surface curvature on sorption film thicknesses in the water-filled region of a partially filled pore system. **The results shown are not transferable in this way** to a system of fibrous silicate hydrates and felted microcrystalline structures corresponding to the schematic representation in **Figure 2.5**. This is due to the fact that the inner surfaces are not concave as assumed at the beginning of this section, but directed into the pore spaces, which are largely convex. These models would be more applicable to mineral materials with other microstructures, such as brick materials, especially sintered products.

On the other hand, especially the film thicknesses in the water-filled pore region are derived on the basis of thermodynamic approaches and investigated or verified, for example, on the basis of AFM. From this, it is concluded that the determined sorption film thicknesses are also applicable to structures with dimensions of only a few nanometers in size, such as the fibrous CSH.

In Section 1.4.6.4, an approach for estimating film thicknesses in the water-filled pore-region is proposed that is pore-radius-independent and therefore applicable.

Section 1.4.4 calculates the inner pore surface area using a proposed method with extensive fitting to the pore shape and is therefore applicable in principle. The computational estimation of moisture storage fractions based on this method, which depends on the internal surface area, is applicable to pore systems of cement-bound products.

In Section 1.4.8, the influence of internal adsorption films on the transfer of edge forces from menisci to pore walls and capillary pressure formation is discussed using cylindrical capillaries. Transferability to other internal surface structures is given.

The informative value of the MIP analyses is, however, limited due to the known reasons. A semi-quantitative comparison of pore systems is, however, well suited. According to the comparative analyses here, the pore volume (porosity) indicated by an MIP analysis largely corresponds to the sum of capillary and gel pores (without the part of air pores). Further conclusions from MIP usually require results of additional observations.

For the transport properties toward liquids and gases, the pore radius r (Entry) denoted by r_{Entry} has a significant influence, so that this radius is used for predictions of the transport properties, compare Section 3.1.7 or, for example, estimating the suitability of HCP layers (membranes) with respect to gas separation resp. gas transport blocking [17].

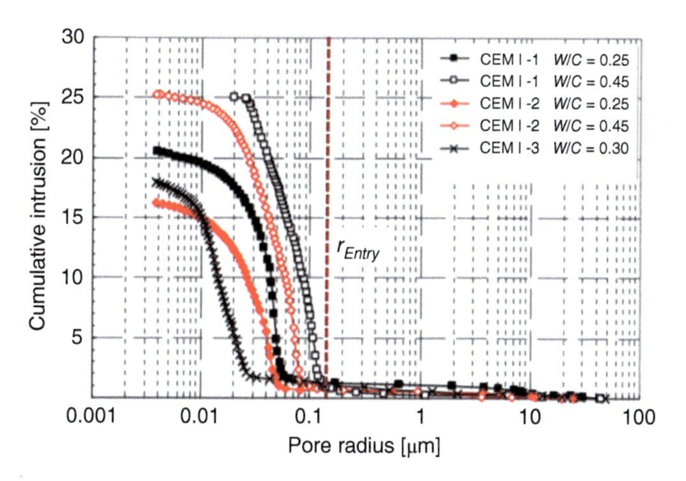

Figure 2.15 $r_{Entry} \approx 0.14$ [μm] as measure for the biggest mean effective capillary pore radius, samples of HCP of three ordinary Portland cements (1), (2), and (3) with different grain size and minimal content of air pores; **90% grain size of cement** (1): <50 [μm], (2): <36 [μm], (3) <19 [μm]. Source: Results from Gluth [17]/Universität Berlin.

The size of this maximum capillary pore radius can be extracted from MIP measurements. **r(Entry)** is the pore radius that must be overcome in the MIP analysis after filling processing pores or air pores at low pressure of mercury when entering the capillary pore system with increasing pressure. Investigation results on pore radius r_{Entry} can be obtained, for example, from Cook and Hover [18], Gluth [17], or Canut and Geiger [19].

Figure 2.15 shows cumulative MIP curves determined on cement paste material from different Portland cements and W/C values with an associated r_{Entry} of 0.14 [μm] = 140 [nm] for an HCP with $W/C = 0.45$. Cook and Hover [18] shows that this entry radius depends on the water–cement ratio and gives a value of $r_{Entry} \approx$ 300 [nm] for 28 days of water-stored HCP with $W/C = 0.70/$.

The applicability of r_{Entry} is highlighted in Section 3.1.7.

Figure 2.16a,b shows the MIP results determined on the concrete C 25/30 with $W/C = 0.60$ and the corresponding $r_{Entry} \approx 300$ [nm].

2.2 Calculations on Porosity, Degree of Hydration, and Material Densities

Powers and Brownyards [20] and Hansen [21] have worked out the well-known procedure for calculating the structural components resulting in the cement paste from Portland cement in the course of progressive hydration. The calculation results provide, depending on the hydration degree m, the water–cement ratio WCR (water–cement ratio) and the type of storage, as well as the volume and weight of cement paste components such as total porosity, capillary pore volume, gel pore volume, hydrated and non-hydrated cement fractions and, for example, bulk densities.

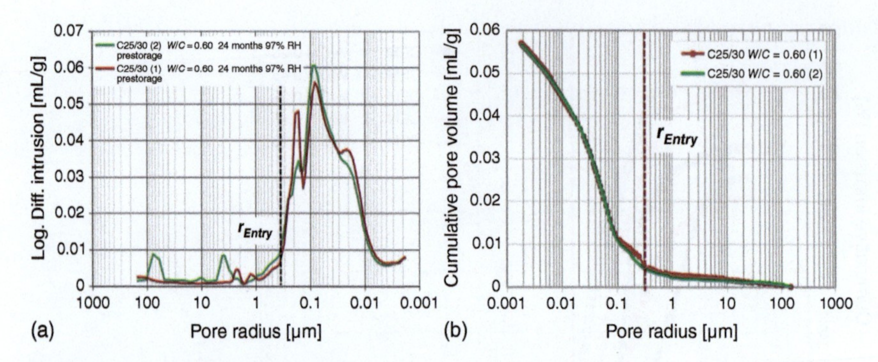

Figure 2.16 Identification of the radius r_{Entry} to concrete C25/30, $W/C = 0.60$ with $r_{Entry} \approx 0.30$ [μm] = 300 [nm] using (a) pore size distribution and (b) cumulative pore volume.

These data are important for a number of considerations, for example in the derivation of sorption isotherms with varying material compositions. The method has so far shown a weakness in the calculation of the proportions of capillary pores and gel pores, which according to Powers/Hansen result from a fixed ratio of physically bound to chemically bound water. Particularly, the previous definition of the gel water by Powers/Hansen can be questioned, compare Espinosa [22]. A proposal for the calculation of gel water is given below.

In the following Section 2.2.1, by way of introduction, the relations for the determination of the concrete mixture components are written down, taking into account production-related air content or air voids, extended to cement paste or concrete of "composite cements," compare [23] or [24].

2.2.1 Calculations of Initial Composition Parameters of Concretes

The total volume V_0 of all added components in the initial state of the mixture is $V_0 = 1\,[\text{m}^3] = 1000\,[\text{dm}^3]$. The designations or abbreviations used are:

Abbreviations:

CEM I	mass of the cement used in [kg/m³],	density $\varrho_{CEMI} = 3.05\text{–}3.1$ [kg/dm³]
W	initial water amount	in [kg/m³]
WC	water–cement ratio	

Other cements alternatively:

CEM II, CEM III …, CEMx in [kg/m³] with density $\varrho_{CEMII/III} = 2.80\text{–}3.05$ [kg/dm³]

Mineral additives:

SF	Silica fume [kg/m³],	density $\varrho_{SF} = 2.2\text{–}2.4$ [kg/dm³]
FA	Fly ash [kg/m³],	density $\varrho_{FA} = 2.2\text{–}2.6$ [kg/dm³]
GGBS	Ground-Granulated Blast-furnace slag [kg/m³],	density $\varrho_{GGBS} = 2.9$ [kg/dm³]
LS	Lime stone [kg/m³],	density $\varrho_{LS} = 2.6\text{–}2.7$ [kg/dm³]

Mass ratios m of the additives:

$$SF = m_{SF} \cdot CEMx \quad FA = m_{FA} \cdot CEMx \quad GGBS = m_{GGBS} \cdot CEMx$$
$$LS = m_{LS} \cdot CEMx \tag{2.1}$$

Total mass B of the binder with several additives:

$$B = CEMx \cdot (1 + m_{SF} + m_{FA} + m_{GGBS} + m_{LS}) \quad [\text{kg/m}^3] \tag{2.2}$$

Mass $CEMxx$ = the hydrating part of B, calculated with the hydrating ratiosk_{add}:

$$CEMxx = CEMx \cdot \left(1 + k_{FA} \cdot m_{FA} + k_{SF} \cdot m_{SF} + k_{GGBS} \cdot m_{GGBS}\right) \quad [\text{kg/m}^3] \tag{2.3}$$

Total volume of a binder with several additives in the initial mixture:

$$V_B = CEMx \cdot \left(\frac{1}{\varrho_{CEMx}} + \frac{m_{SF}}{\varrho_{SF}} + \frac{m_{FA}}{\varrho_{FA}} + \frac{m_{GGBS}}{\varrho_{GGBS}} + \frac{m_{LS}}{\varrho_{LS}}\right) \quad [\text{dm}^3/\text{m}^3] \tag{2.4}$$

Aggregates:
M_{Aggr} mass of the aggregates [kg/m³], density $\varrho_{Aggr} = 2.65\ (2.60\text{–}2.85)$ [kg/m³]

Air voids:
Sum of air voids = Air voids + Compactation voids:

$$V_{Air,sum} = V_{Air} + V_{Compact} = P \cdot V_0 = P \cdot 1000 \quad [\text{dm}^3/\text{m}^3]$$

The value of P in [%] of the initial volume of the sum of all concrete or mortar components

Not all mineral additives can be combined at will, both in terms of quantity and depending on the cement grade CEMx used. Also, as a rule, a minimum cement content must be provided. The background is that there must be sufficient $Ca(OH)_2$ in the concrete for hydration of the pozzolanic admixtures and granulated blast furnace slag, as well as sufficient basicity at the hydration end in the hardened concrete in view of corrosion protection of steel.

The water supply from storage conditions and a special environment (exposure class) can also play a role. Details on this, in particular limit values to be observed, can be taken from the European standard EN 206-1. The application area (building construction or hydraulic engineering) can also play a role. For example, when fly ash FA is used, the coefficient k_{SF}= 0.4 according to EN 206-1; for special applications in hydraulic engineering, k_{SF}= 0.7 can be considered.

For given values $CEMx$ **and** $WC = W/CEMx$ or in case of using mineral admixtures with given $CEMxx$ according to Eq. (2.3), resp. $WCxx = W/CEMxx$ and V_B according to Eq. (2.4), as well as an air pore volume P, the **content of rock aggregate** M_{Aggr} follows from (2.5)

$$M_{Aggr} = \left(1000 - WCxx \cdot CEMxx - V_B - P \cdot 1000\right) \cdot \varrho_{Aggr} \quad [\text{kg/m}^3] \tag{2.5}$$

For a given mass ratio aggregate/$CEMx$ and water-binder value $WCxx$, the cement content $CEMx$ follows from (2.6) to

$$CEMx = \frac{(1-P)\cdot 1000}{\frac{M_{Aggr}/B}{\varrho_{Aggr}}\cdot(1+m_{add}) + \left(\frac{1}{\varrho_{CEMx}} + V_{m_{add}}\right) + WCxx\cdot(1+k\cdot m_{add})} \quad [\text{kg/m}^3] \tag{2.6}$$

with

M_{Aggr}/B = specified Aggr/Binder ratio

$WCxx$ = specified Water/Binder ratio

$V_{m_{add}} = \frac{m_{SF}}{\varrho_{SF}} + \frac{m_{FA}}{\varrho_{FA}} + \frac{m_{GGBS}}{\varrho_{GGBS}} + \frac{m_{LS}}{\varrho_{LS}}$

$m_{add} = m_{SF} + m_{FA} + m_{GGBS} + m_{LS}$

$k\cdot m_{add} = k_{FA}\cdot m_{FA} + k_{SF}\cdot m_{SF} + k_{GGBS}\cdot m_{GGBS},\ k_{LS} = 0$

$k = (k_{FA}\cdot m_{FA} + k_{SF}\cdot m_{SF} + k_{GGBS}\cdot m_{GGBS})/m_{add}$

With $CEMx$ follows B from Eq. (2.2) and $CEMxx$ from Eq. (2.3).

If the binder is $CEMx$ **without further additives and only the water/cement ratio** WC **and the mixing ratio aggregate/cement** $M_{Aggr}/CEMx$ **are given for a concrete mixture** (as, for example, for the determination of the components for the reference mortar REF), then, with a known content P of the introduced air, the cement content is obtained as follows

$$CEMx = \frac{1000\cdot(1-P)\cdot\varrho_{Aggr}}{M_{Aggr}/CEMx + \left(\frac{1}{\varrho_{CEMx}} + WC\right)\cdot\varrho_{Aggr}} \quad [\text{kg/m}^3] \tag{2.7}$$

The required **cement content for the production of 1 [m³] of hardened cement paste HCP at a given** WC **value** and with the air content P of the introduced air results from Eq. (2.7), after $M_{Aggr}/CEMx$ was set to **equal zero.**

Once the cement weight and WC value are given, the aggregate particle size M_{Aggr} is known to be obtained from the following volume balance:

$$M_{Aggr} = \left(1000 - W - \frac{CEMx}{\varrho_{CEMx}} - P\cdot 1000\right)\cdot\varrho_{Aggr} \quad [\text{kg/m}^3] \tag{2.8}$$

The cement paste volume or the **C**ement paste volume (including air pore volume) **of a concrete or mortar** $V_{HCP,M}$ results consequently from the total volume under deduction of the aggregate grain size to:

$$V_{Aggr} = \frac{M_{Aggr}}{\varrho_{Aggr}} \quad V_{HCP,M} = 1000 - V_{Aggr} \quad [\text{m}^3] \tag{2.9}$$

As an example, the composition of the mortar REF is calculated here (compare Section 4.2). Table 4.1 contains the given values: $WC = 0.45$ and M_{Aggr}/CEM I = 3.0.

With these specifications and the value $P = 4.5\%$, $V_{m_{add}} = 0$, $m_{add} = 0$ and $\varrho_{Aggr} = 2.60$ [kg/dm³], the cement content of the REF mortar is obtained to be 495 [kg/m³], $W = 223$ [kg/m³] and the aggregate quantity $M_{Aggr} = 1483$ [kg/m³].

2.2.2 Required Hydration-Related System Parameters for the Method of Powers/Hansen

For the computational description of the internal volume changes associated with the hydration progress of the reactive mineral constituents of the cement and, if applicable, of the additives, the resulting porosity as well as the associated densities as a function of the storage conditions and the raw densities, in addition to the data on the material composition, the following material-related parameters are mainly required when necessarily the method of Powers/Hansen is used:

Wchem0	Necessary water percentage based on cement weight for the complete hydration of the cement clinker present
$\boldsymbol{k_{chem}}$	To describe the volume reduction of the water fraction incorporated during the formation of the hydrate phases, relative to *Wchem*0
$\boldsymbol{k_{gel}}$	Characteristic value to describe the gel pore content in the hydrated microstructure, related to *Wchem*0
$\boldsymbol{k_{phys}}$	Parameter for estimating the physically bound water, related to *Wchem*0
$\boldsymbol{m}$	degree of hydration

In the following, first the relations are given for the determination of the calculable material properties on the basis of the mentioned material parameters. Subsequently, the magnitudes of the parameters k_{chem}, k_{gel}, *and* k_{phys} are discussed, and limit values for the hydration degrees attainable are calculated as a function of the water/cement ratio.

2.2.3 Total Porosity of the Hardened Cement Paste or Concrete After Standard Drying

In the water-filled state of the pore system, the pore system is divided into gel pores V_{gel}, capillary pores V_{cap}, and pores P_{shrink} from chemical shrinkage. **The water fraction referred to as "physically bound water" is assigned to gel pores.**

If the HCP or concrete is dried by an appropriate drying procedure (for example, at 40 °C/1 mbar corresponding to Section 4.3.4), the dry material state is obtained, to which the dry bulk density is related and in which the hydrated cement phases contain only the chemically bound water *Wchem*0.

In contrast to the chemically bound water, the water content possibly contained in the previously mentioned pore groups is removed from the pore system during the drying process mentioned above.

In its dried state, the concrete or the cement paste portion of the material contains the following total porosity V_{pores}:

$$V_{pores} = W + P \cdot 1000 - m \cdot Wchem0 \cdot k_{chem} \cdot CEMx\,[\mathrm{dm^3/m^3}] \tag{2.10}$$

with the initial water content W, the initial air volume $V_{Air,sum} = P \cdot 1000 = P \cdot V_0$, the degree of hydration m, and the increase in volume of the hydrated cement fraction due to the water fraction incorporated during hydrate formation.

In the case of the reference mortar REF, which will be considered in more detail, storage with sealed surfaces of the specimen bodies was also carried out in comparison with water storage, whereby, among other things, the influence of water-filled air voids and the influence of the shrinkage pore volume on the dried water quantity can be observed.

2.2.4 Pore Fractions of Gel and Capillary Pores in the Total Pore Volume V_{pores}

According to Powers/Hansen, the total porosity is divided into gel pores V_{gel} and capillary pores V_{cap}. The extent of the gel pores is thereby calculated from the previously defined "physically bound water fraction." However, establishing a boundary between gel pores and capillary pores is a definitional issue, especially since there is a smooth transition of pore sizes between these pore groups. This issue, which has been discussed in the relevant literature, will be addressed later.

The gel pore volume V_{gel} within the total pore volume V_{pores} according to Eq. (2.10) can be simply written, for example, starting from Eq. (2.18) as follows:

$$V_{gel} = k_{gel} \cdot m \cdot Wchem0 \cdot CEMx \ [\text{dm}^3/\text{m}^3] \tag{2.11}$$

The volume of capillary pores in the water-saturated state of the material can now be calculated as the difference to the total pore volume as

$$V_{cap} = V_{pores} - P \cdot 1000 - V_{gel} \tag{2.12}$$

Using the relations (2.12) and (2.13), it follows:

$$V_{cap} = W - m \cdot CEMx \cdot Wchem0 \cdot (k_{gel} + k_{chem}) \ [\text{dm}^3/\text{m}^3] \tag{2.13}$$

with W the initial water content of the material mixture.

A part of these capillary pores consists of the **shrinking pores, according to Eq.** (2.14).

Powers/Hansen give a quantity for the cement-gel volume contained in the hydrated HCP and for the gel pore volume contained in the cement gel. They integrate into the calculation the physically bound water k_{phys} to the extent $\approx 3.3 \cdot V_m$. V_m corresponds to a mono-layer of water molecules on the inner surface of the cement paste, determined according to BET. The gel water used to calculate the gel pore volume is equated to the physically bound water by Powers/Hansen and related to the chemically bound water according to the procedure used in Section 2.2.3, calculated with the value $W_{gel} = 0.825 \cdot Wchem0$.

Determining the gel pore volume based on the internal surface area proves difficult. Instead, it can be more easily determined on the basis of limiting pore radii for gel pores or on the basis of permeability considerations, preferably taking into account adsorption isotherms.

It should be noted that the physically bound water fraction necessary for hydration is expressed here mainly via k_{gel}, approximately via k_{phys} (compare Section 2.2.7.4). Both parameters are defined here independently: k_{phys} via the derived internal surface area and sorbed water molecule layers (properties that can only be approximated), and k_{gel} by defining a pore-radius boundary between gel and capillary pores.

Even the latter boundary can only be an approximate determination, since there is a smooth transition between the pore groups.

Nevertheless, for formal reasons, a fixed boundary that is as physically plausible as possible should be chosen. According to the IUPAC report (Sing/Everett 1985 and Thommes et al. 2015 [25]), pores with pore **radii** between 1 and 25 [nm] are classified as meso-pores. Gel pores are also in this pore class. A relative air humidity of about 96% is assigned to the pore radius of 25 [nm]. By Espinosa and Franke [26], also based on investigation results of Jennings [27] this pore radius was chosen as formal boundary between gel porosity and capillary porosity. Further extensive investigation results on CSA structure and gel porosity are discussed by Jennings et al. [28].

The chosen pore size limit is justified in more detail in Section 2.2.7.3.

2.2.5 Chemical Shrinkage of the Hydrating Cement Product

The increase in volume of the clinker phases due to hydration is less than the initial volume of water *Wchem*0 incorporated into the hydrate phases. Taylor [29] assumes a volume reduction of water to 74%, Hansen [21] and others generally assume $k_{chem} = 0.75$.

The consequences of this mechanism modeled by the volume reduction coefficient k_{chem} are shown schematically in **Figure 2.17**. There, reference is made to **the effects of "chemical shrinkage" and "autogenous shrinkage."**

In the first hours of hydration in the plastic state, the chemical reaction is converted into an external volume reduction; in the subsequent sufficiently hardened

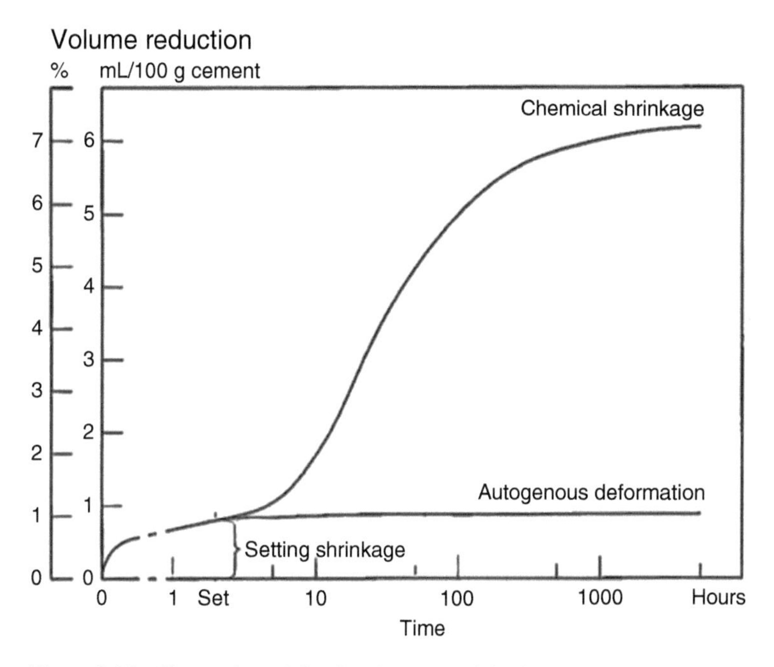

Figure 2.17 Illustration of the development of the internal (empty) volume (chemical shrinkage) and the external volume reduction (autogenous shrinkage) of cement paste. Source: Jensen and Hansen [30]/with permission of Elsevier.

state, the internal shrinkage pores are then formed as a result of the progressive hydration. When the outer surface of the cement paste specimens or concrete is sealed, the shrinkage pores are empty. In the course of the formation of these empty inner pores, liquid menisci are formed there, resulting in capillary pressure (as suction), which can lead to a corresponding external contraction of the material as **"autogenous shrinkage"** and to internal **micro-crack formation.** The resulting capillary suction can be estimated by measuring the internal relative humidity of the air. The process is also described as self-desiccation.

The autogenous shrinkage must be lower the more inert filler is present in the material, or stop when the material hardens sufficiently.

When test specimens are stored in water, the shrinkage pores in the edge zone fill with water, at smaller specimen dimensions, complete filling of the shrinkage pores can also take place. This process can occur synchronously with hydration, so that autogenous shrinkage does not occur.

The chemical shrinkage of the material, the smaller part of which consists of the autogenous shrinkage, continues until the limit hydration degree is reached. The condition of empty shrinkage pores always occurs inside concrete structural elements.

Section 2.1.2 shows that this porosity forms in the inner region (Ip) and especially in the outer region (Op) of hydrating clinker grains. The pores formed by chemical shrinkage can be assigned **proportionally to the gel and capillary pores**. In the following, it will be considered as a capillary pore part. However, at low water/cement ratios, there must be rather large gel pores. For the relationships presented here, however, this classification is of secondary importance.

The resulting pore volume due to chemical shrinkage can be estimated with very good approximation by the following Eq. (2.14).

Consider 1 [m³] of cement paste or fresh concrete with a water–cement ratio *WC*, then in the course of hydration (when water exchange at the surface is prevented) the initial liquid-water content W is reduced by the volume

$$V_{shrink} = m \cdot Wchem0 \cdot (1 - k_{chem}) \cdot CEMx \ [\mathrm{dm}^3/\mathrm{m}^3] \tag{2.14}$$

According to the Eq. (2.14), the total volume of shrinkage pores formed in the reference mortar REF during appropriate storage amounts to 29 $[\mathrm{dm}^3/\mathrm{m}^3]$. They thus amount to about 2/3 of the additional air pores still present in the mortar.

Further discussion of chemical shrinkage can be obtained from Section 2.2.7.2.

2.2.6 Calculation of the Densities ϱ_{wet}, ϱ_{humid}, and ϱ_{dry}

The bulk densities are related to 1 $[\mathrm{dm}^3]$ by applying $V_0 = 1000\,[\mathrm{dm}^3]$ in the Eqs. (2.15) to (2.17). **With surface sealing (foil storage)**, there is no moisture exchange with the environment. Then the density after 1 day or 28 days of storage is:

$$\varrho_{humid} = \frac{CEMx + W + M_{Aggr}}{1000} \ [\mathrm{kg/dm}^3] \tag{2.15}$$

With water storage or possible water supply via the surfaces, the density increases from the initial value Eq. (2.15) due to the filling of the shrinkage pores formed Eq. (2.14) and maybe partial filling ΔP of air pores:

$$\varrho_{wet} = \frac{CEMx + W + M_{Aggr} + P_{shrink} + \Delta P \cdot 1000}{1000} \text{ [kg/dm}^3\text{]} \tag{2.16}$$

The dry density of the material depends only on the degree of hydration and the air content and is thus largely independent of the type of storage, provided the type of storage is taken into account in determining the degree of hydration. The air content does not appear in the equation because it is included in the calculation of the concrete composition, thus influencing cement weight and aggregate weight. It has a relatively large influence on the raw dry density. The influence of the type of drying has already been pointed out before:

$$\varrho_{dry} = \frac{M_{Aggr} + CEMx \cdot (1 + m \cdot Wchem0)}{1000} \text{ [kg/dm}^3\text{]} \tag{2.17}$$

The following **Table 2.1** contains a comparison of calculated and measured characteristic values of the REF mortar and a Portland cement concrete. There is a very good agreement of the values of the products considered.

The Powers/Hansen method, somewhat modified in this chapter, **is the only one that can predict the treated properties with very good accuracy** on the basis of the original composition of a cement-bound product. The calculated results are often more accurate than scattering test results. However, the precondition is that, in addition to the known original composition, the degree of hydration m and, in particular, the initial air content P are entered with good approximation.

Table 2.1 Comparison of pore-volumes and densities of a mortar and a concrete made of Portland cement, calculated with the modified method of Powers/Hansen and determined experimentally.

	Reference mortar REF		CEMI 42.5 R – Concrete $M_{aggr}/CEMI = 5, W/C = 0.55$	
Property	**Calculated values**	**Experimental values**	**Calculated values**	**Experimental values**
V_{pores} [dm^3]	180.1	180.8	158.2	161.1[a)]
V_{gel} [dm^3]	94.8	95[b)]	64.8	67[b)]
V_{shrink} [dm^3]	28.5	[c)]	19.5	[c)]
Air-content P	4.2%[d)]	4.20%	2.5%[d)]	2.5%
ρ_{wet} [kg/dm^3]	2.25	2.25	2.31	2.30
ρ_{dry} [kg/dm^3]	2.103	2.10	2.17	2.16

a) 136.1 [dm^3] measured + 25 [dm^3] P.
b) From the corresponding sorption-isotherm at 92%RH.
c) Not measured.
d) Air-content of fresh mixture measured.

The method is also of importance in the conversion of sorption isotherms when the composition of the material varies. However, the accuracy of the method still needs to be verified **when several mineral additives are used simultaneously**.

2.2.7 Definition of the System Parameters for the Powers/Hansen Method

The system parameters used here are the material parameters mentioned at the beginning:

*Wchem*0, k_{chem}, k_{gel}, and hydration degree m. Their previously used magnitude will be reviewed and partially redefined in the following. The coefficient k_{phys} for estimations is also briefly considered.

2.2.7.1 Water Fraction *Wchem*0 Required for Hydration of Cement Clinker

The relatively simple-seeming theory of the composition of cement-bound materials according to Powers/Hansen [20, 21] as well as Gätje [23] nevertheless yields results that agree very well with reality or measured values, if certain boundary conditions are observed.

Here, first consider the amount of water *Wchem*0 required or chemically incorporated until complete hydration of a Portland cement. This can be determined by measuring the ignition loss of the completely hydrated and predried cement. If the predrying is carried out at 105 °C, the amount of water chemically required until complete hydration is obtained as $Wchem0 \approx 0.23$, based on the weight of the hydrated cement.

This value was also used by Powers and Hansen in their presentation of the fundamentals of the process considered here.

Because of the now generally known microstructure-altering effect of 105 °C drying on cement-bound products, compare Section 4.3.4, calculations based on this value do not lead to property values of the HCP or concrete that agree with tests. When milder drying procedures are used, higher values for *Wchem*0 result. Deviating suggestions are also made below regarding the amount of physically bound water still to be considered in the context of this theory.

Stark/Wicht show in [12] the water-binding of the mineral phases of Portland cement by considering the hydration reactions. After knowing the necessary amounts of water for hydration of the individual phases, the total amount of water required for complete hydration of all phase components of the cement can be determined from the known mineral phase composition of a Portland cement (calculated by Bogue from the chemical analysis of the Portland cement clinker). For the CEMI 42.5 R used as an example, the following phase composition, according to Bogue, was determined as mass fractions of the cement:

$$C_3S = 52.3\%, \quad C_2S = 20.5\%, \quad C_3A = 8.2\%, \quad C_4AF = 7.6\%, SO_3 = 3.0$$

The necessary amount of water for hydration of the C_3S fraction, for example, is $m_W = 0.276 \cdot C_3S$. Related to the CEMI 42.5 R : $m_W = 0.276 \cdot 52.3\% = 0.144$ $[g/g_{CEM\,I}]$.

For the complete hydration of the previously mentioned cement phases of CEMI 42.5 R, consequently, according to Stark/Wicht [12], the following water fraction is required:

$$Wchem0 = 0.144 + 0.043 + 0.072 + 0.028 = 0.287\ [g/g_{CEMI}]$$

Extensive experimental investigations on the maximum water-binding *Wchem*0 were carried out by Gätje/TUHH [23].

In addition to the theoretical calculation of the water binding of the individual clinker phases in the cement, the maximum hydrate-water contents were determined thermogravimetrically and checked. The investigations were carried out on samples from mortar prisms made of Portland cement CEM I 42.5% R with a water/cement ratio of 0.40, 0.50, and 0.60 from two cement plants A and AB up to an age of 3 and 3.5 years, respectively. Storage was continuous at approximately 100% RH and 23 °C. Any carbonated marginal areas were removed beforehand. Drying methods used for comparison were 105 °C and the much milder freeze-drying. Prior to performing thermogravimetric analysis (TG), the ≥3 year-old Portland cement mortar samples were "post-cured" over steam.

Evaluations of the measurements yielded *Wchem*0 values for fully hydrated mortar test specimens of approx. 23 mass% after 105 °C drying and 27% (plant AB) to 30% (plant A) after freeze-drying.

In the following, $Wchem0 = 0.28$ **[g/g$_{CEMx}$]** is assumed when appropriate resp. mild drying procedure is used, $Wchem0 = 0.23$ **[g/g$_{CEMx}$]** at 105 °C drying.

2.2.7.2 Volume-Reduction Coefficient k_{chem} and Chemical Shrinkage of the Hydrating Cement Product

The quality of the calculated estimation depends on the precision of the required parameter hydration degree m for the considered time point as well as on the parameter k_{chem}.

The value of k_{chem} is largely binder-dependent. The values given by Taylor [29] and Hansen [21] for Portland cements of 0.74 and 0.75, respectively, were rechecked against measured values. Suitable measured values are those of chemical shrinkage, which are extensively reported in the literature.

For comparison, among others, measured values of HCP from CEM I 42.5 N, $W/C = 0.50$ from Deschner et al. [31] were used on the shrinkage pore volume as a function of time. Measurements were made according to ASTM Standard C 1608-07 by following the water absorption of small cement paste samples during hydration. Parallel calorimeter measurements and measurements of chemically bound water were performed. The volume of chemical shrinkage is found to be $P_{shrink} = 0.070\,[dm^3]$ after a reaction time of more than 28 days. The chemically bound water *Wchem*0 was determined in [31] to be approximately 0.28 after a hydration time of over 100 days, a value that is also used here.

According to the comparative calculations, which are presented in more detail in Section 2.2.7.3, **for Portland cement, the most probable value is** $k_{chem} = 0.72$ to 0.73.

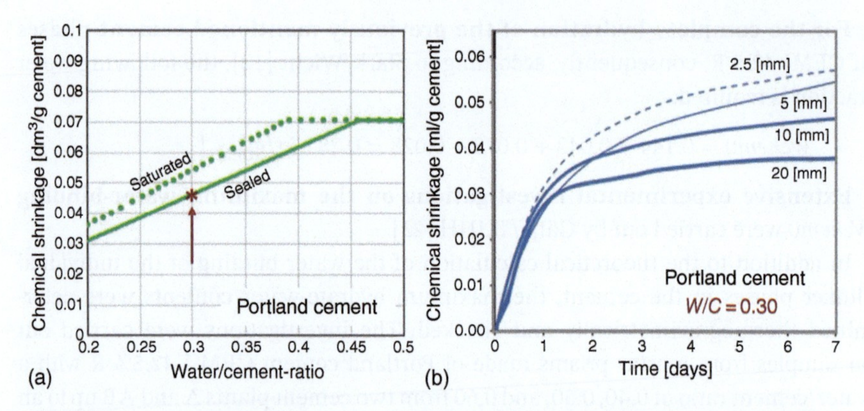

Figure 2.18 (a) Calculated maximum chemical shrinkage of Portland cement samples versus W/C-ratio, sealed and saturated conditions (assumed $m_{max} = 0.90$). The red star indicates the value to be expected in a comparative measurement with $W/C = 0.30$, compare measurement results from [32] in figure b. (b) Results of measurements on chemical shrinkage from Hui Chen et al. [32] with ASTM C 1608-07 procedure ($W/C = 0.30$, saturated samples with size of 2.5–20 [mm]). Source: (a,b) Chen et al. [32]/with permission of Elsevier.

Figure 2.18b shows results of chemical shrinkage measurements according to ASTM Standard C 1608-07 from [32] on Portland cement paste or HCP with $W/C = 0.30$. One can see the strong dependence of the measured values on the material-sample thickness used in a glass vial. According to [32], the sample thicknesses of 2.5 and 5 [mm] give the applicable measurement results. In these standardized tests, the samples are submerged in water. Since they are prepared by stirring, they also inevitably get air pores, which may also affect the measurement result. From these measurements, a shrinkage pore volume limit of approximately 0.055 can be derived for the W/C value of 0.30.

It should be noted that the numerous publications with results on chemical shrinkage give significantly diverging values for comparable boundary conditions. Therefore, it seems more reasonable to use the expected values from the present theory when in doubt.

Figure 2.18a shows the expected volume of shrinkage pores (including autogenous shrinkage) during hydration of Portland cement materials as a function of the water/cement ratio present, assuming a maximum degree of hydration of $m = 0.90$. It can be seen that an upper limit of the pore volume of approx. 0.071 [dm^3/g cement] is reached for water-saturated specimen bodies or in water-saturated areas from a W/C value of approx. 0.39, in the interior of larger bodies or for sealed surfaces only from a W/C value of ≈0.46. Taking $m = 0.87$ as a basis, an upper limit of the pore volume of 0.068 [dm^3/g CEM] results. The relationship to the respective water–cement value is also discussed in Section 2.2.8.

2.2.7.3 Definition of the Characteristic Value k_{gel} to Calculate the Gel Pore Volume

In recent work from 2015, Jennings et al. [33] propose to use desorption isotherms (of cement paste from CSH phases) to characterize pore groups. They suggest that the pore group that desorbs below 85% RH to 25% RH on drying should be referred to as gel pores.

However, this definition does not appear to be generalizable to cement-bound products. A look at the adsorption and desorption isotherms of the material REF. **Figure 4.24** shows that 85% RH related to the adsorption isotherm results in a water content that would give an RH of about 98% related to the adsorption isotherm. This definition would include a significant fraction of capillary pores.

Instead, it is proposed to define the **characteristic** k_{gel} **based on the adsorption isotherms from the water content at** 91–92% RH, corresponding to a **pore radius of 12.5 [nm]** or a pore width of 25 [nm]. This limit for the gel pore volume can be justified on the basis of permeability measurements as a function of the adsorption water content. This relationship will be discussed in more detail when defining the transport coefficients in the context of this book.

Analogous to the calculation of k_{phys} (see Eq. (2.22)), the value k_{gel} is also related to the required chemically bound water *Wchem*0.

Let the water content of the mortar product or concrete product at equilibrium after storage at 92% RH be W_{gel} [kg/m^3], experimentally related to a dry weight that justifies the use of a chemically bound water quantity of 28%. The standard drying used for this purpose is 40 °C and 1mbar air pressure.

With ***CEMx*** **= cement content** of the product and m = degree of hydration, it follows that

$$k_{gel} = \frac{W_{gel}}{m \cdot CEMx \cdot Wchem0} \tag{2.18}$$

The coefficient k_{gel} **is determined below from experimental results** on 3 different mortar or concrete products made from Portland cement:

1. **Reference mortar REF with** M_{Aggr}**/CEMI = 3**, WC = 0.45, CEMI 42.5 R = 497 [kg/m^3] 84d water storage, m = 0.82, $V_{Air,sum}$ = 4.2%.
   Measured W_{gel} = 95 [kg/m^3] from the adsorptions-isotherm at φ = 92% RH
   $$k_{gel} = 95/(0.82 \cdot 497 \cdot 0.28) = 0.832 \tag{2.19}$$
2. **Portland-CEM-concrete with** M_{Aggr}/CEMI = 5, WC = 0.55, CEMI 42.5 R = 349 [kg/m^3] 28d water storage, m = 0.80, $V_{Air,sum}$ = 2.5%,
   Measured $M_{92\%}$ = 3.10% relative to ϱ_{dry} from the adsorptions-isotherm at φ = **92%** RH $W_{gel} = M_{92\%} \cdot \varrho_{dry}$ = 3.10% · 2160 = 67 [kg/m^3].
   $$k_{gel} = 67/(0.80 \cdot 349 \cdot 0.28) = 0.856 \tag{2.20}$$
3. **Portland-CEM-concrete with** M_{Aggr}**/CEMI = 6**, WC = 0.55, CEMI 42.5 R = 307 [kg/m^3] 28d storage with total sealed surface, m = 0.80, $V_{Air,sum}$ = 3.5%.
   Measured $M_{92\%}$ = 2.65% relative to ϱ_{dry} from the adsorptions-isotherm at $\boldsymbol{\varphi}$ **= 92%** RH $W_{gel} = M_{92\%} \cdot \varrho_{dry}$ = 2.65% · 2190 = 58 [kg/m^3].
   $$k_{gel} = 58/(0.80 \cdot 307 \cdot 0.28) = 0.844 \tag{2.21}$$

Figure 2.19 shows the calculated pore fractions of a HCP CEMI as a function of the W/C value. It can be seen, for example, that the relative gel pore volume above a W/C value of 0.35 is independent of the W/C value.

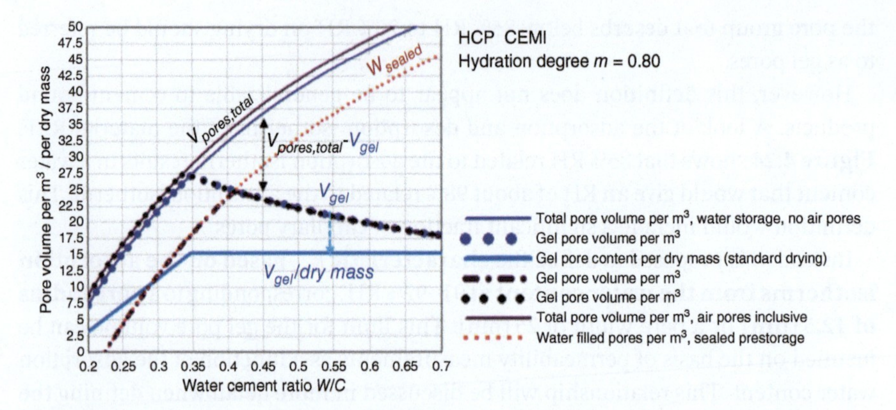

Figure 2.19 Calculated relation of the total pore volume $V_{gel} + V_{cap}$ with and without air pores (= water content after water storage) and the gel pore volume V_{gel} of HPCI (CEMI) to the W/C ratio. Reference values: 1 [m^3] and dry mass.

This result is also confirmed by the NMR measurements of Holthausen and Raupach in [34] on different cement mortars with W/C values between 0.40 and 0.70.

It is particularly interesting that these k_{gel} values almost coincide with the value for k_{gel} of 0.825 estimated by Powers/Hansen assuming an internal sorption film.

Based on these results, k_{gel} = 0.83 is defined to calculate the gel-water content of a cement-bound product, corresponding to a water content up to a pore **radius** of 12.5 [nm] or the **adsorption moisture at φ = 91.5%.**

2.2.7.4 Coefficient k_{phys} for Estimating the Physically Bound Water and the Necessary Water–Cement Value

One way to estimate the necessary water–cement value for complete hydration of the cement component is based on the coefficient k_{phys}.

As measurements have shown, **the physically bound water is necessary for the progress of hydration of the clinker phases.** This is due to the fact that even in the water-filled pore system, some molecular layers of water are bound to the inner surface to such an extent that they are not further available for internal transport and chemical conversion, compare Sections 1.4.7, 1.4.6, and 1.4.4.5.

From Section 1.4.4.5, it can be seen that when the pores are completely filled with water on the inner surface, a water fraction of about 0.8–0.9 [nm] thickness, corresponding to about 2–3 molecular layers, is predominantly solidly adsorbed. Assuming that for a hydration to progress freely, moving water fractions must be present, **the firmly adsorbed water amount can be the lower limit of an approach of physically bound water.** The adsorbed water quantity results from the experimental investigations on the reference mortar REF are as follows:

According to the evaluation in Section 1.4.3, the internal surface area of the mortar or its cement paste fraction is A_{BET} = 72.4[m^2/cm^3]. Assuming a sorbed layer thickness of 0.8 [nm], the adsorbed water fraction per [m^3] of REF mortar is $W_{sorp} = 0.8\,[nm] \cdot A_{BET}$:

$W_{sorp} = 72.4\ [m^2/cm^3] \cdot 10^6 [cm^3] \cdot\ 0.8 \cdot 10^{-9}$[= Layer thickness in m] $\cdot\ 10^3$ $[kg/m^3] = 57.9\ [kg/m^3]$.

W_{sorp} **is assumed here to be the physically bound water required for hydration**. Following the approach of Powers/Hansen or Taylor [29], this water fraction is related to the chemically bound water *Wchem* by **applying the factor** k_{phys} corresponding to

$$W_{sorp} = k_{phys} \cdot Wchem$$

From the hydrated cement fraction of the REF mortar, k_{phys} (with $m = 0.82$) is obtained as follows:

$$k_{phys} = \frac{W_{sorp}}{m \cdot CEMx \cdot Wchem0} = \frac{57.9}{0.82 \cdot 504 \cdot 0.28} = 0.50 \tag{2.22}$$

Therefore, $\boldsymbol{k}_{phys} = 0.50$ can be defined.

In Section 2.2.8, the dependence between achievable hydration degree and water–cement ratio has already been presented using the characteristic values k_{chem} and k_{gel}.

The following water–cement ratio would be at least necessary with the present approach k_{phys} for a complete hydration with $m = 1.0$:

$$WC = m \cdot (1 + k_{phys}) \cdot Wchem0\ = 1.0 \cdot (1 + 0.50) \cdot 0.28 = 0.42 \tag{2.23}$$

Further conclusions are hardly possible on the basis of this coefficient k_{phys}.

2.2.8 Dependence of the Attainable Degree of Hydration on the Water/Cement Ratio

Assuming constant temperature boundary conditions of, for example, 20 °C, the final degree of hydration attainable in a cement paste HCP of Portland cement depends mainly on the supply of water in the pores and on the grain size or grinding fineness of the cement.

At the beginning of cement hydration, all the initial water of the material mixture is available for hydration. As the reaction progresses, the pore space and the water content contained therein decreases. This pore space corresponds to the respective capillary pore volume according to Eq. (2.13), with the water content still available for further hydration of the microstructure therein.

This water fraction differs depending on whether a water supply from the material surface is possible or if a sealed material is present with shrinkage pores according to Eq. (2.14).

The hydration progress must come to a standstill when the volume of the still existing water-filled capillary pores approaches zero.

This follows from the fact that only the water-filled pore space present at the beginning of hydration can be used for the achievable volume increase of the hydrating cement phases, since solidification of, for example, a CEMI 42.5 R can already start after 60 minutes by forming a "coarse-meshed" hydrate phase framework, so that a subsequent increase of the pore space after the end of solidification after a few hours is not possible without structural damage.

The then existing hydration degree m_0 for HCP of Portland cement results from Eq. (2.13) in dependence of the water/cement ratio WC at water supply from the surface resp.

for the water-saturated state to

$$m_{0(WC)} = \frac{W_{(WC)} + \Delta P_{(WC)} \cdot 1000}{CEMx_{(WC)} \cdot Wchem0 \cdot (k_{gel} + k_{chem})} \tag{2.24}$$

In this $\Delta P_{(WC)}$ is the water content in %/m³, which corresponds to the proportion of air pores that were filled when water was supplied from outside, e.g. by water storage.

When the surface is sealed, the shrinkage pores created cannot be refilled from the outside, so the shrinkage pore volume must be subtracted from the initial amount of water available using Eq. (2.14). Furthermore, parts of the air pores cannot be refilled either. Consequently, m is given by the following Eq. (2.25).

$$m_{0(WC)} = \frac{W_{(WC)} - P_{shrink(m,WC)}}{CEMx_{(WC)} \cdot Wchem0 \cdot (k_{gel} + k_{chem})} \tag{2.25}$$

It can be seen that in Eq. (2.25), $P_{shrink(m,WC)}$ itself depends on m.

It is therefore necessary to calculate this pair of Eqs. (2.14) and (2.25) sequentially 2 to 3 times at most.

Example: **sealed surface:** Let an HCP resp. cement paste fraction of a $CEMx$ concrete be made with $WC = 0.40$, $\varrho_{CEMx} = 3.05$ [kg/m³], and an (estimated) air void content of 3%. For an HCP, Eq. (2.7) gives the required cement content as 1332 [kg/m³], and with $WC = 0.40$ the associated water content $W_{(WC)} = 553$ [kg/m³]. To calculate the shrinkage pore content, first choose an m-value of 0.95, which is larger than what was actually expected. Thus, from Eq. (2.14) $P_{shrink(m,WC)} = 99$ [dm³/m³] and for $m_{0(WC)} = 0.75$, the 2nd calculation with the previously determined value $m_{0(WC)}$ gives 0.79, and the subsequent calculation gives 0.78 as the attainable degree of hydration.

Example: water storage: The attainable degree of hydration at this boundary condition for the HCP considered previously is given by Eq. (2.24) (assuming $\Delta P_{(WC)} = 0$) to be $m = 0.92$, i.e. significantly higher than in the sealed condition.

In general, a lower maximum value than $m = 0.92$ is obtained for m at higher WC values:

The dependence of the maximum attainable degree of hydration for sealed and water-saturated areas for the range $W/C = 0.20$ to $W/C = 0.50$ is shown in **Figure 2.20.**

Accordingly, larger degrees of hydration at saturated conditions result when part of the air pores are filled with water, assumed in the diagram to be $\Delta P_{(WC)} = 2\% \cdot 1000$ [dm³/m³]. The gray area in the diagram labeled "max. possible hydration" indicates the range in which the degree of hydration may be

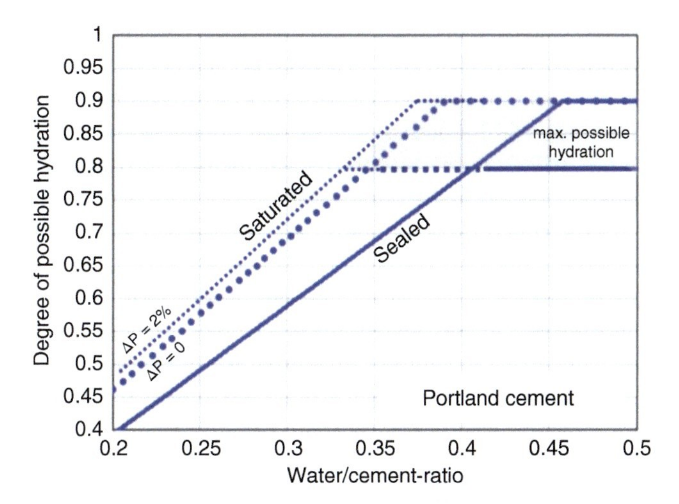

Figure 2.20 Calculated degrees of possible hydration of Portland cement samples, depending on W/C ratio, storage conditions, and effect of air voids.

limited, mainly by different grinding finenesses of the cement. This range, given between about $m = 0.80$ and 0.90, can also be further differentiated in the case of grinding fineness deviating from the norm.

The parallel calculated **limit curve of chemical shrinkage** can be taken from **Figure 2.18a** in Section 2.2.7.2.

2.2.9 Equations to Describe Hydration Kinetics

In Section 2.2.8, the question of maximum achievable hydration degrees was addressed. In a number of question, the rate of hydration degree increase to the maximum value is needed.

On the subject of the time-dependent development of hydration and cement structure, there are a large number of publications, which, however, are largely confined to shorter hardening periods not exceeding 90 days.

Stark provides an overview of theoretical studies on cement hydration and microstructure development published up to 2011 in [35]. Beaudoin and Odler provide a broad presentation on the chemical and mineralogical development of individual phases and cement Beaudoin and Odler [36] and Thomas et al. [37] on hydration modeling and simulation, respectively.

Long-term measurements of the hydration of Portland cement binders with the addition of latent hydraulic or pozzolanic constituents are scarcely available (in contrast to measurements of strength development for example).

In [Bentz.2006], Bentz investigates the possibilities of modeling the hydration rate or the resulting degree of hydration on the basis of selected material parameters. The predictions succeed approximately with deviations from the measured values in the order of 10–20%.

By Gätje/TUHH [23] also extensive **measurements were carried out on the hydration degree development** of HCP from Portland cement CEMI 42.5 R, measurements on the development of CEMI 42.5 R with coal fly ash addition up to 50 mass% up to a total duration of 22 months. In particular, long-term hydration observations were made on HCP from the blast furnace cements CEMIII/A 42.5 with 42.7% blast furnace slag content and from CEMIII/B 42.5 with 71.6% blast furnace slag content up to a total duration of 15 months. Storage was always at 100% RH and 23 °C.

The hydration grades *m* determined on HCP specimens from CEMI 42.5 R of producer A are shown in **Figure 2.21** for the investigated W/C values 0.30–0.60. For this HCP from CEMI 42.5 R, the maximum values after one year of hydration (at 100% RH/23 °C storage) for the W/C values 0.4 to 0.6 were more than 80%. After a curing time of three years, a degree of hydration of about 90% was reached at $W/C = 0.60$.

For modeling the hydration degree increase or related concrete properties it is useful to use mathematical functions.

The course of the hydration degree increase of the HCPI from CEMI 42.5 R is plotted (for the same measured values) in **Figure 2.21** on the left up to an evaluation time of 90 days on a linear scale, in the right picture up to the total measuring time of three years on a logarithmic scale. Only in the right-hand picture can the familiar S-shaped course of curing be recognized.

The formulation of the Eq. (2.26), which goes back to a formulation of Byfors [38] supplemented here and is also applicable for other time-dependent problems with a similar value course, proves to be a very well-adaptable function. $m_{basic}(t)$ **is here**

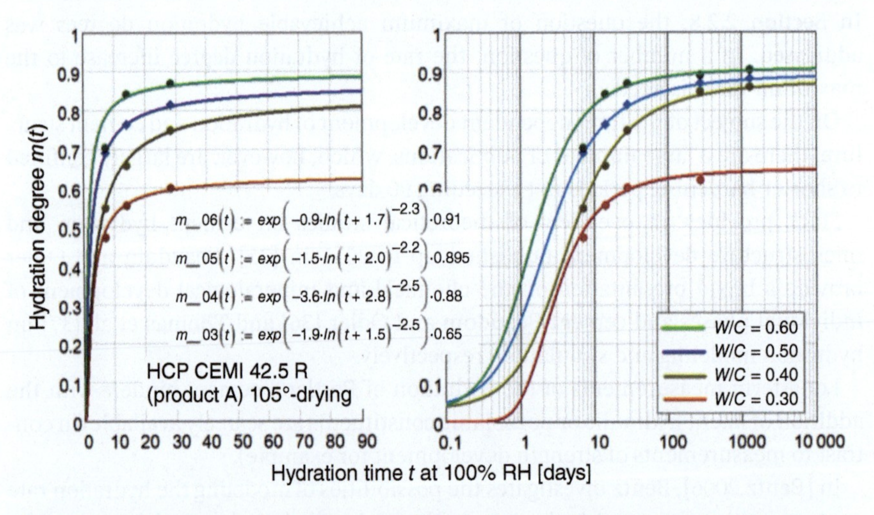

Figure 2.21 Development of the degree of hydration of HCP prisms made of Portland cement CEMI 42.5 R of cement plant A when stored at 100% RH and 23 °C. (right) Representation of the measured values from seven days up to three years for W/C values 0.30–0.60 on a logarithmic scale with the associated functions listed in the figure on the left. (left) Display of the function section from $t = 0$ days to $t = 90$ days in linear scale.

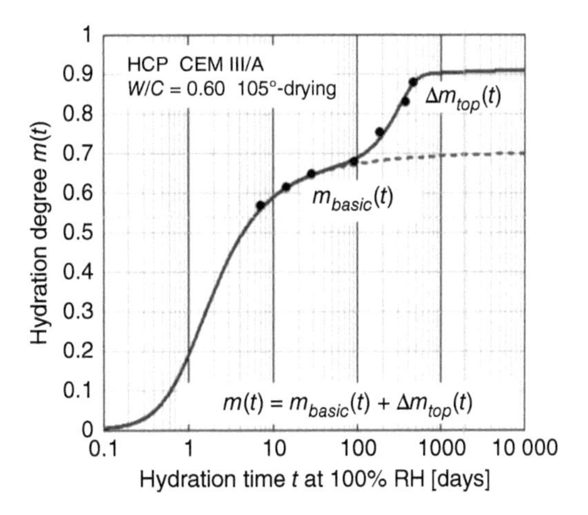

Figure 2.22 Model diagram for the determination of the mathematical functions as regression curves for the measured values in **Figures 2.21** and **2.23a,b** with reference to the Eqs. (2.28), (2.26), and (2.27).

the sought course $m(t)$ **with** t **in days**. The parameter $m_{basic,max}$ is the measured or assumed final degree of hydration for the respective W/C value.

The values $-k_0 \leq 1.0$ and Δt_0 together control the curve values at $t = 0.1$ and $t = 1$ days, respectively. e_0 determines the height position of the curve progression in the range $t = 10$ days. **The parameters** used can be obtained from the equations given in **Figure 2.21**.

$$m_{basic}(t) = exp\left(-k_0 \cdot [ln(t + \Delta t_0)]^{-e_0}\right) \cdot m_{basic,max} \tag{2.26}$$

A different behavior is shown by the determined hydration degree of development of HCP from the mentioned blast furnace cements in Figure 2.23a,b for the investigated W/C values of 0.40 and 0.60. (The mineralogical composition of the blast furnace slag used by the cement plant is given in [23]).

It is clearly visible that after about 100 days of hydration duration, where there seems to be an indication of reaching a limit of maximum hydration, there is actually an accelerated increase in hydration. This concerns both the CEMIII/A 42.5 cement with approx. 43% blast furnace slag content and the CEMIII/B 42.5 with 72% blast furnace slag content at both W/C values investigated.

The cause obviously lies in the interplay of calcium hydroxide supply and a delayed hydration of the blast furnace slag fraction.

$$\Delta m_{top}(t) = \left(1 - exp(-[k_1 \cdot (t - \Delta t_1)]^{+e_1})\right) \cdot \Delta m_{top,max} \tag{2.27}$$

$$m(t) = m_{basic}(t) + \Delta m_{top}(t) \tag{2.28}$$

A functional modeling of the hydration behavior according to Figure 2.22 can be accomplished via Eq. (2.28) by adding the basis function $m_{basic}(t)$ from Eq. (2.26) and the later onset resp. effective partial function $\Delta m_{top}(t)$ from Eq. (2.27). $m_{basic,max}$ **and** $m_{top,max}$ **are the fractions at maximum hydration.**

The value $-\Delta t_1$ in Eq. (2.27) specifies a time Δt (in days) from which this function becomes effective. Sample parameters can be obtained from the equations in **Figure 2.23a,b**.

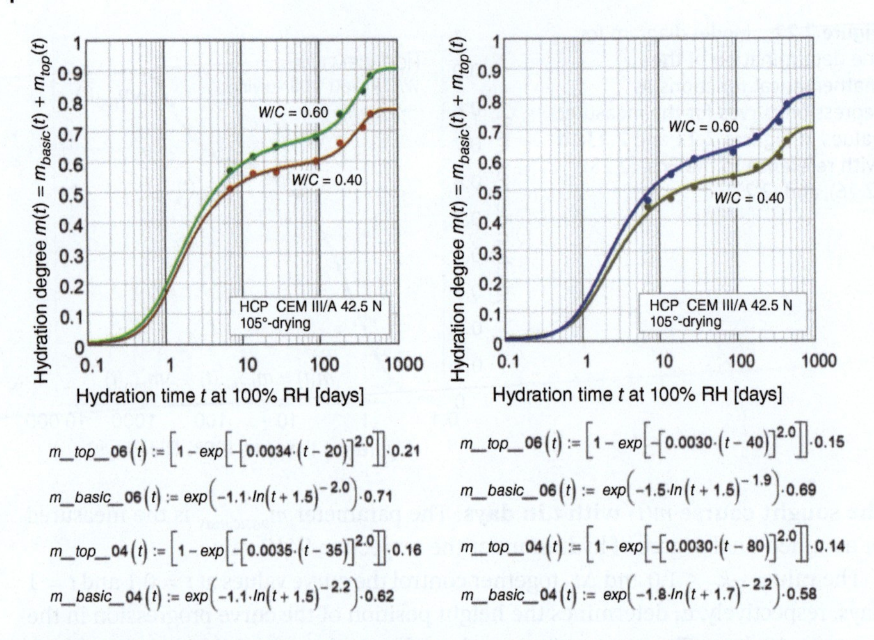

Figure 2.23 Evolution of the degree of hydration of HCP prisms made of blast furnace cement CEMIII/A with 43% blast furnace slag (a) and CEMIII/B with 72% blast furnace slag (b). Measured values from 7 days to 1.5 years are presented on a logarithmic scale, with storage in 100% RH and 23 °C. **Development of the regression functions** with Eq. (2.28). Plot of the functions with the parameters used below each figure.

The investigation results presented previously refer to the hydration degree of development when the material is stored at 100% RH.

By Adam [39], it was experimentally investigated on different concretes on the basis of non-evaporable water, which change of hydration development results, if after few days of storage at 100% RH a storage at constant lower relative humidity takes place.

Figure 2.24 shows results from [39] of measurements on concrete slices of thickness about 12 [mm], which after a few days of hardening at 100% RH were transferred to an air humidity of 50% RH/20 °C. Their water binding was followed up to 90 days. The accompanying rapid drying of the test specimens as a result of the storage transfer resulted in a significantly slowed hydration process. Subsequent storage at high air humidity or subsequent water storage would again cause an increase in the hydration rate.

In [39], **the temperature influence on hydration** is also considered.

2.2.10 Comparison of a Predicted and Measured Composition of a Cement Paste as a Function of the Degree of Hydration

Muller et al. [40] and Valori et al. [41] reported in 2013 on ^{1}H NMR (proton nuclear magnetic resonance) measurements performed on cement paste on the evolution of the composition of hydrate phases and pore evolution as a function of the degree

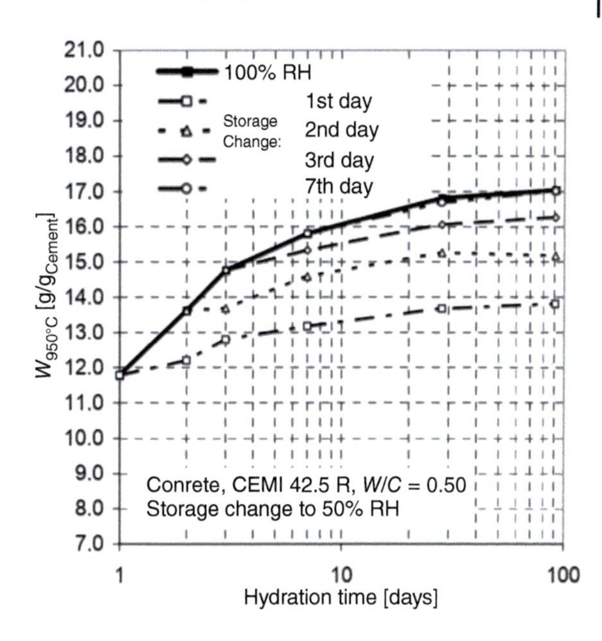

Figure 2.24 Measurements by Adam [39] of the hydration course of concrete slices transferred from 100% RH storage to 50% RH/20 °C after one to seven days after production. Indication of the course over the non-evaporable water, determined at 950 °C. Dimensions of the concrete slices are approximately 100 · 50 · 12 [mm^3]. The slices were previously dried at 105 °C. Source: Adapted from Adam [39].

of hydration of the HCP **of sealed material samples**. The advantage of this measurement method is that no structure-altering material drying method is required. The HCP material was prepared from a low-iron Portland cement (white cement) of 67% C_3S, 20% C_2S, 3.5% C_3A, and 4.4% calcium sulfate, sealed and hydrated. The *WC* values used were 0.32, 0.40, and 0.48.

Figure 2.25b of Muller et al. [40] shows the ^{1}H NMR measurement results resp. evaluations as relative values with respect to the total volume. For comparison, we

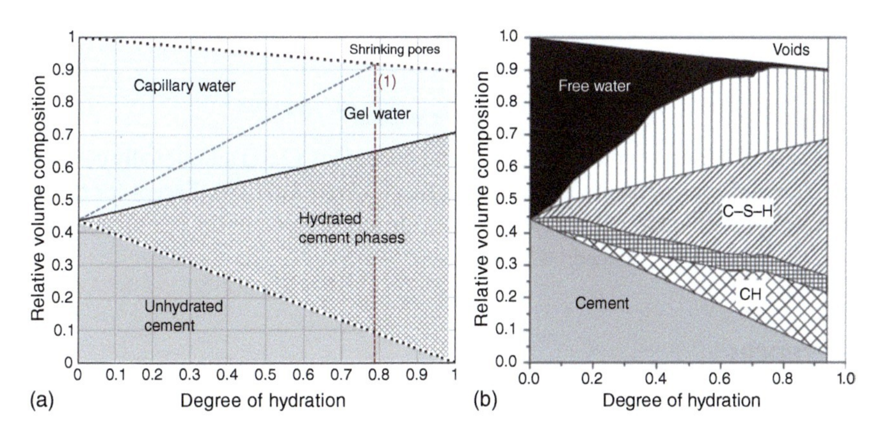

Figure 2.25 Development by volume of the components of hydrating white Portland cement samples as a function of the degree of hydration ($W/C = 0.40$, sealed storage), comparison of calculation and measurements. (a) Calculated with the reviewed method of Powers/Hansen. (1): **The vertical red line indicates the maximum attainable hydration degree m_{max} of HCP at the actual conditions.** (b) Results of ^{1}H NMR measurements from Muller et al. [40] and Valori et al. [41].

now checked which prediction results are obtained for the evolution of the composition as a function of the degree of hydration based on the modified Powers/ Hansen theory, as relative values related to $V_0 = 1000\,[\mathrm{dm}^3]$.

The calculation results, which largely agree with the measurements, are shown for the *WC* value 0.40 in Figure 2.25a. Only the partitioning of the pore content into gel pores and capillary pores shows notable differences in the measurements at hydration degrees below $m \approx 0.7$, where the measurements showed a nonlinear behavior in contrast to the computational estimation. If necessary, this behavior could also be modeled.

In **Figure 2.25a**, a vertical red line is drawn with the **label (1)**, indicating the maximum hydration degree attainable under these conditions at about 0.78, which is consistent with the results from Section 2.2.8.

The homogeneously distributed proportions of hydrated phases and of not yet hydrated cement are given by the following equations. Volume of hydrated cement phases:

$$V_{Hydr(WC,m)} = m \cdot CEMx_{(WC)} \cdot \left(\frac{1}{\varrho_{CEMx}} + Wchem0 \cdot k_{chem} \right) \tag{2.29}$$

Volume of the non-hydrated cement fraction:

$$V_{CEMx,Hydr(WC,m)} = \frac{CEMx_{(WC)} \cdot (1 - m)}{\varrho_{CEMx}} \tag{2.30}$$

The parameters used are: $Wchem0 = 0.28$, $k_{chem} = 0.72$, $k_{gel} = 0.83$.

The gel-water, capillary-water, and shrinkage pore contents were calculated for **Figure 2.25a** using the Eqs. (2.11), (2.13), and (2.14). As air-porosity content was 3.0%, was used and it was assumed that this value was also true in terms of size for the experimental material.

As a check, the sum of the proportions of the cement paste must always give V_0 or 1000 [dm^3] according to the following Eq. (2.31) with the values of m and W/C given:

$$V_0 = 1000 = V_{gel} + V_{cap} + V_{Hydr} + V_{CEMx,Hydr} + 1000 \cdot P \tag{2.31}$$

In the case of a mortar or concrete, the right part of Eq. (2.31) additionally contains the volume of the aggregate $V_{Aggr} = M_{Aggr}/\varrho_{CEMx}$ from Eq. (2.9). Except for the air void volume $P \cdot 1000\,[\mathrm{dm}^3]$, the proportions currently given in Eq. (2.31) then refer to the corresponding reduced HCP volume.

References

1 Reichinger, M. (2007). Poröse Silicate mit hierarchischer Porenstruktur: Synthese von mikro-/mesoporösen MCM-41 und MCM-48 Materialien aus zeolithischen Baueinheiten des MFI-Gerüststrukturtyps. Dissertation. Ruhr-Universität Bochum.

2 Sousa, S.J.G. and de Holanda, J.N.F. (2005). Sintering behavior of porous wall tile bodies during fast single-firing process. *Materials Research* 8: 197–200.

3 Franke, L. and Bentrup, H. (1993). Evaluation of the frost resistance of bricks in regard to long service life. II. *ZI International* 46 (9): 528–536.
4 Bentrup, H. (1992). Untersuchungen zur Prüfung der Frostwiderstandsfähigkeit von Ziegeln im Hinblick auf lange Lebensdauer. Dissertation. Hamburg University of Technology TUHH.
5 Hedenblad, G. (1993). *Moisture Permeability of Mature Concrete, Cement Mortar and Cement Paste*. Division of Building Materials. Forschungsbericht. Lund Institute of Technology.
6 Rossen, J.E. and Scrivener, K.L. (2017). Optimization of SEM-EDS to determine the C–A–S–H composition in matured cement paste samples. *Materials Characterization* 123: 294–306.
7 Richardson, I.G. and Groves, G.W. (1993). Microstructure and microanalysis of hardened ordinary Portland cement pastes. *Journal of Materials Science* 28: 265–277.
8 Richardson, I.G. (1999). The nature of C-S-H in hardened cements. *Cement and Concrete Research* 29: 1131–1147.
9 Richardson, I.G. (2004). Tobermorite/jennite- and tobermorite/calcium hydroxide-based models for the structure of C-S-H: applicability to hardened pastes of tricalcium silicate, h-dicalcium silicate, Portland cement, and blends of Portland cement with blast-furnace slag, metakaolin, or silica fume. *Cement and Concrete Research* 34: 1733–1777.
10 Rossen, J.E., Lothenbach, B., and Scrivener, K.L. (2015). Composition of C–S–H in pastes with increasing levels of silica fume addition. *Cement and Concrete Research* 75: 14–22.
11 Taylor, R., Richardson, I.G., and Brydson, R.M.D. (2010). Composition and microstructure of 20-year-old ordinary Portland cement–ground granulated blast-furnace slag blends containing 0 to 100% slag. *Cement and Concrete Research* 40: 971–983.
12 Stark, J. and Wicht, B. (2000). *Zement und Kalk. Der Baustoff als Werkstoff*. Basel: Birkhäuser.
13 Richardson, I.G. (1997). The structure of the calcium silicate hydrate phases present in hardened pastes of white Portland cement/blast-furnace slag blends. *Journal of Materials Science* 32: 4793–4802.
14 Rößler, C., Stark, J., Steiniger, F., and Tichelaar, W. (2006). Limited-dose electron microscopy reveals the crystallinity of fibrous C–S–H phases. *Journal of the American Ceramic Society* 89: 627–632.
15 Fylak, M.J. (2011). Untersuchungen zum frühen Hydratationsverhalten von Portland- und Portlandkompositzementen. Dissertation. Martin-Luther-Universität Halle-Wittenberg.
16 Schaan, G., Rybczynski, S., Ritter, M. et al. (2019). Transmission electron microscopy investigations of fatigue induced changes in the cement structure of ultra-high performance concrete. *41st Internatinal Conference on Cement Microscopy 2019*, San Diego, CA.
17 Gluth, G.J.G. (2011). Die Porenstruktur von Zementstein und seine Eignung zur Gastrennung. Dissertation. Technischen Universität Berlin.

18 Cook, R.A. and Hover, K.C. (1999). Mercury porosimetry of hardened cement pastes. *Cement and Concrete Research* 29: 933–943.

19 Canut, M.M.C. and Geiker, M.R. (2011). Pore Structure in Blended Cement Pastes. *DTU Civil Engineering Report; No. R-268*. Technical University of Denmark.

20 Powers, T.C. and Brownyards, T.L. (1948). Studies of the physical properties of hardened cement paste. Research Laboratories of the Portland Cement Association, Chicago Bulletin 22.

21 Hansen, K.K. (1986). *Sorption Isotherms. A Catalogue*. Technical University of Denmark. Building Materials Laboratory.

22 Espinosa, R.-M. (2005). Sorptionsisothermen von Zementstein und Mörtel. Dissertation. Technischen Universität Hamburg.

23 Gätje, B. (2004). Nachträgliche Ermittlung betontechnologischer Parameter in Zementstein, Mörtel und Betonen unbekannter Zusammensetzung. Dissertation. Technischen Universität Hamburg.

24 Schlüßler, K.H. and Mcedlov-Petrosjan, O.P. (1990). *Der Baustoff Beton*. Berlin: VEB Verlag für Bauwesen.

25 Thommes, M., Kaneko, K., Neimark, A.V. et al. (2015). Physisorption of gases, with special reference to the evaluation of surface area and pore size distribution (IUPAC Technical Report). *Pure and Applied Chemistry* 87: 160.

26 Espinosa, R.M. and Franke, L. (2006). Influence of the age and drying process on pore structure and sorption isotherms of hardened cement paste. *Cement and Concrete Research* 36: 1969–1984.

27 Jennings, H.M. (2000). A model for the microstructure of calcium silicate hydrate in cement paste. *Cement and Concrete Research* 30: 101–116.

28 Jennings, H.M., Bullard, J.W., Thomas, J.J. et al. (2008). Characterisation and modeling of pores and surfaces in cement paste: correlations to processing and properties. *Journal of Advanced Concrete Technology* 6 (1): 5–29.

29 Taylor, H.F.W. (1997). *Cement Chemistry*. London: Thomas Telford Publishing.

30 Jensen, O.M. and Hansen, P.F. (2001). Autogenous deformation and RH-change in perspective. *Cement and Concrete Research* 31: 1859–1865.

31 Deschner, F., Winnefeld, F., Lothenbach, B. et al. (2012). Hydration of Portland cement with high replacement by siliceous fly ash. *Cement and Concrete Research* 42: 1389–1400.

32 Chen, H., Wyrzykowski, M., Scrivener, K., and Lura, P. (2013). Prediction of self-desiccation in low water-to-cement ratio pastes based on pore structure evolution. *Cement and Concrete Research* 49: 38–47.

33 Jennings, H.M., Kumar, A., and Sant, G. (2015). Quantitative discrimination of the nano-pore-structure of cement paste during drying: new insights from water sorption isotherms. *Cement and Concrete Research* 76: 27–36.

34 Holthausen, R.S. and Raupach, M. (2018). Monitoring the internal swelling in cementitious mortars with single-sided ^{1}H nuclear magnetic resonance. *Cement and Concrete Research* 111: 138–146.

35 Stark, J. (2011). Recent advances in the field of cement hydration and microstructure analysis. *Cement and Concrete Research* 41 (7): 666–678.

36 Beaudoin, J. and Odler, I. (2019). Hydration, setting and hardening of Portland cement. *Lea's Chemistry of Cement and Concrete.*

37 Thomas, J.J., Biernacki, J.J., Bullard, J.W. et al. (2011). Modeling and simulation of cement hydration kinetics and microstructure development. *Cement and Concrete Research* 1257–1278.

38 Byfors, J. (1980). *Plain Concrete at Early Ages.* Stockholm: Swedish Cement and Concrete Institute.

39 Adam, Th. (2006). Ein Modell zur Beschreibung der Hydratation von Beton in Abhängigkeit vom Feuchtegehalt. Dissertation. Technischen Universität Darmstadt.

40 Muller, A.C.A., Scrivener, K.L., Gajewicz, A.M., and McDonald, P.J. (2013). Densification of C-S-H measured by ^{1}H NMR relaxometry. *Journal of Physical Chemistry C* 117: 403–412.

41 Valori, A., McDonald, P.J., and Scrivener, K.L. (2013). The morphology of C–S–H: lessons from ^{1}H nuclear magnetic resonance relaxometry. *Cement and Concrete Research* 49: 65–81.

3

Basic Equations for the Description of Moisture Transport

3.1 Moisture Flows at the Volume Element

For the calculation of the transport of liquids and gases in porous materials, Darcy's transport equations are used as basic equations. They provide the relationship between the amount of mass $\dot{m}$ transported per unit time in $[kg/(m^2\ s)]$ and the potential difference across the material body length Δx [m].

Depending on the driving potential, the appropriate transport coefficients must be used.

In the following, the hydraulic conductivity and the permeability derived from it are first discussed in more detail for the case of capillary pressure as the driving potential. In addition, the transport coefficient diffusivity is also illuminated for the case where water content is chosen as the driving potential.

In this book and in the following, the term moisture is used. This is an imprecise term for the water content which is the sum of liquid and vapor water, partially also in adsorbed form, in the pore system. In some cases, only the liquid water content is meant.

k_f = Hydraulic conductivity, dimension [m/s]: transport coefficient determined on a material of given porosity and given temperature, for instance, $\theta = 20\,°C$, in an experiment according to **Figure 3.1** resp. **Figure 3.2 with total water saturation** of the material.

K_l = Permeability, dimension $[m^2]$ is derived from the conductivity k_f, compare Eq. (3.8). The viscosity and density of the fluid used in the experiment (according to **Figure 3.1**) are taken into account at the temperature of the experiment.
This makes permeability a material property (**for the liquid-saturated state**) that is independent of the liquid used and is therefore often referred to as "**intrinsic permeability.**"
According to (3.7), the transport coefficient *ksat*0 can then also be easily derived from the permeability.

D_k = Transport coefficient, dimension [s], for a (capillary) pore system for liquids or liquid water according to Darcy, **for the saturated and the unsaturated states** of the pore system, compare Eq. (3.1).

Moisture Storage and Transport in Concrete: Experimental Investigations and Computational Modeling, First Edition. Lutz H. Franke.

ksat0 = Transport coefficient [s], corresponds to D_k for the **saturated state** of the pore system. Temperature dependence: see Eq. (3.7).

D_{w0} = Diffusivity, dimension [m²/s]: Transport coefficient for liquid water as driving potential according to Eq. (3.13) **for the saturated state** ("intrinsic diffusivity"), $D_w(w)$ or $D_w(S_w)$ **for the unsaturated state** of the pore system, where w = water content in [kg/m³] and S_w= saturation (value 0–1.0).

3.1.1 Capillary Pressure as Transport Potential

With exceptions, the capillary pressure Δp in [Pa] is chosen as the potential difference in this book.

The transport properties of the given material for liquids are considered in this case via the transport coefficient of conductivity D_k resp. the permeability derived from it or the vapor diffusion coefficient D_v:

$$\dot{m}_k = D_k \cdot \frac{\Delta p}{\Delta x} \quad [\text{kg}/(\text{m}^2 \cdot \text{s})] \tag{3.1}$$

Except in exceptional cases, the material-dependent transport coefficient must be determined experimentally. It is usually dependent on the moisture saturation level of the material. With respect to liquid transport, the transport coefficient is usually determined as shown schematically in **Figures 3.1 and 3.2**. The filter coefficient k_f (hydraulic conductivity), however, is determined when the material sample is completely saturated with water. The transport in non-water-saturated materials is determined accordingly via the Darcy equation, although there is a reduction in the transport coefficients compared to the reference value of permeability at complete

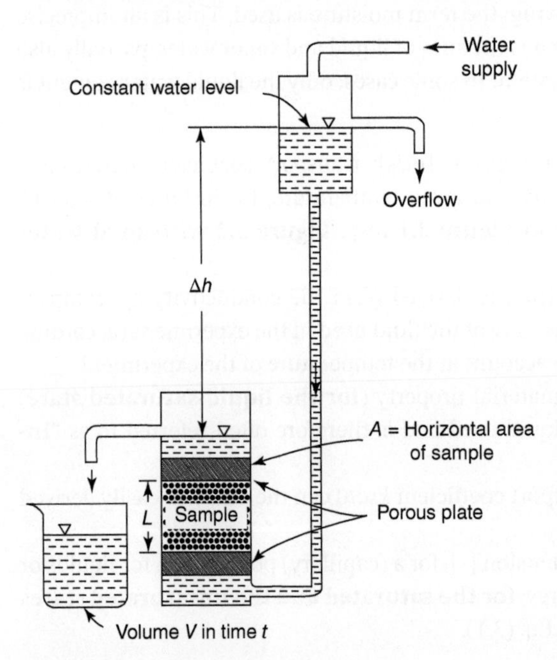

Figure 3.1 Determination of the hydraulic conductivity, compare Todd [1].

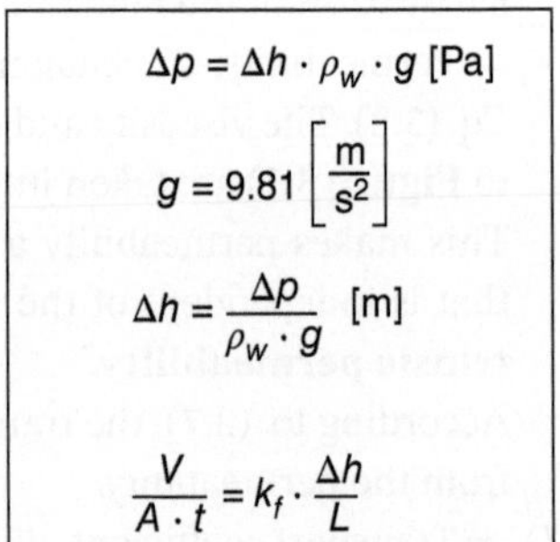

$$\Delta p = \Delta h \cdot \rho_w \cdot g \text{ [Pa]}$$

$$g = 9.81 \left[\frac{\text{m}}{\text{s}^2}\right]$$

$$\Delta h = \frac{\Delta p}{\rho_w \cdot g} \text{ [m]}$$

$$\frac{V}{A \cdot t} = k_f \cdot \frac{\Delta h}{L}$$

k_f = Hydraulic conductivity

Figure 3.2 Hydraulic conductivity k_f [m/s].

material saturation. The derivation of these reduced coefficients will be discussed in Section 5.2 of this book.

From Figure 3.2, with $\dot{m}_k = V \cdot \varrho_w \cdot g/(A \cdot t)$ and with $\Delta x = L$, we get the relation

$$\dot{m}_k = \frac{k_f}{g} \cdot \frac{\Delta p}{\Delta x} \tag{3.2}$$

The capillary transport coefficient D_k for water-saturated condition results in

$$D_k = \frac{k_f}{g} \quad [s] \tag{3.3}$$

where k_f has the unit [m/s], gravity g the unit [m/s^2], and **thus D_k the unit [s].**

(According to SI Unit system: 1 [Pa] = 1 [N/m^2] = 1 [kg/(m $\cdot$ s^2)]), giving $\dot{m}_k$ in Eq. (3.2) with units [kg/(m$^2 \cdot$ s)].

3.1.2 Consideration of the Temperature Influence on the Capillary Transport Coefficients

Before using k_f, it must be known at what temperature k_f was determined, usually 20 °C. The magnitude of **moisture transport in the water-saturated state** under capillary pressure difference (compare Eq. (3.25)) in a volume element (VE) can be described by the material-specific coefficient *ksat*0 for a temperature of 20 °C:

$$D_k = ksat0 \ [s], \quad \dot{m}_k = ksat0 \cdot \frac{\Delta p}{\Delta x} \tag{3.4}$$

From the pressure plate test at 20 °C corresponding to **Figures 3.1** and **3.2,** $ksat0_{20\,°C}$ can be derived using the acceleration due to gravity g in [m/s^2]:

$$ksat0_{20\,°C} = k_{f_{20\,°C}}/g \tag{3.5}$$

For varying temperatures, $ksat0_{20\,°C}$ can be fitted via the temperature dependence $\eta(\vartheta)$ of the viscosity (with ϑ in °C between 0 °C and 100 °C according to **Figure 3.3**).

$$\eta(\vartheta) = [0.1 \cdot (273 + \vartheta)^2 - 34.335 \cdot (273 + \vartheta) + 2472]^{-1} \quad [\text{Pa} \cdot \text{s}] \tag{3.6}$$

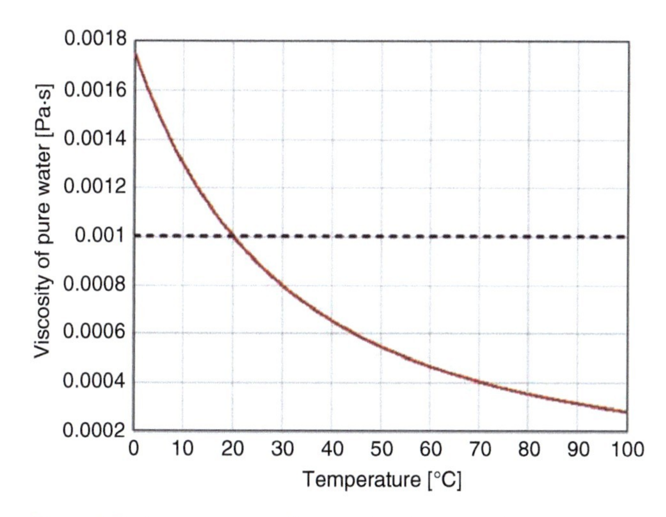

Figure 3.3 Temperature effect on the viscosity of pure water.

At 20 °C, $\eta_{20\,°C} = 1.0 \cdot 10^{-3}$ in [Pa · s].

A temperature adjustment of *ksat*0 results accordingly to:

$$ksat0_{(\vartheta)} = ksat0_{20\,°C} \cdot \frac{\rho_{w(\vartheta)}}{\rho_{w20\,°C}} \cdot \frac{\eta_{20\,°C}}{\eta_{(\vartheta)}} \approx ksat0_{20\,°C} \cdot \frac{\eta_{20\,°C}}{\eta_{(\vartheta)}} \tag{3.7}$$

Instead of the transport coefficient k_f/g, the so-called permeability resp. the permeability coefficient K_l in [m²] can be given.

In K_l, the viscosity η and the fluid density ϱ_w are considered with the experimental temperature prevailing in the determination of k_f/g. Thus, the quantities k_f [m/s], η, and ρ_w must be related to the same temperature.

This makes the permeability formally temperature-independent. It is defined in Eq. (3.8)

$$K_l = \frac{k_{f(\vartheta)} \cdot \eta_{(\vartheta)}}{g \cdot \rho_{w(\vartheta)}} \quad [\text{m}^2] \tag{3.8}$$

*ksat*0 can now also be derived from the permeability K_l in [m²] by reintroducing $\eta_{(\vartheta)}$ *and* $\rho_{w(\vartheta)}$ for the desired temperature, here 20 °C.

Using K_l, this then gives the intrinsic transport coefficient (complete capillary saturation) at 20 °C to

$$ksat0_{20\,°C} = K_l \cdot \frac{\rho_{w20\,°C}}{\eta_{20\,°C}} = K_l \cdot 10^6 \quad [s] \tag{3.9}$$

The coefficient $ksat0_{20\,°C}$ for the mortar REF amounts to $3.0 \cdot 10^{-13}$, dimension [s] (the corresponding intrinsic hydraulic conductivity to $3.0 \cdot 10^{-12}$ [m/s] and the corresponding intrinsic permeability to $3.0 \cdot 10^{-19}$ [m²]).

3.1.3 Note on the Difference Between Capillary Pressure and Hydraulic External Pressure

Figures 3.1 and 3.2 show the classic experiment with hydraulic external pressure. This produces the desired flow in the material sample. This flow takes place mainly in the porous material in the larger pore fractions in the capillary pores above θ_{gel} resp. in the superhygroscopic region, if present.

The superhygroscopic region is the pore region that can no longer be filled by external vapor absorption at relative humidities above 100%. Defining the transfer to the superhygroscopic pore region at 98.5% RH, the corresponding pore volume can account for 70–80% of the total pore volume for certain materials, such as ceramic bricks. In this pore region, the capillary pressure is small at complete water saturation.

Let us now **consider a capillary suction test** on a material only partially or non-saturated with water **with additional hydraulic external pressure influence**.

With regard to the effectiveness of an external hydraulic pressure on the fluid transport in the pore system concerned, it is useful to remember here that the driving potential is a suction stress (which is usually set with a positive sign in the transport calculations) caused by the effect of the surface energy.

A hydraulic external pressure is superimposed from the external surface as a compressive stress (with opposite sign (−)) on the capillary tensile stress, which is thereby lowered. Therefore, at a volume element of the material, between the side facing the hydraulic external pressure and the opposite side of the VE, there is an increase in the potential difference at the VE, causing an acceleration of the transport.

The resulting effects are evident, for example, from experimental **results of Rucker-Gramm and Beddoe [2], Figure 3.4a,b**.

Influencing the water absorption resp. the liquid transport in a pore system by an **external hydraulic** pressure is in principle limited to pores that develop a capillary suction tension in the order of magnitude of the external pressure.

Figure 1.16 a,b shows that for the material REF at a water saturation of 0.10 [m^3/m^3] corresponding to approx. 92% RH, the associated capillary pressure is at least 10^7 [Pa]. An external hydraulic pressure of 1 bar equal to 10^5 [Pa] would therefore mean a pressure difference of 1% for pores of this size and would then be negligible for such and smaller pores. Accordingly, only pores that are filled with water at relative humidities above 92% RH, according to the moisture storage function, could be influenced by external hydraulic pressure.

Section 5.2.1 shows that this pore area corresponds to the capillary pore fraction of the material. The results in **Figure 3.4** show an example of the acceleration of capillary water absorption into a free-pore system due to relatively low external hydraulic pressure.

Of interest is the question of the extent to which these explanations can be reconciled with the explanations in Section 3.1.7, where the determination of the capillary transport coefficient is explained, which is determined via **measurements of the hydraulic conductivity [m/s] with hydraulic pressure**. These tests to determine

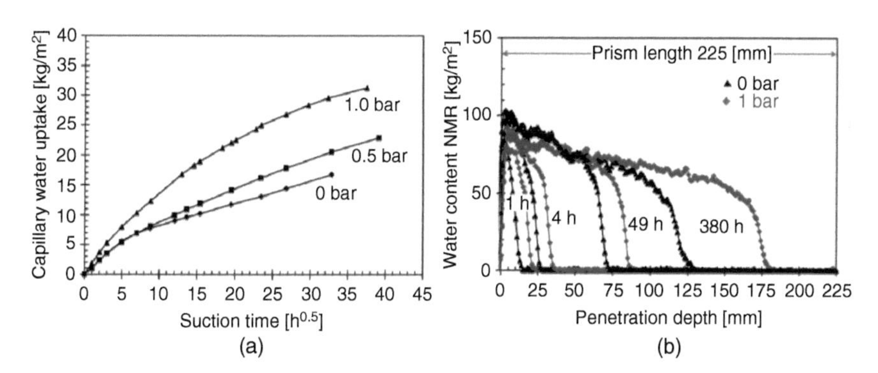

Figure 3.4 Measurements of the influence of external hydraulic pressure on water uptake and water distribution in Portland-cement-mortar prisms. Source: Rucker and Beddoe [2]/John Wiley & Sons. Mortar prism of 45 · 45 · 225 [mm^3], mortar of quartz sand and 515 [kg/m^3] cement CEM I 42.5 R, *W/C* ratio of 0.60; storage: three months water storage plus one month storage at 50 °C followed by storage at 23 °C/50% RH until equilibrium. **(a)** 1D water uptake tests with one-sided external hydraulic pressure of 0, 0.5, and 1 bar. **(b)** Comparative NMR measurements of the water content distribution within the prisms in the capillary water uptake test without and with 1 bar pressure after different test times.

ksat are carried out on water-saturated test specimens with hydraulic pressures of up to 10 bar. In these tests, transport takes place via the capillary pores, i.e. not all of the water content is transported synchronously. Due to the pressure gradient within the test specimens, the mean value formation, and the influence of time, these tests can also only be an approximation for the required value of *ksat*.

In the author's opinion, the initial value of *ksat* entered in computational transport calculations should therefore be corrected in accordance with Section 3.1.7.4.

In transport calculations, usually the same transport coefficients are used for both capillary pressure and hydraulic external pressure. Although hydraulic external pressure generates all-round pressure inside the material, capillary suction, on the other hand, generates all-round tensile stresses. Permeability parameters derived from hydraulic pressure tests can therefore only be used for both effects if the fluid used is largely incompressible and the material tested proves to be sufficiently rigid. For solid or high-strength mortars and concretes and water as liquid, this is usually assumed to be given.

In the following, the influence of changing temperatures and the possible salt content of an aqueous solution on the liquid transport coefficient will be discussed.

3.1.4 Influence of Salinity and Temperature on Fluid Transport Coefficients

Verify that liquid-water transport coefficients also allow reliable modeling for saline aqueous solutions. Petra Rucker [3] addresses this issue in more detail, drawing on results of H. Garrecht [4]. Furthermore, capillary suction experiments with sodium chloride solutions by various authors, for example, Lunk and Wittmann [5] are addressed by Rucker [3]. In these experiments, "retardation" of the dissolved ions was observed. Rucker describes the cause of the "retardation" is to be found in the chemical and physical binding of chloride components mainly to calcium aluminate phases of the cement paste under formation of Friedel's salt. Therefore, in the case of binding processes not taking place, such "retardation" resp. transport slowing down is not to be expected.

Nevertheless, a change in the transport velocity of salt-containing solutions compared to pure water may take place due to the change in viscosity. Rucker [3] cites the results of Garrecht [4], who makes the change in transport coefficient dependent on the change in viscosity and the change in surface energy resp. surface tension. However, these results refer to **diffusivity** in [m^2/s].

In the case of capillary pressure as the driving potential, a change in surface tension affects capillary pressure, but not the associated **conductive transport coefficients**, which are affected only by a change in viscosity.

The simultaneous influence of changing temperature and dissolved salts on viscosity is complex. Recent studies on the change in viscosity of alkalis, acids, and especially salt solutions are presented in [6] and [7] and by RuizLlamas and Macías-Salinas [8]. The viscosity influence depends on the solute resp. varies from salt to salt. Laliberté [6] and [7] provides the following procedure elaborated by him on the basis of an extensive literature review.

In Eq. (3.10), w_i gives the mass fraction of the solute and w_w gives the water fraction of the solution. Equation (3.11) provides the sought viscosity of the solution. η_w is the viscosity of pure water at the given temperature, and η_i the theoretical viscosity of the solute. This is to be determined via Eq. (3.12). The parameters $v1$ to $v6$ are substance-dependent. T in [Kelvin] is again the given temperature.

Figure 3.5 shows the course of the viscosity of calcium chloride, calculated over the given parameters $v1$ to $v6$, for the temperatures given in figure up to a chloride mass fraction of 30% (investigation results of Laliberte [7]).

$$w_i + w_w = 1 \tag{3.10}$$

$$\eta_m = \eta_w{}^{w_w} \cdot \eta_i{}^{w_i} \tag{3.11}$$

$$\eta_i = \frac{exp\left(\frac{v1 \cdot w_i{}^{v2} + v3}{v4 \cdot (T-273.15)+1}\right)}{v5 \cdot w_i{}^{v6} + 1} \tag{3.12}$$

The parameters to be used from [6] for calcium chloride are as follows:

$v1 = 32.03 \quad v2 = 0.7879 \quad v3 = -1.1495 \quad v4 = 0.002699 \quad v5 = 780860 \quad v6 = 5.844$

the parameters for sodium chloride

$v1 = 16.22 \quad v2 = 1.3229 \quad v3 = -1.4849 \quad v4 = 0.0074691 \quad v5 = 30.98 \quad v6 = 2.0583$

and, for example, the parameters for sulfuric acid

$v1 = 7.77 \quad v2 = 1.2344 \quad v3 = -1.9164 \quad v4 = 0.0056363 \quad v5 = 46.007 \quad v6 = 2.8307$

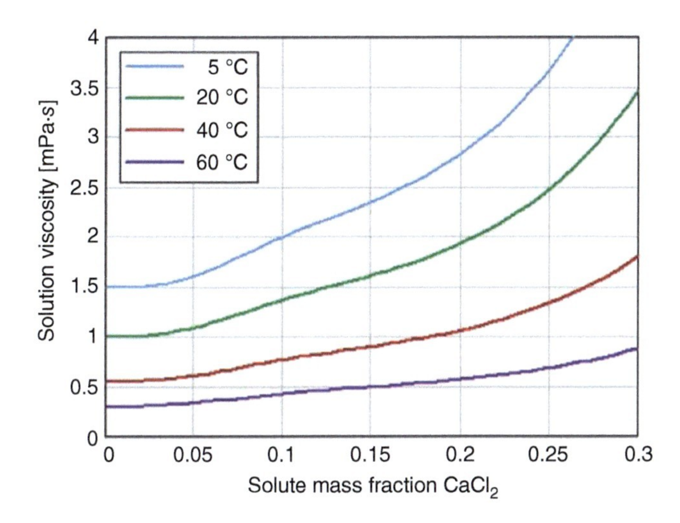

Figure 3.5 Viscosity dependence of solutions of calcium chloride on temperature and chloride concentration.

3.1.5 Diffusivity as Transport Coefficient for Water Saturation as Driving Potential

The water flow is also formulated for 1D diffusive transport according to Eq. (3.13)

$$\dot{m}_w = D_w(w) \cdot \frac{\Delta w}{\Delta x} \quad [\text{kg}/(\text{m}^2 \cdot \text{s})] \tag{3.13}$$

where $\Delta w = \theta \cdot \rho_w$ (with θ as relative water content and the water density ρ_w [kg/m^3]) as driving potential given by the difference of the water contents θ at the volume element VE of width Δx.

The transport coefficient **diffusivity** $D_w(w)$ (units [m^2/s]) is water content-dependent.

For complete saturation $\theta = 1.0$, the value of the **intrinsic diffusivity** D_{w0} results.

The use of the relative water content as driving potential in transport calculations for largely inert porous materials with suitable pore size distribution has been "state of the art" for decades, also in commercial computational programs. Nevertheless, definitions and basic considerations will be discussed again here, since they will be referred to in Section 3.1.7 (even though mainly capillary pressure is considered as the driving potential in this book).

The determination of values for diffusivity is explained by Carmeliet et al. [9] and [10], among others.

First, the moisture distributions from uniaxial capillary water absorption tests within the test specimens must be determined experimentally, for example, by NMR measurements, usually for at least 2 test times t.

An evaluation procedure now consists of using the so-called Boltzmann transformation $\lambda = \frac{x}{t^{1/2}}$ to convert the $w(x,t)$ distributions of the different time points t into function distributions $w(\lambda)$. The diffusivity $D(w)$ can now be determined using the following equation:

$$D_w(w) = -\frac{1}{2} \cdot \left(\frac{\partial \lambda}{\partial w}\right)_w \cdot \int_{w_0}^{w} \lambda(w) dw \tag{3.14}$$

In Figure 3.6a,b, as an example from [10], various measurements have been summarized via the Boltzmann transformation of the measurement results. **Figure 3.6b** schematically represents the averaged function obtained from them. In Eq. (3.14), the inverse function of $w(\lambda)$ is integrated from the lowest moisture content w_0 at the end of the moisture distribution to the moisture content w for which the diffusivity is to be determined. Usually this is first the water absorption surface, i.e. at $\lambda = 0$ resp. $x = 0$. For this moisture content w, the still needed slope of the function $\left(\frac{\partial \lambda}{\partial w}\right)_w$ is then to be determined.

The coefficient for moisture transmission within the material concrete can only insufficiently be determined on the basis of the Boltzmann transformation due to the deviation from the root-time law for longer suction times).

Therefore, a second way to determine the diffusivity coefficient is taken by Krus [11] or by Rucker [3] for concrete/mortar using NMR, compare Section 3.1.7.

They therefore evaluate at least 2 measured moisture distributions using the Eq. (3.13) as a basis. They determine by integration of these moisture distributions

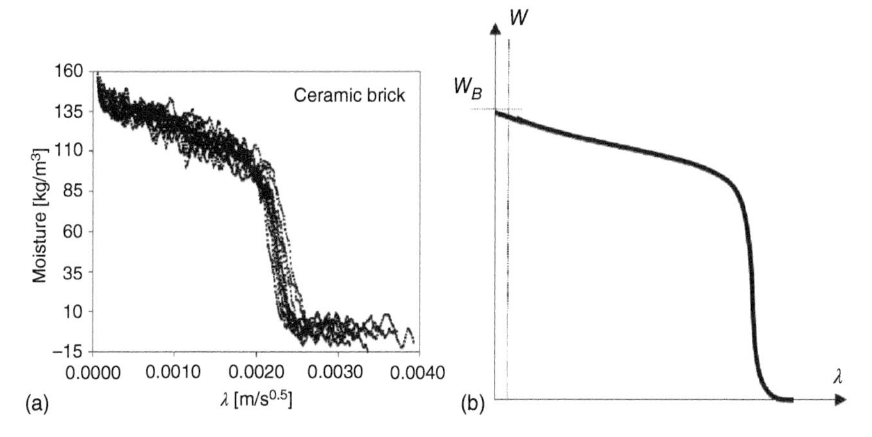

Figure 3.6 Moisture distributions of a ceramic brick material according to the Boltzmann λ-transformation. (a) Measured water content profiles after transformation. (b) Possible function after fitting by averaging. Source: Carmeliet et al. [10]/Sage Publications.

for the 2 associated time points t for a selected reference point x the associated moisture flow $\dot{m}_w$ and divide this moisture flow by the mean value of the moisture gradients in the reference point x.

In experimental determination of diffusivity, the scatter of the moisture distribution curves is principally problematic, compare Carmeliet et al. [10] and Rucker [3].

There is also a problem in specifying the moisture boundary conditions when integrating the moisture distribution curves. Using concrete or mortar material, the moisture content of the test specimens at the beginning of the determination of the distribution curves may also have an influence on the accuracy of the coefficient determination. When determining moisture distribution curves, especially with NMR in prismatic concrete bodies that had a partial saturation corresponding to, for example, 50% RH before the start of the suction test, the moisture distribution curves may deviate from the curves that would have been obtained in the dry initial state. This must be taken into account when determining the diffusivity values [3].

3.1.6 Comments on the Issue of Moisture Transport Modeling Based on Diffusivity or Permeability

The modeling of moisture transport in porous materials is very often formulated on the basis of relative water saturation as the potential resp. diffusivity as the associated transport coefficient.

However, it should be noted that, on the basis of diffusivity, the physically effective potential of moisture transport in porous systems, i.e. capillary pressure or vapor pressure, is not taken into account.

For materials that have separate moisture storage functions for adsorption and desorption, there are different driving capillary and vapor pressures for a given water saturation, and thus different transport responses for adsorption and desorption. Obviously, these relationships can only be modeled in a more complex way by modeling based on diffusivity.

For a given vapor pressure φ, there are (for different desorption and adsorption curves) different water contents with different transport characteristics for water vapor and capillary water but identical potentials from vapor pressure and from capillary pressure, **compare also Section 4.6.7**. Using the water content as transport potential again results in modeling complications.

A number of other authors, for example, Roels [12], Scheffler [13], and Funk [14], believe that capillary pressure is the more appropriate method as a driving potential, if different moisture storage functions for adsorption and desorption are to be considered.

The modeling and investigations described below in this publication are therefore made only on the basis of capillary and vapor pressure as driving potentials. Considerations based on diffusivity are made where appropriate for comparison with investigation results of other authors.

3.1.7 Diffusivity and Permeability Coefficients for Mortars and Concrete

For a reliable calculation of the moisture transport, correct transport coefficients are necessary, which must be adapted to the material in question.

In the following, **the main calculation and modeling methods are addressed** and partially discussed.

Between the diffusivity and the permeability exists a mathematical relation given below, which in principle also allows a determination of the intrinsic permeability from the diffusivity D_{w0}. These relations are used in some of the procedures described below.

For the diffusivity that can be written, compare Carlier/Burlion [15] or Feng Yang et al. [16]:

$$D_w(S_w) = -K_l \cdot \frac{krl(S_w)}{\overline{V} \cdot \eta} \cdot \frac{dp_c}{dS_w} \tag{3.15}$$

where $krl(S_w)$ describes the dependence of permeability on the degree of saturation, $\overline{V}$ [m³/m³] is the total porosity of the material, η is the viscosity [kg/(10³·m·s)] at the chosen 20 °C, dp_c/dS_w is the slope of the dependence of the capillary pressure on the water saturation.

The intrinsic permeability after transformation of the previous equation is then

$$K_l = -D_{w0} \cdot \overline{V} \cdot \eta \cdot \frac{dS_w}{dp_c} \ [\mathrm{m}^2] \tag{3.16}$$

A large number of publications exist on the subject of the section title, which can only be partially addressed here.

The majority of the methods refer to the determination of diffusivity [m²/s]. Other methods deal with the determination of the permeability [m²] or the corresponding hydraulic conductivity [m/s], as well as the derivation of the permeability from the diffusivity.

For HCP, concretes and mortars with W/C values of at least 0.38, the permeabilities range between 10^{-21} and 10^{-17} [m^2]. The corresponding intrinsic conductivities [m/s] are larger by a factor of 10^7 for 20 °C, i.e., range between 10^{-14} and 10^{-10}.

The diffusivity values range for HCPI/mortar/concrete between approx. 10^{-10} and 10^{-6}.

Measurement results for permeability [m^2] are given in [17] [18], measurements for (intrinsic) permeability in [19] [20] [21] [22] [23] [24] [25], and for diffusivity in [3], for example.

Overviews of methods for the determination of transport coefficients investigated in research work or applied in modeling are presented, for example, by Scheffler [13], Liu [25], Zhang and Scherer [26], and Patel et al. [27].

3.1.8 Methods for Determining the Transport Parameters

3.1.8.1 Measurement of the Hydraulic Conductivity [m/s] via Hydraulic Pressure

The measurement methodology shown in principle in **Figure 3.1**, where water is forced under pressure through a material sample and where conductivity is derived from the generated relationship pressure versus flow (**Figure 3.2**), is considered as a classical reference method for the determination of the transport coefficient k_f [m/s].

However, the numerous reports show that these experiments are laborious and the result depends not only on the material but also on the experimental parameters.

For example, sample age or degree of hydration and W/C value (as also with other measurement methods) have the expected influence, but also sample shape, test duration until flow equilibrium, and especially the level of hydraulic pressure have a significant influence on the result.

At high pressures above approx. 4 bar, the results usually change and lower conductivities are measured. For W/C values lower than approx. 0.40, higher pressures and thinner samples are generally required. Therefore, even with this reference method, significant scattering is to be expected for a given material in principle and depending on the laboratory.

The cause of the decrease in permeability during a corresponding compression experiment at high pressures has not yet been determined. From the investigations in Section 1.4.7, measurement results are now available, which indicate a clear increase in the solidification of the surface films on the pore walls of water-filled pores when mutual pressure is exerted on the adsorbed films via the pore walls. It should therefore be investigated whether the application of high pressures to the pore solution can also lead to a solidification of sorbed layers and thus to a narrowing of the pore cross-sections associated with a reduction in permeability during the test.

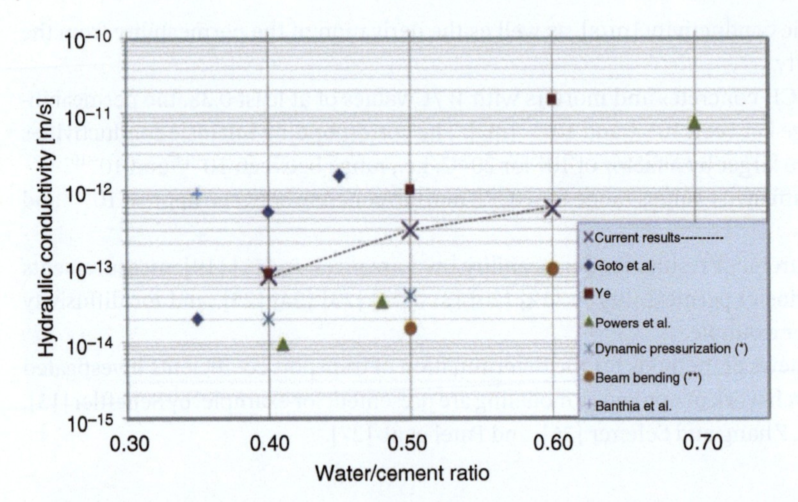

Figure 3.7 Effect of W/C value on hydraulic conductivity of HCP. Measured values of Quoc Tri Phung et al. [23]/with permission of Elsevier compared to results from other authors.

From Phung et al. [23], measurement results are presented, which were determined after 28d water storage on HCP specimens made of CEMI 52.5 N (EN 197) of thickness 25 [mm] at W/C values of 0.40, 0.50, and 0.60. The method used by the author, which also works via hydraulic pressure, differs from previous ones in that a constant water flow input was used.

According to the authors, these steady-state conditions allow much higher measurement accuracies to be achieved, especially at extremely low water flows.

The results are presented in Figure 3.7. The measured hydraulic conductivities [m/s] are compared with the measurement results on HCP (from also varying W/C values but partly other cements) from other authors. The aim was to check the magnitude of the dependence of the hydraulic conductivity on the W/C value with otherwise unchanged composition.

Measurement results of the **dependence of the hydraulic conductivity [m/s] on temperature** are given for different concretes by Jooss and Reinhardt in [28].

Also, because of the overall effort of the hydraulic conductivity test method, some of the methods discussed below are preferable or should be used for comparison.

3.1.8.2 Transport Coefficients by Measurements of the Electrical Conductivity

An overview of methods for predicting "effective diffusion coefficients" of hardened cement paste (HCP), mortars, and concrete in the pore water-saturated state on the basis of electrical measurement methods can be obtained from Patel et al. [27], which considers the entire pore solution, i.e., pore water with the ions effectively dissolved in it. The transport characteristics considered are generally referred to as diffusivity in this paper, although they are also used to measure hydraulic conductivity or permeability.

Results from different authors are evaluated together and presented as relative diffusivity De/Do. Do is the conductivity of the pure pore solution, and De is the conductivity of the pore solution through the material. All methods that use electrical measurement techniques are examined. For example, measurement of ion migration in the above materials by applying electric fields and using the Nernst–Planck equation is also discussed. Experiments using the addition of tritiated water (HTO) as a tracer in the transported water will also be addressed.

In the methods using the electrical conductivity or the electrical resistivity of the material samples (for example, the Katz–Thompson method described in more detail below), the conductivity through the material De contains the influence of the existing tortuosity or porosity of the material and further possibly transport-reducing influences, through which a measure for the permeability [m^2] or the relative diffusivity and of diffusivity changes can be derived.

The following general results may be derived from Patel's work:

With the help of electrical measurement techniques, it is obviously possible to predict the transport characteristics at least approximately.

It is deduced that the diffusivity or permeability of concrete can be calculated with good accuracy based on the diffusivity/permeability of the HCP fraction:

$$D_{concrete} = V_{HCP(concrete)} \cdot D_{HCP}$$

where D_{HCP} is the separately determined transport coefficient with the composition of the HCP fraction present in the mortar or concrete, $V_{HCP(concrete)}$ is the volume fraction of the HCP in the concrete, and $D_{concrete}$ is the sought transport coefficient.

This also means that if the diffusivity of a reference mortar or concrete is known, the diffusivity of other concrete compositions (with unchanged W/C but varying cement contents) can be derived from the diffusivity of the reference material.

Patel also gives a relationship for the influence of changes in W/C on relative diffusivity. Furthermore, he finds that the interface transition zone (ITZ) exerts only a subordinate influence on the transport characteristics.

3.1.8.3 Prediction of Permeability [m^2] According to Katz–Thompson

The following relationship was developed by Katz–Thompson to determine the permeability [m^2] of rock using electrical conductivity measurements. In the past, there have been a large number of publications that have assessed this procedure as suitable and as less suitable. The suitability of the method for the application of it to building materials, especially HCP and concrete, has been extensively tested in the meantime and confirmed in principle, although one can expect only an approximate value.

In the following, we will check whether this method could have been used in advance to determine the permeability of the material REF for use in the computational program, at least approximately.

$$K_l = \frac{1}{226} \cdot d_c{}^2 \cdot \frac{\sigma}{\sigma_0} \tag{3.17}$$

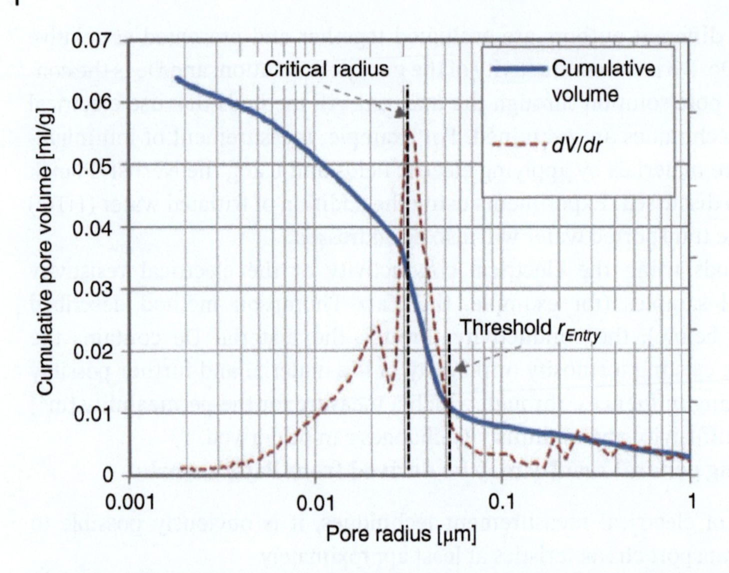

Figure 3.8 Definition of the relevant pore radii (determined via the MIP measurement results) for the alternative calculation of σ/σ_0 according to Eq. (3.18), in view of the calculation of the permeability of the material REF according to Katz–Thompson with Eq. (3.17).

where d_c **is the critical pore diameter = 2· critical radius** according to **Figure 3.8** and σ_0 = electrical conductivity of the pore solution and σ = the material-through conductivity measured on the sample.

Here, the electrical conductivity σ of the material is set in relation to the measurement of the conductivity of the extracted pore solution σ_0. The measurement of the conductivity σ or the electrical resistance is relatively easy, but the measurement of the pore solution itself is more complex because the solution has to be extracted from the sample material by pressure resp. squeezing. For concrete with W/C values below 0.40, a squeezing out procedure can be difficult to realize. An alternative is to use the conductivity σ_0 of pore solutions of comparable composition. However, these conductivities can vary significantly despite similar solute contents [27].

Because of the necessary effort to determine the relation σ/σ_0 by electrical measurements, the following determination equation is proposed alternatively by Katz and Thompson:

$$\frac{\sigma}{\sigma_0} = \frac{d_{max}}{d_c} \cdot \theta_{total} \cdot \theta_{d_{max}} \tag{3.18}$$

The meaning of the pore diameters or pore radii is defined in more detail by Nokken and Hooton in [20].

In **Figure 3.8**, the corresponding radii are given in relation to the measured MIP curves for the material REF.

Here, d_{max} or r_{max} corresponds to the radius r_{Entry} defined in more detail in Section 2.1.3.

θ_{total} corresponds to the total porosity [m^3/m^3], and $\theta_{d_{max}}$ [m^3/m^3] is the pore volume fraction above r_{Entry}.

In the following, it will be checked whether the Katz–Thompson method can provide a suitable permeability value for the material REF. Thereby, the ratio σ/σ_0 shall be determined according to Eq. (3.18). The porosity values contained in it are calculated alternatively from the MIP measurements on the one hand and from the sorption isotherm of the material REF on the other hand.

The pore radii r_{Entry} and $r_{critical}$ are taken from Figure 3.8 to 0.051 μm = $0.51 \cdot 10^{-7}$ [m] and 0.031 μm = $0.31 \cdot 10^{-7}$ [m].

Permeability using the porosity volumes determined by the MIP measurements:

From the original MIP measurements of material REF:
$\theta_{total} = 0.135$ and $\theta_{r_{Entry}} = 0.0246$ [m^3/m^3] are taken. This gives
$\sigma/\sigma_0 = 1.02/0.62 \cdot 0.135 \cdot 0.0246 = 0.00405$
$K_l = 1/226 \cdot (0.62 \cdot 10^{-7})^2 \cdot 0.00405 = 0.69 \cdot 10^{-19}$ [m^2]

Permeability using the porosity volumes from the adsorption isotherm:

From the adsorption isotherm $\theta(r)$ in **Figure 1.23b**, the following is obtained:
$\theta_{total} = 0.165$ and $\theta_{r_{Entry}} = 0.025$ [m^3/m^3]. Then follows
$\sigma/\sigma_0 = 1{,}02/0.62 \cdot 0.165 \cdot 0.025 = 0.00679$
$K_l = 1/226 \cdot (0.61 \cdot 10^{-7})^2 \cdot 0.00679 = 1.15 \cdot 10^{-19}$ [m^2]

These two previously calculated results have to be compared with the transport coefficient $ksat0_{20\,°C}$ used of the water-saturated material REF:

According to Eq. (3.9)
$ksat0 = K_l/10^6 = 3.0 \cdot 10^{-13}$ from which the permeability results to $3.0 \cdot 10^{-19}$.

For a mortar with similar composition to material REF, nearly the same permeability ($1.4 \cdot 10^{-19}$) was determined in a 3-axial pressure apparatus (Section 4.6.8).

It can be seen that the value of permeability determined is very **close to the "nominal value" when taking into account the sorption isotherm**, while the value determined with the porosity volumes from MIP data is too small by a factor of approx. 0.5. However, it must be taken into account in the evaluation that the limiting radii determined via the MIP analysis were used for both permeability values. Only the values of the porosity volumes taken from the MIP data in the relevant calculation lead to the deviation.

Nevertheless, in this case, the calculation procedure is rather classified as positive, especially also because of the simplicity of the procedure and because the majority of other methods do not give more accurate values.

It should also be noted that the coefficient 1/226 in the Katz–Thompson equation has an empirical background, see also Thompson [29]. In other publications, alternative quotients up to 1/180 are proposed, which, however, do not lead to any fundamental change.

3.1.8.4 Determination of Diffusivity [m²/s] via NMR According to Krus and Rucker-Gramm

The procedure for measuring the diffusivity coefficients has already been indicated in Section 3.1.5. For the determination of $D_w(w)$, the two equations at Eq. number (3.19) are applied for chosen points x and times t. They are solved for $D_w(w)$ by setting equal after inserting the corresponding gradients, taken from the measured water content distributions, into the left equation and after integrating the measured distributions of the right equation.

$$\dot{m}_{W(x,t)} = -D_w(w) \cdot \frac{dW(x,t)}{dx} \left[\frac{\text{kg}}{\text{m}^2 \cdot \text{s}}\right]$$

$$\dot{m}_{W\left(x,\frac{t_2-t_1}{2}\right)} = \frac{1}{2 \cdot (t_2 - t_1)} \int_{a=x}^{\infty} (W(a,t_2) - W(a,t_1))\, da \tag{3.19}$$

In the one-sided water uptake experiment (NMR measuring device; **see Figure 3.9**), the equations apply to a homogeneous initial saturation w of the prismatic bodies. The measured moisture distributions shown in **Figure 3.4b** are an example of the type of distributions.

A part of the diffusivities measured on mortars and concrete according to this method can be taken from **Table 3.1**. A commentary on the values is given below.

Further details on the measurement procedure can be obtained from Krus [11], in particular for the situation of water redistribution within the prisms after suppression of the unilateral water supply.

3.1.8.5 Diffusivity [m²/s] and Permeability [m²] via Capillary Sorptivity

A number of scientists have investigated in the past whether and, if so, how the transport parameters diffusivity [m²/s] and permeability [m²] can be derived from the capillary sorptivity S or the capillary water absorption coefficient ww without having to perform difficult measurements of the moisture content distributions occurring in the suction tests.

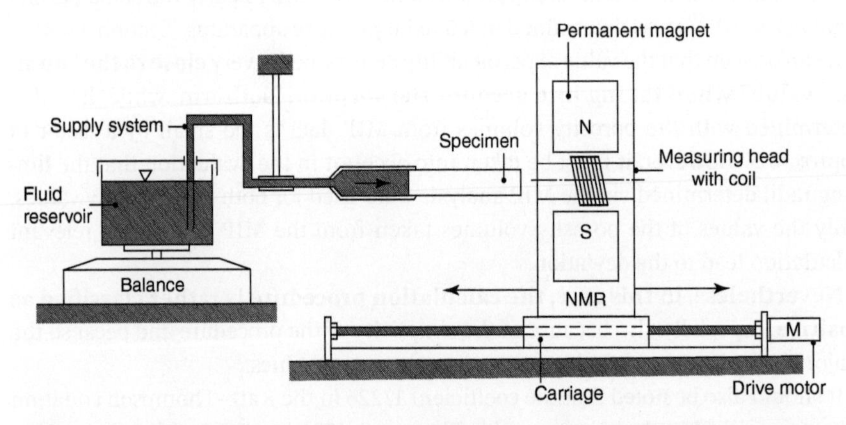

Figure 3.9 NMR measuring device from Krus [11] for measuring the water content distribution in the longitudinal direction in prismatic mortar specimens during capillary sorption tests. The NMR device is guided stepwise over the test specimens.

Table 3.1 Diffusivities Dw(θ1) and Dw(θ2) [m^2/s] of mortars (cement contents 516–650 [kg/m^3]) and concretes (cement contents 320–409 [kg/m^3]) as a function of the degree of water saturation θ1 and θ2 (θ= relative water content to capillary water saturation).

Cement Type	Cement *W/C* [kg/m³]	Dw(ϑ1)	Dw(ϑ2)	ϑ_1	ϑ_2
CEM I 42.5 R	0.60 E 516	1.5E-1Q	6.4E-08	0.23	0.75
CEM I 42.5 R	0.50 E 578	1.0E-10	7.0E-08	0.30	0.71
CEM I 42.5 R	0.40 E 658	2.0E-10	1.0E-08	0.26	0.38
CEM I 42.5 R	0.50 E 778	1.5E-10	7.0E-08	0.31	0.72
CEM I 42.5 R	0.60 E 580	8.0E-10	1.0E-07	0.30	0.75
CEM I 42.5 R	0.50 E 650	3.9E-10	1.3E-07	0.34	0.78
CEM I 42.5 R	0.60 E 320	3.6E-10	4.6E-08	0.21	0.69
CEM I 42.5 R	0.50 E 359	1.5E-10	1.0E-08	0.18	0.50
CEM I 42.5 R	0.40 E 409	6.0E-11	3.2E-08	0.26	0.60

Source: Measured values adapted from Rucker-Gramm [30] derived by NMR measurements corresponding to **Figure 3.9**.

They start from the associated well-known equations (3.20):

$$\text{Sorptivity } i_{(t)} = S \cdot t^{0.5} [\text{m}],\ S\left[\frac{\text{m}}{\text{s}^{0.5}}\right];\ \text{Absorption } m_{(t)} = ww \cdot t^{0.5}\left[\frac{\text{kg}}{\text{m}^2}\right],\ ww\left[\frac{\text{kg}}{\text{m}^2 \cdot \text{s}^{0.5}}\right] \tag{3.20}$$

Lockington et al. [31] and Zhou [32] show that the derived (intrinsic) diffusivity D_{w0} can be described approximately as follows:

$$D_{w0} = \left(\frac{S}{\theta_{sat} - \theta_0}\right)^2 \cdot \tau(n) \tag{3.21}$$

Therein is θ_{sat} = water content at saturation [m^3/m^3], θ_0 = homogeneous initial water content, for example, ≈ 0. $\tau(n)$ is a shape factor for the water content distributions within the material during the one-sided sorption test, additionally characterized by the parameter n.

The time-dependent distribution curves are normalized by means of the Boltzmann method (compare Section 3.1.5) and by subsequent adjustment.

After complex derivatives, $\tau(n)$ is given by

$$\tau(n) = \frac{n^2}{(2n-1) \cdot exp(n) - n + 1} \tag{3.22}$$

n additionally characterizes the shape of the distributions and can be chosen between 4 and 8 depending on the material. The authors suggest $n = 6$ for concrete-type materials.

$n = 8$ provides bulbous distribution curves, $n = 4$ more sloping curves. In Zhou [32], the n-dependent model curves are presented.

The diffusivity D_{n0} is calculated below for the material REF according to this method:

The required sorptivity coefficient S can be transferred after conversion from the capillary absorption coefficient ww by Eq. (3.20), since it is the same experiment.
$ww = 0.0175$ at the dry initial state. $S_{(REF)} = ww/\varrho_w = 0.0175/1000 = 1.75 \cdot 10^{-5}$.
According to dry initial state of the prisms: $\theta_{sat} - \theta_0 = 0.15$ (excluding the air void volume part).
With $n = 5.5$ follows $\tau(n) = 0.0124$. This gives $D_{w0}(REF) = 1.69 \cdot 10^{-10}$.
Since a measured value for the material REF does not exist, the accuracy of the calculation cannot be specified directly, but only for similarly composed materials with similar W/C values.

Furthermore, it was examined to what extent the permeability K_l can be derived from the diffusivity D_{w0}. For this purpose, the Eq. (3.16) is used.
In the following, this method is verified for the material REF:

The total porosity is $\overline{V} = 0.17$ [m³/m³], and the viscosity at 20 °C $\eta = 10^{-3}$ [kg/(10³·m·s)].
A much discussed difficulty is to determine the required gradient dS_w/dp_{cap} of the water content–capillary pressure curve for the state near water saturation.
The function in **Figure 1.16b** is used for this purpose. The slope can be determined incrementally or by using the analytical function of the curve. Then

$$\frac{dS_w}{dp_{cap}}(\theta_s) = 1.01 \cdot 10^{-8}$$

D_{w0} and the previously determined values inserted into Eq. (3.16) yields $K_l = 1.94 \cdot 10^{-21}$.

This value is about 2 powers of 10 lower than the value actually determined for the material REF. Based on this result, it must be questioned whether the conversion method according to Eq. (3.16) using the diffusivity determined from sorptivity is suitable for determining the permeability. The reason for this discrepancy is attributed to the fact that the diffusivity value is clearly too low.

From the measurements in the **Table 3.1**, it can be seen that there are diffusivity values of the order of $5.0 \cdot 10^{-8}$ [m²/s] consistently for mortars of similar composition to the reference mortar REF (despite being well below complete water saturation). **Using this diffusivity value for the previously performed conversion** from diffusivity to permeability, we obtain a value $K_l \approx 5.0 \cdot 10^{-19}$, which correctly corresponds to the true permeability value of the REF.

3.1.8.6 Transport Parameters from Reverse Calculations

In a number of publications, attempts are made to determine permeability or diffusivity backwards using calculation programs. For this purpose, it is necessary to determine the parameters sought in such a way that experimental test results, determined on the same material, can be correctly simulated by the program with the boundary conditions of the test.

Some publications attempt to determine permeability or diffusivity characteristics via evaporation tests and vapor adsorption tests:

These methods can only provide correct water transport coefficients by chance, because, for example, the modeling of the boundary conditions of the vapor transitions for moisture storage via vapor and for drying by evaporation uses additional different parameters at the material surface, which can influence or distort the value of the liquid transport coefficients to be determined, in addition to the influences of the sorption isotherms and hysteretic behavior present.

Zhidong Zhang and Ueli Angst [33] report on corresponding experiments on HCP to determine permeabilities via a computational program. They conclude: "When fitting the anomalous mass change curves, we could not find close values of permeability for different drying (or wetting) conditions. Therefore, the development of a universal dual-permeability model for anomalous moisture transport needs to consider more factors, such as chemical interactions of water with hydration products and the microstructural changes during drying/wetting." More successful are parameter determinations for fluid transport, if experimental investigation results for the liquid transport in the considered material are taken as a basis. An example for this is the use of water content distributions, determined, for example, with NMR, see Krus [11] and Rucker et al. [30], which describe the liquid transport directly and depend predominantly on the water transport coefficients in the calculation program.

However, a major disadvantage is that obtaining the distribution curves for the particular material under consideration is a laborious task.

The intrinsic permeability of the material REF used here for experimental investigations was also specified by back-calculation, in which an initial value is first entered for the permeability resp. the liquid transport coefficient *ksat0* [s]. At the start of the program, the liquid transport coefficient is then specified in such a way that the water absorption test measured for the material is accurately recalculated with an adjusted coefficient.

3.1.8.7 Relative Diffusivity Modeled from Morphology resp. Structure Data

Patel et al. [27] also examine the transport models of a number of authors related to both HCP and mortar and concrete. Purely empirical modeling as well as modeling based on material structure theories are examined.

They summarize: "A large number of models have been proposed to predict effective diffusivity of hardened cement paste, mortar and concrete. None of these models are devoid of calibration of parameters. For cement paste, even the most advanced models require parameter associated with diffusivity in the gel pores of C–S–H. For mortar and concrete models, whether or not they are empirical, there are always parameters that need to be calibrated with relevant experiments. All empirical models once calibrated to the data compiled in this study show similar predictive behaviour. Models accounting for cement paste morphology in simplified way and advanced models show a similar predictability." It can be concluded that morphology-based modeling can be expected to predict diffusivity correctly if it is related to the material class under investigation and if the user has the special data needed for the modeling.

3.2 Base Modeling of Moisture Transport

The following equations describe one-dimensional transport but could be extended to multidimensional transport accordingly (see, e.g. [34]). The following description, however, uses other size labels and takes a different descriptive route of the basic equations.

The discretization is done using VEs. For the mass balance of the transported water, the well-known equation applies:

$$\frac{dW_{(p_k)}}{dt} - \dot{W}_S = \frac{\partial}{\partial x}\left(J_{W_{(p_k)}} + J_{V_{(pV)}}\right) \tag{3.23}$$

That is, the time-dependent change in capillary pressure-dependent water content W corresponds to the local change in moisture fluxes into and out of the volume element (VE) per unit time, taking into account the moisture fluxes $J_{W_{(p_k)}}$ and $J_{V_{(pV)}}$ from capillary pressure $p_{(k)}$ and vapor pressure $p_{(V)}$.

To solve the problem, discretization is necessary. This is done on the basis of the following equations for moisture transport, referring to **Figures 3.10** and **3.11**.

The driving potentials are supposed to be the capillary pressure as well as the vapor partial pressure. For simplicity, the discretization is formulated below **first without the vapor fraction.**

The balance equation for a time step 0–1 is obtained considering a moisture source term $\Delta \overset{0-1}{W}_{s_i} / \Delta t$ (one-dimensional) to:

$$\left(\frac{\overset{1}{W}_i - \overset{0}{W}_i}{\Delta t}\right) \cdot \Delta x_i \cdot A - \frac{\Delta \overset{0-1}{W}_{s_i}}{\Delta t} = \left[\left(\overset{0}{j}_{i-1,i} - \overset{0}{j}_{i,i+1}\right) + \left(\overset{1}{j}_{i-1,i} - \overset{1}{j}_{i,i+1}\right)\right] \cdot \frac{1}{2} \cdot A \tag{3.24}$$

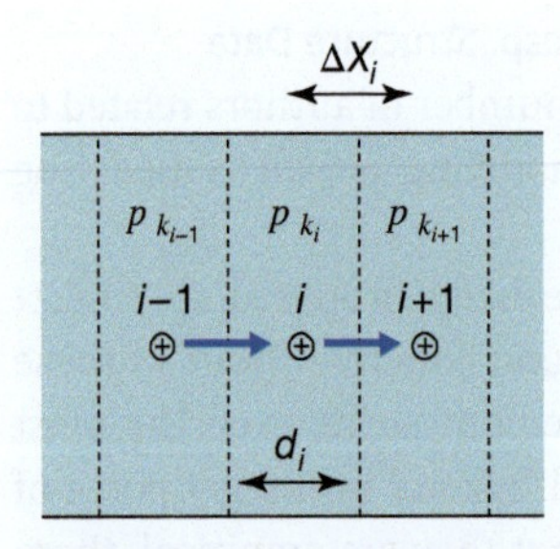

Figure 3.10 Section of a 1D net discretization with capillary moisture flow due to pressure difference acting on element i.

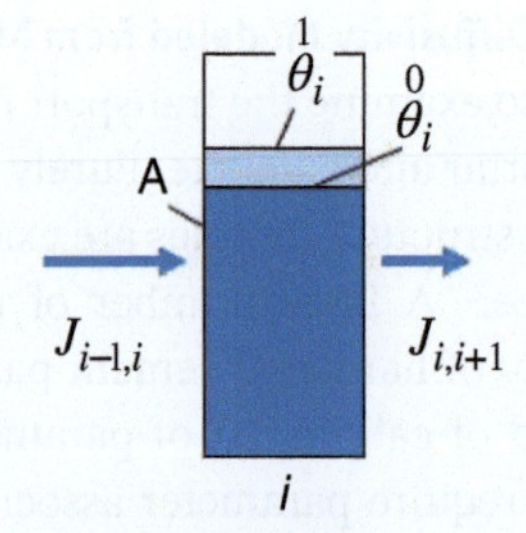

Figure 3.11 Single volume element with inflow and outflow and water contents θ_0 and θ_1 to $t0$ and $t1$.

The water flows contained therein at the times $t0$ and $t1$ are as follows:

$$\overset{0}{j}_{i-1,i} = \frac{\overset{0}{D}_{i-1,i}}{\Delta x_{i-1,i}} \cdot \left(\overset{0}{p}_{k_i} - \overset{0}{p}_{k_{i-1}} \right) \tag{3.25}$$

$$\overset{1}{j}_{i-1,i} = \frac{\overset{0}{D}_{i-1,i}}{\Delta x_{i-1,i}} \cdot \left(\overset{1}{p}_{k_i} - \overset{1}{p}_{k_{i-1}} \right) \tag{3.26}$$

The initial version of the balance Eq. (3.24) for the volume element VE_i is one of n equations of the chosen n nodes resp. n VE of the material body. A linear system of equations results, for the n values Pk_i of the capillary pressure in the VE at time $t1$. **The capillary pressures at time $t0$ as well as the other incoming parameters are considered to be known.**

This equation is correspondingly applicable for heat and mass transfer.

The formulation of the Eq. (3.24) corresponds to the Crank–Nicolson formulation with the coefficient $f = 0.5$ there, assuming a linear change of the acting currents from the value at $t0$ to $t1$. The solution was tested here. The method can apparently provide accurate solutions at larger time steps, as heat transport calculations have shown. However, for more complex initial parameters (moisture transport), the solution procedure tends to produce oscillatory resp. completely erroneous results.

In contrast, an always stable solution is obtained if the following modification of the Eq. (3.24) is made.

In this case, only the fluxes at time $t1$ are taken into account in the balance portion of the fluxes. However, the reliability of the method is bought by shorter time steps. The introduced control calculation for checking the arithmetical accuracy of the result will be discussed later.

The basic Eq. (3.24) must now be prepared for the actual solution of the system of equations. The fluxes at time $t1$ correspond to the formulation Eq. (3.26) and contain the unknown capillary pressures as well as the location- and moisture-dependent transport coefficients, which have to be calculated from the state at time $t0$ (Eq. (3.25)).

Instead of W_i, the relative water content θ_i was introduced, and instead of Δx_i, the element width d_i was used. If the source term here is initially assumed to be 0, it remains to define the term $(\overset{1}{\theta}_i - \overset{0}{\theta}_i)/\Delta t$ with respect to the possibility of solving the equations (3.27). In it, $\overset{1}{\theta}_i$ is the (relative) still unknown water content of the VE_i at time $t1$, cf. also **Figure 3.11**.

$$\left(\frac{\overset{1}{\theta}_i - \overset{0}{\theta}_i}{\Delta t} \right) \cdot \rho_W \cdot d_i \cdot A - \frac{\overset{0-1}{\Delta\theta}_{S_i}}{\Delta t} = \left(\overset{1}{j}_{i-1,i} - \overset{1}{j}_{i,i+1} \right) \cdot A \tag{3.27}$$

A change in the relative water content θ_i in the VE_i leads to a change in the pore filling degree of the pore system of the VE and the corresponding change in the associated mean radius of the pores filled until then, as well as the associated capillary pressure. This results in a physical dependence between the relative water content θ and the capillary pressure pk_i.

The functional dependence between θ and pk is obtained from the experimentally determined adsorption isotherms $\theta(\varphi)$ by applying the Eq. (1.39).

Figure 1.16b shows the moisture storage function generated from the adsorption isotherm for material REF as a function of capillary pressure $\theta(pk)$.

When applying the moisture storage function $\theta(pk)$, according to the explanations in Sections 1.4.4.1 and 1.4.6.3, split the storage function into the output function $\theta(pk) = \theta_{cap+vap}(pk)$ and the net storage function $\theta_{net}(pk) = \theta_{cap}(pk)$ without the sorptive vapor storage part, according to **Figures 1.23a and 1.32**.

This results in **the separate storage curves shown in Figure 3.12**.

The net storage function can be formulated according to Eq. (1.66) as follows:

$$\theta_{net}(\varphi) = \theta(\varphi) \cdot \varphi^{0.5} \tag{3.28}$$

The relationship between the known water content at time $t0$ and the unknown water content at time $t1$ in VE_i can now be expressed according to **Figure 3.13** as a function of the capillary pressure p_k and the slope of the net water storage function (governing capillary transport) at point θ_i by Eq. (3.29):

$$\overset{1}{\theta}_i = \overset{0}{\theta}_i + \left(\frac{\overset{0}{d\theta}}{dp_k}\right)_i * \Delta_{pk(i)} = \overset{0}{\theta}_i + \left(\frac{\overset{0}{d\theta}}{dp_k}\right)_i * \left(\overset{1}{p}_k - \overset{0}{p}_k\right)_i \tag{3.29}$$

Substituted into Eq. (3.27), the prepared Eq. (3.30) **for the moisture balance of capillary transport and vapor transport** (compare subsequent remarks) of VE_i follows in the now solvable system of equations.

$$\left(\frac{\overset{0}{d\theta}}{dp_k}\right)_i * \left(\overset{1}{p}_{k_i} - \overset{0}{p}_{k_i}\right) * \frac{\rho_w \cdot d_i}{\Delta t} - \frac{\overset{0-1}{\Delta\theta_i}}{\Delta t} = \frac{D_{i-1,i}}{\Delta x_{i-1,i}} * \left(\overset{1}{pk}_i - \overset{1}{pk}_{i-1}\right) - \frac{D_{i,i+1}}{\Delta x_{i,i+1}} * \left(\overset{1}{pk}_{i+1} - \overset{1}{pk}_i\right) \tag{3.30}$$

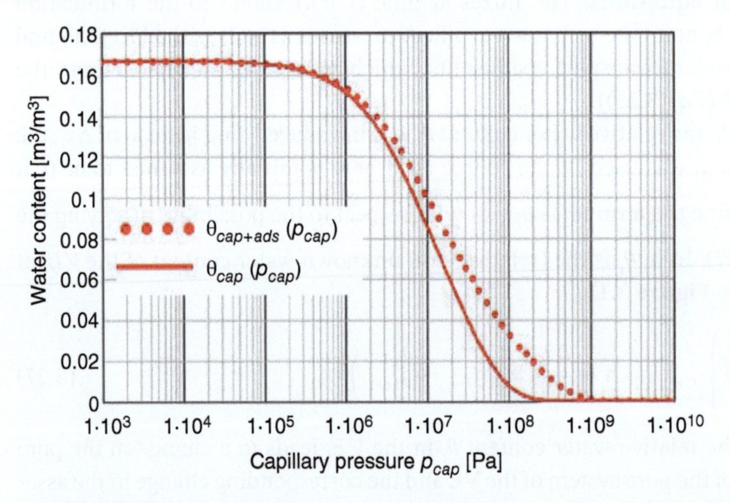

Figure 3.12 Basic water storage function $\theta(pk) = \theta_{cap+ads}(pk)$ of mortar REF **and derived net storage function** $\theta_{net}(pk) = \theta_{cap}(pk)$ (without adsorbed water fraction in the non-water-filled pore fraction).

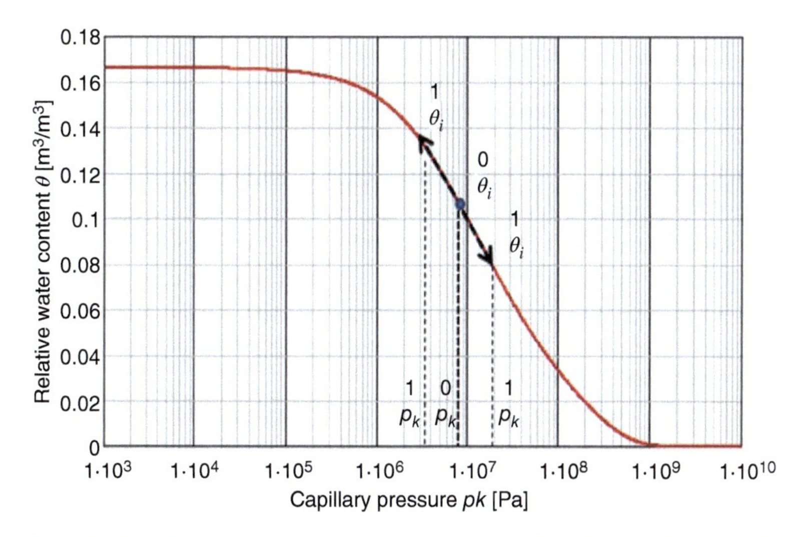

Figure 3.13 Water storage function $\theta(pk)$ of mortar REF, showing the dependence of the unknown value $\overset{1}{\theta}(\overset{1}{pk})_i$ on the slope $(\theta/dpk)_i$ at the known point $\overset{0}{\theta}(\overset{0}{pk})_i$, leading to Eq. (3.29).

The mathematical derivative of the storage function to *dpk* (for example, the Eq. (1.51) of the adsorption isotherm for the material REF) can be expressed directly as a function using the correlation from Eq. (1.39):

$$\varphi = exp\left(-\frac{p_{cap}}{Rd \cdot T \cdot \varrho_W}\right)$$

In general, the mathematical derivative of the storage function can also be calculated on the basis of the discretization for an element i as follows, and the results practically coincide with the analytical solution:

$$\overset{0}{d\theta dpk_i} = \Delta\theta\Delta pk(pk)_i = \frac{\Delta\theta(pk)}{\Delta pk(pk)} = \frac{\theta(pk * 1.01) - \theta(pk)}{pk \cdot 0.01} \tag{3.31}$$

The simultaneously proceeding vapor transport is considered as follows:

The water flow transported in vapor form during the time step as a function of the vapor pressures applied to VE_i is now not to be converted into a capillary pressure fraction via Eq. (1.39). Instead, the water fraction transported from vapor during the time step is **assigned to the source term** and thus taken into account in the moisture balance of Eq. (3.30). Among other things, this has the great advantage that the transported fractions from liquid water and vapor can be calculated and reported separately. The source term in VE_i then corresponds to the difference between the amount of liquid water transported in and out of vapor in the time step Δt and is considered as the slope of the corresponding water fraction in (3.30).

The following Eq. (3.32) **provides the basis for the corresponding vapor fraction.** The vapor pressures required for this are obtained, for example, from Eq. (3.33).

$$\left(\overset{0}{J}_{V_{i-1,i}} - \overset{0}{J}_{V_{i+1,i}}\right) = \frac{D_{Vi-1,i}}{\Delta x_{i-1,i}} * \left(pV_i - pV_{i-1}\right) - \frac{D_{Vi,i+1}}{\Delta x_{i,i+1}} * \left(pV_{i+1} - pV_i\right) \tag{3.32}$$

$$pV_i = \overset{0}{\varphi_i} \cdot \overset{1}{pV\,sat_i} \tag{3.33}$$

The vapor transport coefficients are considered in more detail in the Sections 5.3.1 and 5.

The coefficients in Eq. (3.30) **for capillary water transport** depend on the degree of pore filling in the volume element VE_i and its neighboring elements. The transport takes place according to **Figures 3.10 and 3.11** in the computational model from VE center to center at different moisture contents in each case. Therefore, for higher or lower moisture content, the magnitude of the transport coefficients is different from one VE to the neighboring VE and is homogeneously increasing or decreasing for unchanged material.

For the transport from one center to the next over the half widths of the VEs, the mean value of the corresponding *kx*-values is used, in which the element thicknesses (Δx-values) as well as the water content and, if necessary, time-dependent local diffusion coefficients D_i are taken into account:

$$kx_{i-1,i} = \left(\frac{D_{i-1}}{\Delta x_{i-1}} + \frac{D_i}{\Delta x_i}\right) \cdot \frac{1}{2} \tag{3.34}$$

instead of the coefficients $D_{i-1,i}/\Delta x_{i-1,i}$ formally inserted in Eqs. (3.25) or (3.32).

For different adjacent values Δx_{i-1} and Δx_i should be calculated with $h_{(i-1)} = \Delta x_{i-1}/2$ and $h_i = \Delta x_i/2$ as distances from the midpoints to the common boundary of the neighboring volume elements, the **weighted average** via Eq. (3.35) should be used as the transport coefficient.

$$kx_{i-1,i} = 2 \cdot \frac{\left(D_{i-1} \cdot \Delta x_{i-1} + D_i \cdot \Delta x_i\right)}{\left(\Delta x_{i-1} + \Delta x_i\right)^2} \tag{3.35}$$

The procedure in case of material change from one VE to the next VE is considered in Chapter 5.

If a volume element VE is not water-saturated, D_i must be expressed as a function of the water content, e.g. as follows:

$$D_{i(\theta)} = ksat \cdot \left(\frac{\theta_i - \theta\min}{\theta sat - \theta\min}\right)^{expo1} \tag{3.36}$$

with the pore filling ratio θ and an exponent to be determined "expo1."

The more detailed formulation of the diffusion coefficients as a function of water content and a possible change as a function of time is discussed in more detail in Section 5.2.

The vapor transport coefficients are defined accordingly as a function of the degree of pore filling, see also Chapter 5.

After rearranging the moisture balance equation applicable to the volume element *i* (3.30) and introduction of the abbreviations:

$$\overset{0}{d\theta pk_i} = \left(\overset{0}{\frac{d\theta}{dpk}}\right)_i \qquad \overset{0}{Fk_i} = d\theta pk_i \cdot \frac{\rho w \cdot d_i}{\Delta t} \tag{3.37}$$

as well as the abbreviations for the transport coefficients:

$$k_0 = kx_{0,1} \quad k_{i-1} = kx_{i-1,i} \quad k_{n-1} = kx_{n-1,n}$$

result in **following elements of the linear system of equations for the unknown quantities** $pk(i)$ **at time** $t1$.

The elements i of the left-hand side $\overline{L}$ of the system of equations (3.39) with source term $\overset{0-1}{\Delta Q_i}/\Delta t$ of the time step Δt are for time $t0$:

$$\overset{0}{L}_i = -\overset{0}{p}_{ki} \cdot Fk_i - \frac{\overset{0-1}{\Delta Q_i}}{\Delta t} \tag{3.38}$$

The system of equations in matrix notation:

$$\overline{L} = \overline{M}k \cdot \overline{p}k \tag{3.39}$$

The elements of the vector $\overline{L}$ as well as the vector $\overline{p}k$ of the $n + 1$ sought capillary pressures at time $t1$ as well as the corresponding coefficient matrix $\overline{M}k$ correspond to the representation in **Figure 3.14**. The corresponding new moisture distribution θ_0 to θ_n in the material body for time $t1$ is then obtained with the new pk_i values from the moisture storage function (Figure 3.13).

The known boundary conditions ka, ke, pka, and pke are discussed in Chapter 5.

The previously mentioned transport coefficients are moisture-, temperature-, and partly time-dependent. Furthermore, there are dependencies on the boundary conditions and the moisture storage functions. Therefore, the transport coefficients must be further refined with regard to a realistic description of the moisture transport or the moisture states in the material body. This will be done in Chapter 5, based on the results of our own experimental investigations and investigation results from the literature.

$$\overline{L} = \begin{bmatrix} \overset{0}{L}_0 + ka \cdot p_{ka} \\ \overset{0}{L}_1 \\ \vdots \\ \overset{0}{L}_i \\ \vdots \\ \overset{0}{L}_{n-1} \\ \overset{0}{L}_n + ke \cdot p_{ke} \end{bmatrix} \qquad \overline{p}k = \begin{bmatrix} \overset{1}{p}_{k0} \\ \overset{1}{p}_{k1} \\ \overset{1}{p}_{ki} \\ \overset{1}{p}_{kn-1} \\ \overset{1}{p}_{kn} \end{bmatrix}$$

$$\overline{M}k = \begin{bmatrix} (ka + k_0 - Fk_0) & -k_0 & 0 & 0 & 0 & 0 & 0 \\ -k_0 & (k_0 + k_1 - Fk_1) & -k_1 & 0 & 0 & 0 & 0 \\ 0 & \ddots & \ddots & \ddots & 0 & 0 & 0 \\ 0 & 0 & -k_{i-1} & (k_{i-1} + k_i - Fk_i) & -k_i & 0 & 0 \\ 0 & 0 & 0 & \ddots & \ddots & \ddots & 0 \\ 0 & 0 & 0 & 0 & -k_{n-2} & (k_{n-2} + k_{n-1} - Fk_{n-1}) & -k_{n-1} \\ 0 & 0 & 0 & 0 & 0 & -k_{n-1} & (ke + k_{n-1} - Fk_n) \end{bmatrix}$$

Figure 3.14 Elements of the solution system.

3.3 Structure of the Simulation Program

The simulation program for moisture and heat transport used for the calculations was developed by the book author in the programming language Fortran on the basis of the explanations in Section 3.2, the corresponding Windows interface was created by Deckelmann, compare also [34].

The following characterizes the structure of the program:

Inputs and pre-processing

- Component dimensions and definition of the element mesh
- Read all data sets to be processed successively with time duration in hours via a txt file with the corresponding boundary conditions of the surfaces concerned (surface temperatures, heat transfer coefficients, RH, v_{Air}, definition of moisture exchange: capillary water absorption only, vapor exchange only, sealed)
- Parameters for the adsorption and desorption isotherm(s) to be used
- Parameter $ww[\mathrm{kg/(m^2\ h^{0.5})}]$
- Initial water content in $\mathrm{m^3/m^3}$ and parameters for its distribution
- Height of possible water external pressure
- Initial temperature distributions + initial heat transport parameters
- Default moisture storage functions (ads-isotherm and des-isotherm) for the start of the calculations if necessary under consideration of the boundary conditions for the moisture exchange.
- Choice if scanning should be calculated
- Moisture storage function-dependent moisture parameters θ_{min}, θ_{gel}, θ_{sat}, and θ_{sat0}
- Input parameters *ksat*0 [kg/(m² s Pa)] and *fksat* [–].
- Basic parameters for vapor transport (dry cup)
- Inputs for calculating the function(s) fkt for decreasing of the moisture transport parameters of degree of hydration, drying, and time
- Selection of the VEs for the plot of the associated moisture storage curves in post-processing
- Choice of calculation models and values of moisture transport coefficients

Calculation of heat and moisture transport for the read-in data sets

- Read in and assign all data sets with possibly changing boundary conditions and time durations from file "Input."
- Definition of initial time step size Δt + specification of the measure of increase of time steps in the course of calculation
- Initial position characterizations in the $\theta(Phi)$ plane, ($\theta(Phi)$ values of all VEs)
- Initial distribution of capillary pressure and vapor pressure
- Calculation of boundary conditions θ_{a0}, θ_{e0}, $p_{cap_{a0}}$, $p_{cap_{e0}}$, $p_{vap_{a0}}$, $p_{vap_{e0}}$
- Intermediate outputs on the monitor
- Calculation of all temperature and moisture changes over a time step Δt up to the total duration of the data set (i) in the following order:
- Heat capacity and heat transport coefficients

- Capillary transport coefficients according to Sections 5.2.1 and 5.2.2 as a function of t and θoo
- Vapor transport coefficients
- Scanning management of all VEs in the $\theta(Phi)$ diagram [storage of all VE positions = $\theta(Phi)$ values before and after each δt, checking for direction reversal, calculation of new scannings and associated slopes.....], conversion of scanning results to the $\theta(p_{cap})$ plane
- Condensation/evaporation heat
- Temperature distribution T_{new}
- Slopes $d\theta/dp_{cap}$ at all VEs (along the isotherms and scannings) for capillary transport with/without vapor or for vapor alone
- **Balance calculation of capillary pressure for capillary pressure distribution $p_{cap}new$**
- Calculation of new moisture distributions $\theta(p_{cap})$ and new scannings
- New vapor pressure distribution p_{vap}
- New positions of all θ values of all VEs in the $\theta(p_{cap})$ plane and conversion to $\theta(Phi)$ plane
- Storage of step number and time totals
- Processing of result files
- Storage of the results as a function of t and the x-dependent variables for all time periods Δt of all data sets also for the output
- Output of the scanning traces in the $\theta(p_{cap})$ diagram relative to the moisture storage functions for the associated graphical representation
- Storage of all intermediate results for the start of the next time step or data set

Post-processing:

- Additional control output at monitor
- Output of results to files or output files with adjusted number of data points.

References

1 Todd, D.K. (1959). *Ground Water Hydrologie*. London, New York: Wiley.

2 Rucker, P. and Beddoe, R.E. (2007). Transport von drückendem Wasser in Betonbauteilen. *Beton-und Stahlbetonbau* 102: 414–426.

3 Rucker, P. (2008). Modellierung des Feuchte-und Salztransports unter Berücksichtigung der Selbstabdichtung in zementgebundenen Baustoffen. Dissertation. Technischen Universität München.

4 Garrecht, H. (1992). Porenstrukturmodelle für den Feuchtehaushalt von Baustoffen mit und ohne Salzbefrachtung und rechnerische Anwendung auf Mauerwerk. Dissertation. Universität Karlsruhe.

5 Lunk, P. and Wittmann, F.H. (1996). Feuchtigkeits-und Salztransport in Beton I. Versuchsergebnisse und analytisches Modell. *Werkstoffwissenschaften und Bauinstandsetzen (4. Intern. Kolloqu., Esslingen).*

6 Lalibert, M. (2007). Model for calculating the viscosity of aqueous solutions. *Journal of Chemical and Engineering Data* 52: 321–335.

7 Lalibert, M. (2009). A model for calculating the heat capacity of aqueous solutions, with updated density and viscosity data. *Journal of Chemical and Engineering Data* 54: 1725–1760.

8 Ruiz-Llamas, A. and Macías-Salinas, R. (2015). Modeling the dynamic viscosity of ionic solutions. *Industrial and Engineering Chemistry Research* 54: 7169–7179.

9 Carmeliet, J., Descamps, F., and Houvenaghe, G. (1999). A multiscale network model for simulating moisture transfer properties of porous media. *Transport in Porous Media* 35: 67–88.

10 Carmeliet, J., Hens, H., Roels, S. et al. (2004). Determination of the liquid water diffusivity from transient moisture transfer experiments. *Journal of Thermal Envelope and Building Science* 27: 277–305.

11 Krus, M. (1996). Moisture transport and storage coefficients of porous mineral building materials: theoretical principles and new test methods. Reprint Dissertation. (Stuttgart University) by Fraunhofer IRB Verlag.

12 Roels, S. (2000). Modelling unsaturated moisture transport in heteogeneous limestone. Dissertation. Leuven: Katholiek Universiteit Leuven.

13 Scheffler, G.A. (2008). Validierung hygrothermischer Materialmodellierung unter Berücksichtigung der Hysterese der Feuchtespeicherung. Dissertation. Technischen Universität Dresden.

14 Funk, M. (2012). Hysteresis der Feuchtespeicherung in porösen Materialien. Dissertation. Technische Universitt Dresden.

15 Carlier, J.-Ph. and Burlion, M. (2011). Experimental and numerical assessment of the hydrodynamical properties of cementitious materials. *Transport in Porous Media* 86: 87–102.

16 Yang, F., Lyu, B., and Xu, S. (2021). *Water Sorption and Transport in Shales: An Experimental and Simulation Study*. Water Resources Research.

17 Andrade, C. and Coppens, E. (2020). Quantifying water permeability and pore size through capillary absorption. *4th International RILEM conference on Microstructure Related Durability of Cementitious Composites (Microdurability2020)*, 485–492.

18 Zhou, C., Chen, W., Wang, W., and Skoczylas, F. (2016). Indirect assessment of hydraulic diffusivity and permeability for unsaturated cement-based material from sorptivity. *Cement and Concrete Research* 82: 117–129.

19 Christensen, B.J., Mason, T.O., and Jennings, H.M. (1996). Comparison of measured and calculated permeabilities for hardened cement pastes. *Cement and Concrete Research* 26: 1325–1334.

20 Nokken, M.R. and Hooton, R.D. (2008). Using pore parameters to estimate permeability or conductivity of concrete. *Materials and Structures* 41: 1–16.

21 Ye, G. (2005). Percolation of capillary pores in hardening cement pastes. *Cement and Concrete Research* 35: 167–176.

22 Yonghao, F., Zhongli, W., and Yue, Z. (2008). Time-dependent water permeation behavior of concrete under constant hydraulic pressure. *Water Science and Engineering* 1 (4): 61–66.

23 Phung, Q.T., Maes, N., De Schutter, G. et al. (2013). Determination of water permeability of cementitious materials using a controlled constant flow method. *Construction and Building Materials* 47: 1488–1496.

24 Tumidajski, P.J. and Lin, B. (1998). On the validity of the Katz-Thompson equation for permeabilities in concrete. *Cement and Concrete Research* 28(5): 643–647.

25 Liu, X. (2009). Transport properties of concrete with lightweight aggregates. Dissertation. University of Singapore.

26 Zhang, Z. and Scherer, G. (2018). Determination of water permeability for cementitious materials with minimized batch effect. *physics.app-ph.*

27 Patel, R.A., Phung, Q.T., Seetharam, S.C. et al. (2016). Diffusivity of saturated ordinary Portland cement-based materials: a critical review of experimental and analytical modelling approaches. *Cement and Concrete Research* 90: 52–72.

28 Jooss, M. and Reinhardt, H.W. (2002). Permeability and diffusivity of concrete as function of temperature. *Cement and Concrete Research* 32: 1497–1504.

29 Thompson, A.H. (1991). Fractals in rock physics. *Annual Review of Earth and Planetary Sciences* 19: 237–261.

30 Rucker, P., Beddoe, R.E., and Schießl, P. (2006). Wasser-und Salzhaushalt im Gefüge zementgebundener Baustoffe –Modellierung der auftretenden Mechanismen. *Beton-und Stahlbetonbau* 101 (6): 402–412.

31 Lockington, D., Parlange, J.-Y., and Dux, P. (1999). Sorptivity and the estimation of water penetration into unsaturated concrete. *Materials and Structures/Materiaux et Construction* 32: 342–347.

32 Zhou, C. (2014). General solution of hydraulic diffusivity from sorptivity test. *Cement and Concrete Research* 58: 152–160.

33 Zhang, Z. and Angst, U. (2020). A dual-permeability approach to study anomalous moisture transport properties of cement-based materials. *Transport in Porous Media* 135: 59–78.

34 Deckelmann, G. (2014). *Handbuch für das Programm ASTRA (Allgemeines Stoff-Transport Simulations-Programm).* Hamburg University of Technology TUHH.

4

Experimental Investigations with Regard to the Modeling of Moisture Transport in Mortars and Concrete

4.1 Preliminary Remarks on Moisture Storage

Moisture storage in porous materials is known to occur by hygroscopic water uptake from the surrounding vapor atmosphere or by liquid water uptake via the capillary transport mechanism when the material under consideration does not yet have water-saturated pores. In the material, moisture transport processes result, which have the tendency to assume stationary moisture distributions. The speed resp. the temporal course of these processes depend on many parameters, for example, on the initial moisture distribution in the material, on the pore structure of the material, the resulting storage capacities and transport parameters, as well as on the nature and temporal change of the hygric resp. climatic boundary conditions.

In order to model these processes, it is therefore necessary to know the extent of the moisture storage capability of the material under consideration and its moisture transfer properties within the material when unevenly distributed moisture contents resp. concentrations are present.

The following chapter therefore deals first with the moisture storage properties of the materials considered here at boundary conditions leading to an increase in moisture content and to their decrease with decreasing ambient moisture contents and, if necessary, drying. Finally, however, results on the capillary transport properties and the observed dependencies of the vapor transport coefficients are also presented.

In the present chapter, the explanation of sorption and desorption isotherms occupies an important place. Their different characteristics have a lasting effect on the storage and, in particular, transport behavior of water in porous systems. Measurement results on sorption isotherms have been available in the literature for many years. They differ from material to material in form and degree of expression. The IUAPC Techn. Report 1985, reviewed in 2015 [1] attempts a material-dependent classification.

Moisture Storage and Transport in Concrete: Experimental Investigations and Computational Modeling, First Edition. Lutz H. Franke.

The existence of pronounced differences between adsorption and desorption isotherms has also long been known for concrete, for example, Feldman and Sereda 1968 [2]. **Figure 1.2a**, or Ahlgren 1972 [3]; **Figure 1.2b**, for the field of soil mechanics, for example, Mualem 1974 [4]. The presence of "paths" between these curves (here referred to as scanning isotherms, or scannings for short) can also be inferred from these publications, where the markedly different expression becomes apparent. These correlations have a considerable influence on moisture transport behavior and will therefore also be discussed in more detail below. Explanations of the causes of this behavior can be obtained from Section 4.4.2.

Results on similar behavior of other building materials can be taken, for example, from Funk [5]. It is now state-of-the art that even non-cementitious materials can exhibit hysteretic wetting and drying behavior. Some of these results will be shown below.

4.2 Concrete Data for the Experimental Investigations

For a detailed investigation of the moisture storage and moisture transport behavior, hardened cement paste (HCP), mortars, and concrete using different cements were employed. A central role is played by the so-called reference mortar (REF), which has been used for various projects at the institute over a period of about 20 years, and whose properties are very well known with regard to different issues. This is the test mortar according to DIN EN 196-1, with a W/C value of 0.45. For the production of the mortar, the cement CEMI 42.5 R-SR 0 (Sulfadur Doppel of the manufacturer Dyckerhoff/Buzzi Unicem) with high sulfate resistance ($C_3A = 0$ according to Bogue) and low alkali content (Na_2O-equivalent = 0.50) was always used, compare **Table 4.1**.

To answer the various questions, mortar series of the REF of different sizes were produced several years in succession in each case. After 24 hours of 98% RH air storage, all series were stored mainly under water in comparison, series were also stored above water (≈98% RH) or sealed with different storage times from 28 days to 420 days.

The required 15 [cm]-long prisms, the cubes of dimensions $4\cdot4\cdot4\,[cm^3]$ and the mortar disks of thickness 7 [mm] resp. 8 [mm] were then produced from the prisms $4\cdot4\cdot16\,[cm^3]$ by precise sawing a few days before the start of the experiment. In sealed storage, the prisms were wrapped with double foil after 24 hours and further stored at about 20 °C until the start of the experiment. Compaction of the mortar- and HCP-prisms at 120 seconds by vibration table.

Table 4.1 contains the compositions of the various cement-bound materials with which experiments cited here were carried out at TUHH in various projects.

Table 4.1 Composition of the cement bonded materials used for reported tests of TUHH in various projects.

Mixtures		Cement	Cement/ manufactory	W/C	Ratio cement/ aggregate
Hardened cement paste	HCPI	CEMI 42.5 R	AB	0.40 0.50 0.60	1/0
	HCPII	CEMII/B-S 32.5 R	F	0.40 0.60	
	HCPIII	CEMIII/B 42.5 R	F	0.40 0.60	
	REF	CEMI 42.5 R-SR 0	Sulfadur Doppel[a)]/ Dyckerhoff-Buzzi Unic.	0.45	1/3 2 [mm] max. aggregate size
	MA	CEMI 42.5 R	A[b)]	0.40 0.50 0.60	
Mortar	MAB	CEMI 42.5 R	AB[c)]	0.40 0.50 0.60	
	MIII	CEMIII/B-42.5 R	F	0.40 0.60	
Concrete	C35/45	CEMI 42.5 R	350 [kg/m^3]/$F1$	0.45+plasticizer	8 [mm] max. aggregate size
	C25/30	CEMI 42.5 R	350 [kg/m^3]/$F1$	0.60	8 [mm] max. aggregate size

For material MA/MAB with $W/C = 0.40$ plasticizer was used.
a) High resistance to sulfates ($C_3A = 0$ by Bogue) and low alkali content (Na_2O-equivalent = 0.50).
b) Producer Ennigerloh-Nord/HeidelbergCement (Germany).
c) Producer Alsen AG/Holcim (Germany) AG.

4.3 Data on Porosity of the Considered Materials and Influence of Treatments on Porosity

4.3.1 MIP Results for Pore Size Distribution and Pore Volume

Mercury pressure porosimetry analyses were performed at various ages of all materials used in the experimental studies. Characterizing mercury porosimetry curves and porosity characteristics of the REF mortar and the normal concretes listed in **Table 4.1** are shown below.

To clarify the influence of the drying conditions, a 105 °C drying series was performed on material REF in addition to the 40 °C/1 mbar drying defined as the standard drying, and its influence on the moisture retention was investigated. **Figure 4.2a,b** shows for comparison the cumulative curves as well as the pore size distribution for the drying conditions mentioned after 84 days of initial water storage.

It is confirmed that a higher water content is expelled by the 105 °C drying. There is a pore increase in capillary pores in the pore size range above 12.5 [nm] and a slight decrease in gel porosity. The consequences of 105 °C drying for water storage and water transport are reported in Section 4.3.4.2.

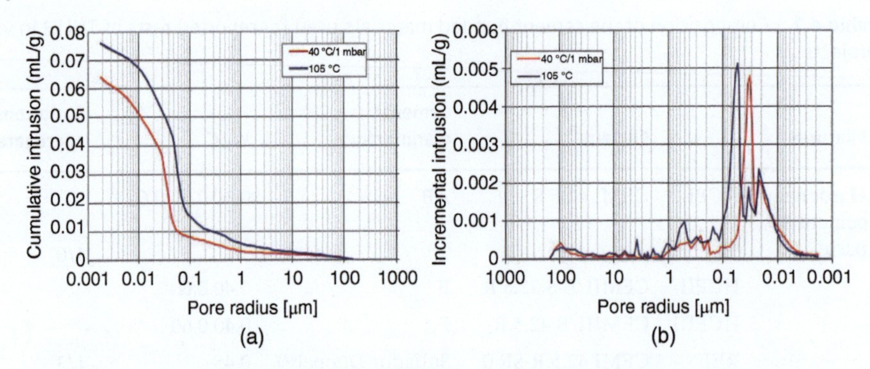

Figure 4.1 Comparison of MIP-results after standard-drying at 40 °C/1 mbar and 105 °C-drying of mortar REF. (a) Cumulative pore volume curves. (b) Pore size distribution.

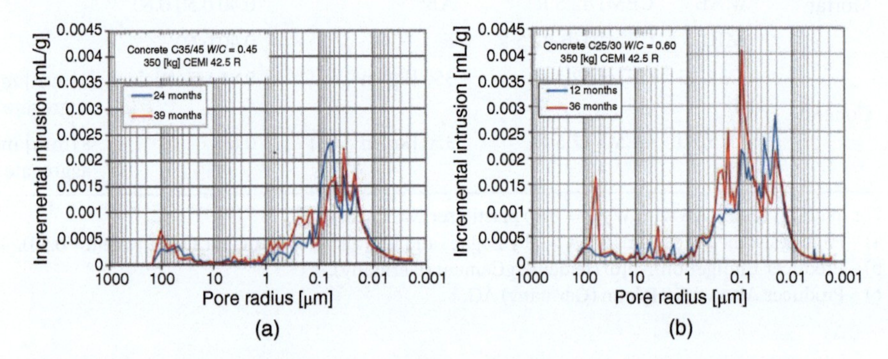

Figure 4.2 Comparison of MIP-results after 105 °C-drying of concrete. (a) Pore size distribution of **C35/45**. (b) Pore size distribution of **C25/30**.

The influence of curing age (56 days and 84 days) on pore-size distribution was checked on REF mortar. According to this, between 56 days and 84 days there is still a pore content decrease of about 10% in the pore size range from 10 to 100 [nm].

Porosimeter measurement results for the concretes C25/30 **and** C35/45 are shown in **Figures 4.2a and 4.1b**. These measurements were made after storage times ≥ 12 months after 105 °C drying. Both concretes have the same cement content of 350 [kg/m^3], the same aggregate grain type, and the same date of manufacture. **Figure 4.2a** shows the cumulative pore volume curves for the C35/45 with $W/C = 0.45$, and **Figure 4.2b** shows the cumulative volume curves for the C25/30 with $W/C = 0.60$, each for 2 ages of the concrete. Among other things, a significantly higher pore volume of the C25/30 in the **capillary pore range from 12.5 [nm] to approximately** 200 **[nm]** is recognizable. Furthermore, the concrete with the higher W/C value indicates a slightly higher proportion of gel pores in the pore region below 12.5 [nm], which is consistent with the behavior of the adsorption isotherms.

4.3.2 Control of the Carbonation Behavior of the Test Specimens

Carbonation leads to densification of the pore system resp. to a significantly lower water absorption capacity. Espinosa [6], for example, writes on the basis of her experimental results "the pore size distribution of a carbonated cement paste specimen is characterized by decrease in pore volume over the entire pore size range." Furthermore, the effect of porosity change on sorption is significant. The moisture content decreases at 92% RH by 60%.

The influence of carbonation on the REF mortar was therefore checked. **Figure 4.3** shows a cube cross section after treatment with phenolphthalein. The cross-section in **Figure 4.3** refers to cubes that were stored under water for 84 days, exposed to 55% RH for four weeks (i.e. not dried), and also stored at 20 °C/65% RH for more than two years after subsequent water saturation. For these **cubes**, the carbonation depth evident in **Figure 4.3** was obtained over the study period.

Carbonated edge samples and non-carbonated core samples were taken from corresponding cubes and tested in the mercury porosimeter, **Figure 4.4**. A decrease in capillary porosity in the carbonated zone and an increase in gel-pore size (pores with radii $\lessapprox$ 12.5 [nm]) are clearly visible. Furthermore, the water absorption after 55% RH storage was investigated. It amounts to about 50% of the carbonated edge compared to the non-carbonated condition. The results confirm the basic finding of a decrease in storage capacity compared to non-carbonated samples. The saturation water content of the carbonated areas also decreases significantly. This means that moisture storage modeling functions of carbonated samples must lead to significantly lower water content values.

In addition, a series of **prisms of mortar REF** were tested for carbonation under different storage conditions after about 2.5 years, compare **Figure 4.5**. Based on the phenolphthalein test, the observed carbonation was lower than that in **Figure 4.3**.

A REF slice series was subjected to drying and wetting storage for 600 days, with the last part of the cycle consisting of 235 days of water storage (**Figure 4.27**). Subsequently, fracture surfaces of the slices were tested with phenolphthalein (**Figure 4.6b**).

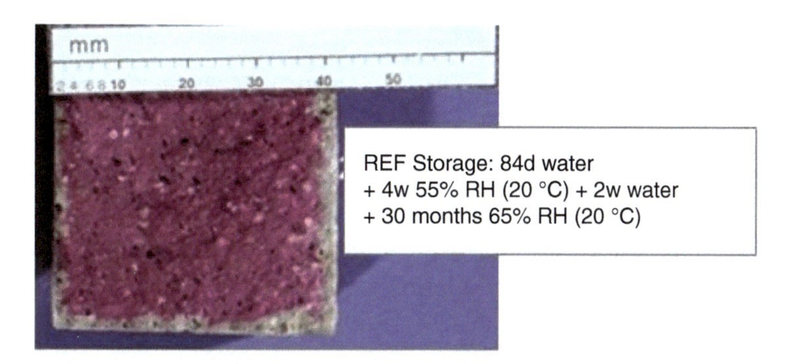

Figure 4.3 Carbonation of the edge of a REF-Mortar cube 4 × 4 × 4 [cm^3] after long-time exposure to 4 weeks at 55% RH, 2 weeks of water storage, and 30 months 65% RH.

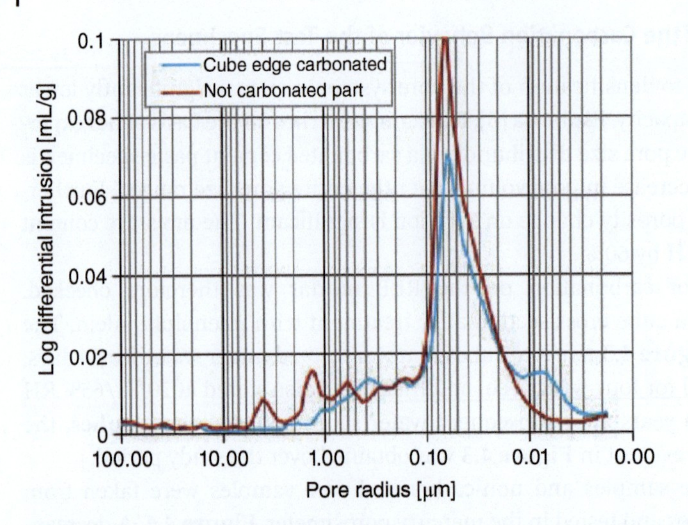

Figure 4.4 MIP pore distribution of the non-carbonated part and the carbonated edge of the REF-mortar cube in **Figure 4.3.**

Figure 4.5 Part from a **prism** (REF series after 84 days of initial water storage, 2 months drying at 40 °C/1 mbar, 2 months subsequent water storage, and 2 years further storage at 55% RH without partial sealing by foil) tested for carbonation. The specimen exhibits a thin carbonated surface layer.

Another series of slices was stored according to **Figure 4.7b**. **Figure 4.7a** shows one of the final tested fracture surfaces. **Figure 4.8,** on the other hand, shows the fracture surface of a slice from a series that was **not finally water-stored.**

These tests make it clear that a reliable prediction of the speed of carbonation and the carbonation depth is difficult. As is well known, **the moisture content of the samples plays a decisive role,** as do the drying speed and duration of the

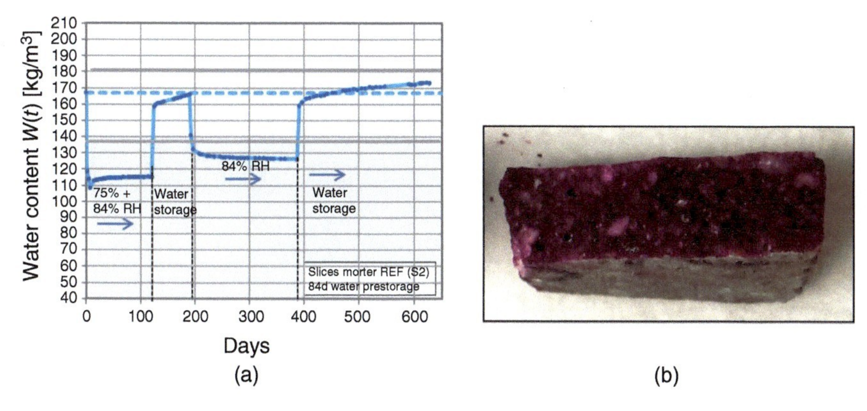

Figure 4.6 Mortar REF, Slice (series S2), thickness ≊7 [mm]. (a) Water content W of the slice series during several storage periods before carbonation testing (84d water prestorage + 120d 75%/85% RH (20°C) + 70d water storage +197d 85% RH (20°C) + 235d water storage); Water content relative to dry mass: $m_W = W/(\varrho_{dry,REF} \cdot 1000)$; $\varrho_{dry,REF} \approx 2.10$.
(b) Test with phenolphthalein of a broken cross section of one of the slices at the end of the multi-storage.

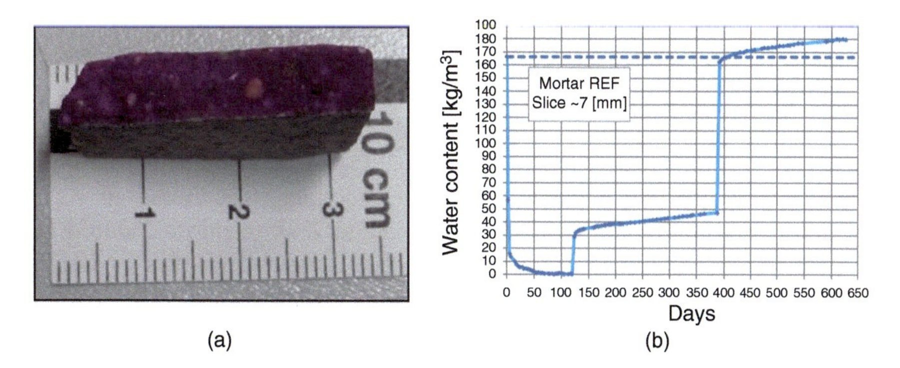

Figure 4.7 Mortar REF, Slice (series S3), thickness ≊ 7 [mm]. (a) Test with phenolphthalein of a broken cross section. (b) Water content W of the slice-series during several storage periods before testing (84d water prestorage + 120d standard drying + 267d 54% RH (20°C) + 240d water storage); $m_W = W/(\varrho_{dry,REF} \cdot 1000)$; $\varrho_{dry,REF} \approx 2.10$.

dryness. For a given material, the available "water reservoir" of the material sample is important for maintaining the highest possible moisture content near the sample surface through water replenishment from the sample interior. According to this, the carbonation rate is greater for slices than for cubes and greater there than for prismatic bodies. It decreases with higher cement content and decreasing WC values.

Further, results on the influence of carbonation on moisture storage (for example, the sorption isotherms) and the change in transport properties are given in Section 4.6.8.

The influence of thin carbonate surface layers on the moisture exchange, in particular the drying rate, is further substantiated in this book on the basis of experimental investigations and comparative calculations in Section 5.3.

Figure 4.8 Slice (thickness ≊3.5 [mm]), cut from the bottom of a prism at the end of a 6.5 months unidirectional capillary water imbibition test. Bottom surface and lateral face have been treated with phenolphthalein. Storage before imbibition test: 84d water, 7d 40 °C/85% RH + 350d 20 °C/85% RH without coating the lateral prism faces. The surface of the bottom is carbonated.

4.3.3 Air-Porosity Content of the Materials Used

In [7], Gätje/TUHH reports, among other things, on the digital figure analysis system she has developed for fully automatic determination of the air and compaction pores via light microscopy with a stepper motor-controlled sample stage on specially prepared thin sections of the material to be examined. According to the resolution limit of light microscopy, which is about 5 [μm], pores with pore diameters ≥ 10 [μm] are detected by the method. **Figure 4.9a and b** are measured examples. The blue shapes correspond to the processing that caused air voids and pores present. On the Portland-cement mortar MA (compare **Table 4.1**), the pore contents and their size distribution were measured for W/C values of 0.40, 0.50, and 0.60, reproduced in **Figure 4.10**.

On specimens made of prisms of material REF, **the air void characteristics of manufacturing caused air content was examined** according to DIN EN 480-11

Figure 4.9 Thin-sections with air voids filled with blue resin, materials without air-entraining agent (a) Mortar MA, max. aggreg. size = 2 [mm] (≈1.5% of aggreg. mass = 4 [mm]) (b) Concrete C30/37, $W/C = 0.45$, $CEMI = 370\,[kg/m^3]$, max. aggregate size = 16 [mm]. Source: Results from Gätje [7]/Hamburg University of Technology.

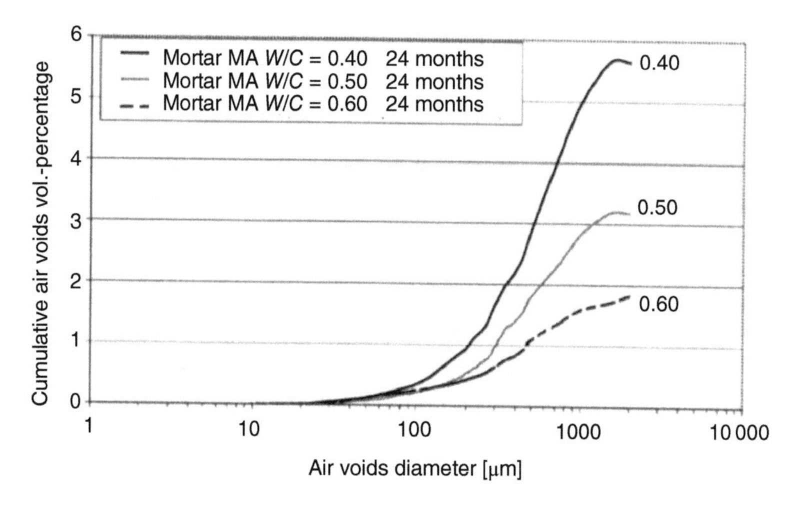

Figure 4.10 Air voids content of mortar MA (made without air-entraining agent) dependent on W/C-value, measured by digital image analysis of thin-sections. Source: From Gätje/TUHH [7]/Universität Hamburg.

Table 4.2 Air voids content of mortar REF made without air-entraining agent, measured by chord length counting method according to European Standard EN480-11 (measured at 1 mortar prism).

Air-voids total volume [%]	Micro air content A_{300} vol. [%]	Distance factor L [mm]
4.81	0.40	0.68

(German/European Standard) on the hardened mortar (by the chord method). The determined air void contents and the spacing factor can be taken from **Table 4.2**. According to this, the total air void content is at a lower measurement limit of 5[μm] is 4.8% in total, while the air void content A300, which is relevant from the point of view of frost resistance, as the air void content between 5 and 300[μm] shows only 0.4% vol., **as expected, since no air-entraining agent was used** in the production of the REF mortar. For admixture-free mortars with different water-cement ratios, Gätje [7] determines about 10[μm] as the smallest pore radius using the 2D image analysis method, compare **Figure 4.10**. Molendowska et al. [8] also finds these smallest pore radii for different concretes in the measurement procedure according to EN 480-11 and with the 2D image analysis method used.

The total air void content of the mortar REF with $W/C = 0.45$ can be derived from this diagram to ~4.5%, which approximates the total pore volume of 4.8% (**Table 4.2**) measured according to standard EN 480-11 by an external institute on another sample.

The air void contents of the C25/30 **and** C35/45 concretes used here (**Table 4.1**) are 1.0% resp. 2.50%, respectively, according to the manufacturing concrete plant. However, the actual air-entrained content of C25/30 concrete is estimated to be more like 1.5–2.0%.

From Liu and Hansen [11] and, for example, Molendowska et al. [8], where a larger number of concretes containing air-entraining agents were studied, it can be concluded that in such concretes there is an accumulation of air voids present at pore-**radii** between about 30 and 120 [μm]. In the case of simultaneous pronounced presence of manufacturing caused air voids, further accumulation is present at pore radii of approx. 300 to 1000[μm]. **When superplasticizers are used, the air-porosity content** decreases significantly. Therefore an initial total air-void volume of approx. 5% to 6% is necessary depending on aggregate size.

Figures 4.11 and 4.12 show measurement results from Liu and Hansen [11] on the pore size distribution of air voids within a concrete, that was installed as road pavement concrete on the one hand and as laboratory concrete on the other. With regard to the highway application, the concrete produced according to ASTM C 192 with a W/C value of 0.45 was produced using a polycarboxylate-based superplasticizer and a synthetic air-entraining agent. The pore size distributions were measured by the linear traverse method according to ASTM C457 on concrete cores taken from the highway and on corresponding samples of the concrete produced in the laboratory. The measurement results of four samples of the highway concrete are shown in **Figure 4.11**, and the corresponding results of laboratory concrete are shown in **Figure 4.12b**. Shown are the air void content distributions as a function of pore size (**Figures 4.11a and 4.12a**) and figures of polished concrete sections (Figures **4.11b and 4.12b**). The diameter-dependent pore fractions are shown as vertical columns, and the air void content sums as curves. The laboratory concrete in **Figure 4.12** shows a pore sum of approx. 2.0% below 300 [mm] and is thus higher than the minimum required of 1.8%.

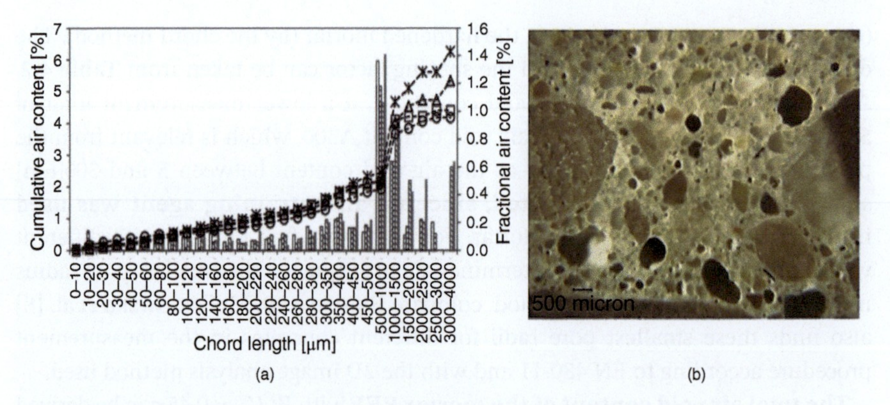

Figure 4.11 Size distribution of air voids in concrete samples from a Michigan highway, obtained by drilling cores. (a) Measured by chord length counting method according to ASTM C457 or EN480 – 11. (b) Microscopic picture of a polished sample surface. Source: (a,b) Liu et al. [9]/with permission of Elsevier.

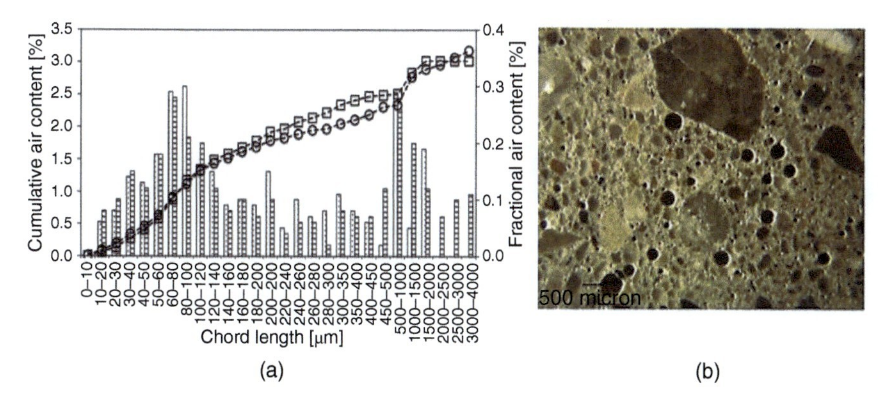

Figure 4.12 Size distribution of air voids in lab concrete mixtures with air entraining agent $W/C = 0.45$, 1 year air curing. (a) Measured by chord length counting method according to ASTM C457. (b) Microscopic picture of a polished sample surface of this concrete. Source: (a,b) Liu et al. [9]/with permission of Elsevier.

In the 60–100 [μm] range, pore contents of up to approx. 2.5% can be seen. The average pore spacing was measured at 0.13 [mm] and is therefore lower than the required minimum spacing of 0.2 [mm]. However, the total pore volume is only 3% (including the processing-related pore fraction) and is therefore significantly lower than the total pore fraction of approx. 5–6% required for safety. Deviating results are obtained for the highway concrete. This concrete only has a pore volume of approx. 1.5% up to 300 [μm] and a lower air content at smaller pores. The pore spacing was measured at 2.3 [mm]. The total air content, on the other hand, shows the desired 5% to 6%, mainly due to large manufacturing pores over 1 to 4 [mm] in diameter. Overall, the possible change in air/porosity content in the course of pavement construction can be clearly seen. Nevertheless, it can be expected that the highway concrete with a pore content of A300 of 1.5% still has an "acceptable" frost resistance.

Figure **4.13a,b** shows the internal interfaces of air voids to the concrete matrix using SEM images. With regard to the stability of the contained air at water saturation, the permeability of these interfaces for molecules is important.

4.3.4 Drying Methods and Influence of Drying

4.3.4.1 Drying Methods

The present chapter is intended to deal primarily with the moisture storage capacity of REF mortars after drying and after different initial water contents. The results of measurements on slices, usually 7 [mm] thick, as well as cubes and prisms of 15 [cm] length are used, occasionally also of half prisms. A very important question for the whole complex of water absorption is the question of which drying conditions may be chosen for establishing the dry state of the material. This condition includes the value of the dry bulk density ϱ_{dry}, to which certain water contents can be related as % of the dry weight, as kg-water content/[m^3], or dimensionless as [m^3/m^3].

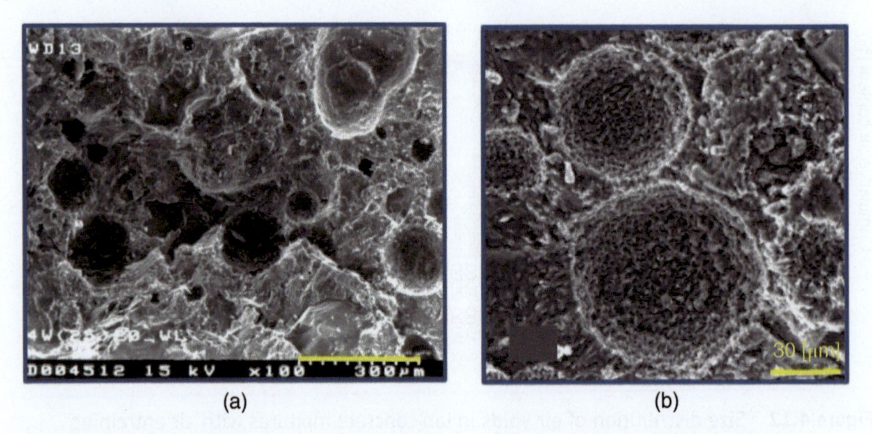

(a) (b)

Figure 4.13 SEM images of inner surfaces of air voids in concrete mixtures with air entraining agent . (a)$W/C = 0.45$, 340 [kg] CEM I, 5.5% air pores (b) HCP CEM I, $W/C = 0.50$, pore diameter 300 [μm]. Source: (a,b) Bollmann [10]/Bauhaus University Weimar.

In the past, it was common practice to dry cement-bound materials at 105 °C as well, including the concrete tests with C25/30 and C35/45, the investigation results of which are used at least partially in the present project.

Different drying methods can be used to achieve the desired drying condition. These drying processes must be evaluated with respect to their intensity and, if necessary, their dehydrating effect on the chemically bound water. A number of comparative tests have been carried out for this purpose. Furthermore, experience from research projects is available. Compare also [12].

On this basis, a list of methods according to their effectiveness is given below.

- **105 °C to mass equilibrium:**
  This method has the advantage of rapidly removing evaporable water from porous systems, but it is less suitable for materials that already lose chemically bound or crystalline water below this temperature. It is therefore only suitable for special cases for cement-bound products, compare the explanations in Section 4.3.4.2.
- **freeze-drying:**
  First, the sample is frozen at −40 °C in an external freezer, which freezes the pore water into pores >1[nm]. The material must therefore be frost-resistant when freeze-drying is used. In subsequent main drying, below the triple point of the phase diagram of water (0.13 °C and 6.1 mbar) external pressure under vacuum, water is largely expelled by sublimation. The equivalent relative air humidity is less than 0.04% RH at 23 °C, Espinosa [6]. The measured values in **Table 4.3** show that more water is dried out by freeze-drying alone than by standard drying plus subsequent vacuum and pressure treatment.
- **40 °C/1 mbar-drying in a vacuum oven:** This drying method, which has since been extensively tested, is defined here as the **standard method** (see also Section 4.6.9). It requires comparatively short drying times and prevents carbonation influences due to the low air pressure. For comparison, for example,

Table 4.3 Hardened cement paste HCPIII (mixture see **Table 4.1**), prestorage 12 weeks at 98%RH/20 °C: Influence of the drying methods on the evaporable water content measured after prestorage, [%] relative to dry mass.

HCP III	$W/C = 0.40$		$W/C = 0.60$	
drying method	**Drying [%]**	**Drying + pressure [%]**	**Drying [%]**	**Drying + pressure**
3% RH (silica-gel)	22.40	23.60	37.40	
40 °C/1 mbar	23.50	24.50	38.50	38.7
Freeze-drying	24.90	26.90	40.80	

sorption isotherms on cement-mortar samples were measured using the desiccator method after drying with 40 °C/1 mbar as well as at 40 °C over silica gel. The results are shown in **Figure 4.14a,b**. One can see the small difference between the congruent adsorption isotherms for the two mortars MIII with W/C values of 0.60 and 0.40.

- **40 °C-drying over silica gel: Figure 4.15** shows measurement results of the drying influence of 40 °C-drying over silica gel and of 40 °C/1 mbar-drying in a vacuum oven for comparison.
- **23 °C-drying over silica gel alone at atmospheric pressure:** The tests carried out on this method here gave the following results:
  Using a large desiccator (height 40 [cm] and inner diameter 33 [cm]) with a weight of 260 [g] of silica gel, a humidity of 2.4% RH was established at 21 °C. Using a medium-size desiccator (height 28 [cm], inner diameter 20 [cm]) and 630 g silica gel, a humidity of 1.7% RH was established at 21 °C, in each case without moist material to be dried. Since a moderately higher internal moisture occurs during material drying, the air humidity must be controlled with a measuring device. The silica gel should be renewed at least every 14 days. In the work of Espinosa/TUHH [6], silica gel drying (defined there as 3% drying) was also used in addition to the previously defined freeze-drying. Information on the

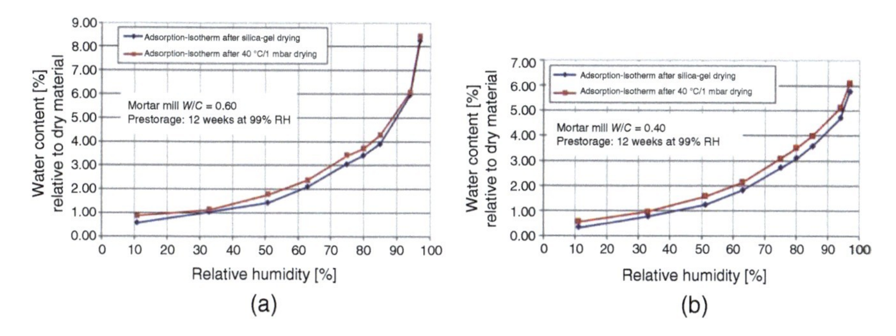

Figure 4.14 CEMIII-Mortar MIII, prestorage 12 weeks at 98% RH/20 °C: Comparison of Influence on the adsorption isotherms of drying procedures 40 °C/1 mbar drying and 40 °C-silica-gel drying. (a) $W/C = 0.60$; (b) $W/C = 0.40$.

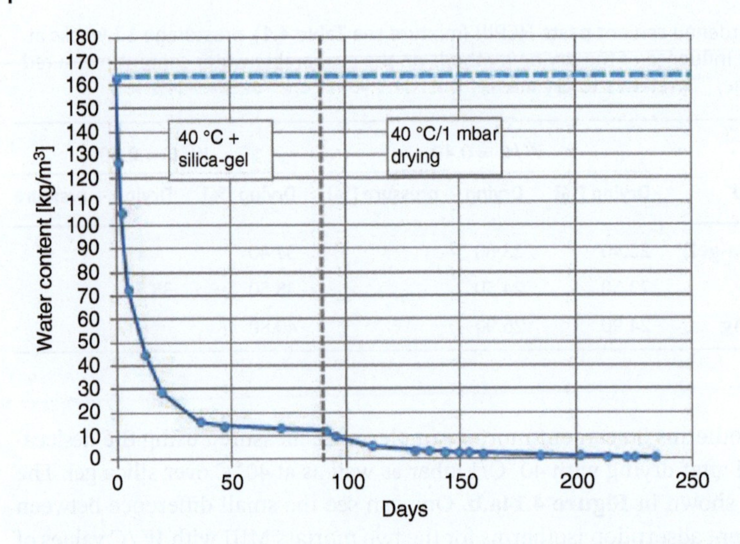

Figure 4.15 CEMI-Mortar REF, prestorage 12 weeks at 98% *RH*/20 °C: Influence on the adsorption isotherms of drying procedures 40 °C/1 mbar drying and 40 °C/silica-gel drying.

frequency of silica-gel renewal and on the RH values resulting in the desiccators was not provided.

The residual moisture remaining in the material after drying by silica gel-drying compared to freeze-drying (or even 40 °C/1 mbar-drying) can also lead to differences in the generation of sorption isotherms [6].

- **Water removal over special salt solutions or alkalis in desiccator:** The saturated salt solutions and other solutions that can be used can be obtained from EN ISO 12571 or in ISO 12570 and EN ISO 483 also for higher evaporation temperatures. The results of control measurements can be obtained from **Table 4.4.**

For very mild, stable temperature-insensitive drying at 11.3% and 20 °C or 23 °C, a $LiCl{\cdot}H_2O$ solution can be used advantageously.

The definition of the moisture equilibrium to be achieved by drying also plays a decisive role for the degree of drying, in particular for the respective drying time. According to EN ISO 12571, desorption tests or adsorption tests can be terminated resp. the moisture equilibrium can be considered as reached, if after each two following exposure days, the change of the total mass of the respective sample is smaller than 0.1%. What this means for reaching sorption equilibrium for mortar slices of dimension 40·40·7 [mm^3] is shown by the measured values in **Figure 4.16a,b**, which were determined in the course of measuring an adsorption isotherm. The red curves show the total mass change in % at the respective measurement times, in each case related to two past days according to the definition of EN ISO 12571. It can be seen that for the 35% RH exposure, the limit values according to ISO 12571 are already reached after about one week and for the 85% RH after about

Table 4.4 Salt solutions to generate defined relative air humidities for isotherm experiments.

Salt solution	RH(%) at 20 °C
K_2SO_4	97.7
KNO_3	93.7
KCI	85.0
$(NH_4)_2SO_4$	81.0
$NaCl$	75.4
$Mg(NO_3)_2$	55.7
K_2CO_3	44.0
$MgCl_2$	33.1
$CaCl_2$	33.3
$LiCl$	11.5

Values from Michael Steiger, University Hamburg and from TUHH.

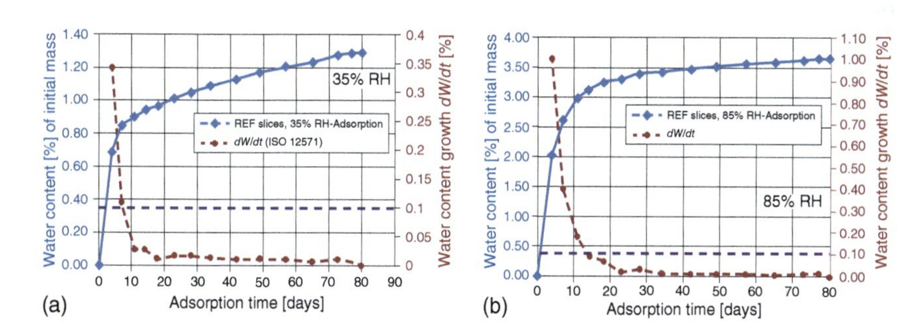

Figure 4.16 CEMI-Mortar REF, 84d water prestorage, slices 40 · 40 · 7 [mm^3]: After 40°C/1 mbar drying, exsiccator exposure to 35% RH (Figure (a)) and 85% RH (Figure (b)) until equilibrium state (blue curves). The red curves dW/dt show the slope of the mass decrease, the horizontal dashed lines show the equilibrium states according to ISO 12571, allowing hypothetically a measuring stop.

14 days (time until the red dashed line crosses below the horizontal dashed line). From the course of the blue curve of the mass loss is evident that, up to this time, the equilibrium has clearly not been reached. Therefore, the drying process was prolonged up to approx. 80 days.

4.3.4.2 Possible Influence of Drying on Capillary Water Uptake

The influence on capillary water absorption of 105 °C drying was compared with drying at a temperature of only 40 °C, accelerated by simultaneous application

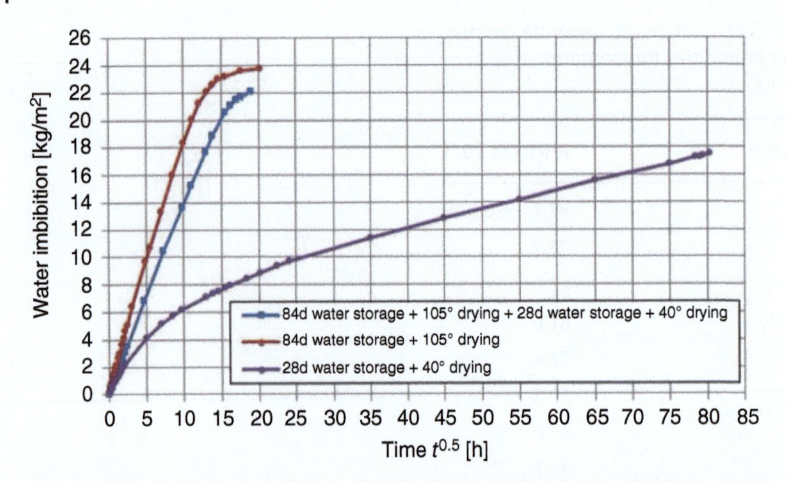

Figure 4.17 Influence of drying procedure on imbibition tests of CEMI-Mortar REF prisms 40·40·160 [mm^3] after 28d water prestorage : Imbibition curves after 105 °C, after 105 °C + 28d water storage + 40 °C/1 mbar drying and after 40 °C/1 mbar drying. Long time water storage does not repair the initial harmful 105 °C drying effect.

of a negative pressure of 1 mbar. Last method used as standard drying corresponds approximately to drying over silica gel, compare **Figure 4.14a,b**.

Figure 4.17 contains the mean curves from three prisms each of the unidirectional capillary water uptake of the REF mortar after 105 °C drying (red curve), after 105 °C drying plus subsequent 28 days of water storage of the prisms and drying at now 40 °C/1 mbar **to check whether structural regeneration can be expected** by long-term water storage after 105 °C drying. A slightly lower capillary water uptake rate (blue curve) can be seen compared to 105 °C drying alone, but a likewise straight-line behavior of the curve up to its curvature, caused by reaching the top (back) side of the prisms and by slight evaporative influence. On the other hand, by drying the same REF material only at standard drying conditions of 40 °C/1 mbar, we obtain a curved curve with a much slower capillary water uptake rate and a pronounced curvature behavior in the first two days of water uptake. These results show high reliability, recognizable by the individual curves (not reproduced here) from which the mean curves shown were formed.

In addition, MIP investigations were performed on the question of the regeneration ability of the hydrated microstructure through by extended water storage. **Figure 4.18a,b** contains the measurement results. It can be seen first (**Figure 4.18b**) that after 105 °C drying, compared to 40 °C drying, there is a significantly higher pore volume, predominantly of capillary pores, in the range between about 40 nm and 4 [μm]. The 28-day water storage after the 105 °C drying leads (according to the subsequent MIP analysis) to a limited pore size change in the capillary pore area but not to a sustainable regeneration of the porosity to the lower total pore volume before the 105 °C drying.

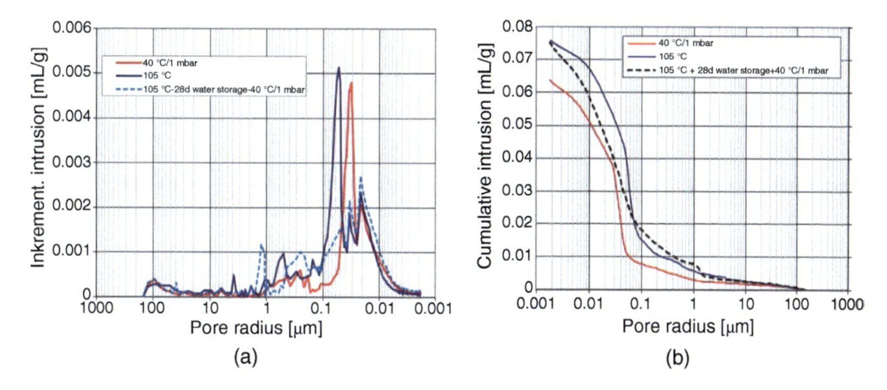

Figure 4.18 CEMI-Mortar (REF), 84d water prestorage: Influence of drying procedure on porosity. Comparison of MIP pore distribution curves (a) and cumulative pore volume (b) after 105 °C-drying and following 28d water storage and drying with standard-drying procedure.

The result of these comparative measurements is that with a 105 °C-drying of cement-bound products a change of the cement gel takes place with a subsequent change in moisture storage and transport behavior. The swelling capacity of the cement-gel can obviously also be changed by a long 105 °C-drying,

It follows that, as a standard drying method for cement-bound products, 105 °C drying should not be used if other moisture engineering tests or parameters are to be based on it. Also, the Powers/Hansen method presented in Section 2.2 and modified there for the advance calculation of properties of a hardened concrete would not provide sufficiently reliable results without adaptation of parameters used.

Gluth [13] states in connection with results of mercury pressure porosimetry that the gel pores of the HCP are particularly severely damaged during 105 °C drying. Further discussion of the effect of drying can also be obtained from Jennings et al. [14].

The measured water contents after 84d water storage after manufacture are given for the REF mortar used as a reference in **Table 4.5** for standard drying at 105 °C. **Table 4.5** contains water storage results of REF measured after standard drying in [kg/m^3] and 105 °C drying in comparison, further an example of calculation of the 84d value.

$$Water\ content = \varrho_{dry} \cdot mass = 2100 \cdot 7.80\% = 163.8 \quad [kg/m^3] \tag{4.1}$$

For example, a water content $m_{105\,°C}$, which was related to the drying temperature of 105 °C, can be converted to the standard drying temperature of 40 °C/1 mbar as follows:

$$m_{(40\,°C/1\,mbar)} = \frac{(1 + m_{105\,°C}) \cdot \varrho_{105\,°C}}{\varrho_{(40\,°C/1\,mbar)}} - 1 \tag{4.2}$$

This conversion possibility refers only to water contents that were not determined after possibly "harmful" drying, such as the drying of the samples.

Table 4.5 Measured water content and water adsorption values of mortar REF after 84d water storage and following drying, measured on cubes 40 [mm^3].

Mortar REF 84d water prestorage	Standard-drying (40 °C/1 mbar)		Drying at 105 °C	
	mass-% [a)]	kg/m³	mass-% [b)]	kg/m³
Initial water content after 84d water-prestorage	**7.80**	**163.8**	**9.13**	**190.2**
Equilibr. water adsorption (by water imbibition after drying)	6.84	143.8	8.00	166.8
Max. water adsorption (by water storage plus 150 bar pressure)	**8.58**	**180.8**	**9.77**	**203.5**
Bulk density of dried material (mean values) [kg/dm³]	**2.105**		**2.085**	

a) In relation to standard-drying-mass.
b) In relation to 105°-drying-mass.

Thus, the conversion refers to adsorption and desorption isotherms and scannings, for example, in **Figure 4.60a,b** where the **water content** in the adsorption tests was determined after drying to 3% RH, or to desorption isotherms where the initial value of the curve (lowest φ–value) is not sharper than standard drying resp. the aforementioned values were 3% RH, and where at the end only all values were related to the 105 °C dry mass.

4.4 Hysteretic Moisture Storage Behavior as Important Issue with Respect to Modeling

In Sections 1.1 and 4.1, it was already clearly pointed out the by now long-known peculiarity in the moisture storage of cement-bound products, namely the clear difference in the functions describing the water uptake and the water release. If this material behavior is to be modeled as accurately as possible, extensive additional knowledge of the storage and transport behavior is required, based on Chapter 3 "Basic equations." In Section 4.4.1, the sorption isotherms for the reference material REF are first described again in more detail. Subsequently, in Section 4.4.2, the causes of the hysteretic storage behavior are illustrated schematically.

In Section 4.4.3, questions are formulated that must first be answered in order to model the processes.

4.4.1 Adsorption and Desorption Isotherms of the CEMI Reference Material

First, the adsorption and desorption isotherms of the reference REF mortar are shown in Figure 4.19, measured at 20 °C. The measurements were made on

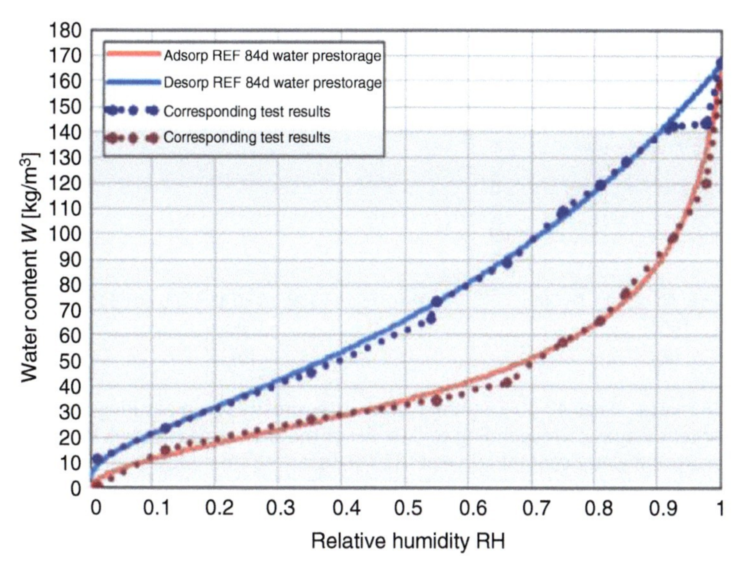

Figure 4.19 CEMI-Mortar REF, 84d water prestorage, $W/C = 0.45$: Measured basic-isotherms, standard-drying. Water content relative to dry mass: $m_W = W/(\varrho_{dry,REF} \cdot 1000)$; $\varrho_{dry,REF} \approx 2.10$.

slices of dimensions $40 \times 40 \times 7.5$ [mm], which were cut out of the water-stored prisms 14 days before the start of the sorption measurements and then stored again under water until the start of the experiment. Previously, the prisms had been stored in the humidity box for 1 day after manufacture and then under water for 83 days.

The measurement points shown in **Figure 4.19** are mean values of three test specimens each. All desorption experiments were started at the same time and stored in desiccators over the associated salt solutions for each relative humidity selected. Some of the selected salts and associated relative humidities are noted in **Table 4.4**. Measurements were continued on all test specimens until no further change in weight in mg was recorded between two measurement times at least three days apart. Regarding the **definition of moisture equilibrium, reference should be made to the Section 4.3.4. Figure 4.20a,b** shows exemplary measurement curves (mean values of three samples in each case) for relative air humidity 81% RH . It is evident that a period of 80 days was necessary until the moisture equilibrium of the adsorption slices was reached after standard drying at 40 °C/1 mbar.

The measurement results resp. sorption isotherms for the material REF can be taken from **Figure 4.19**. The discrete measurement points were then fitted using a suitable function corresponding to the solid lines in **Figure 4.19**.

For the analytical description of these sorption isotherms, Eq. (4.3) was chosen as the relative water content in [m³/m³] related to 1 [m³] of mortar or concrete. The function contains four parameters A, a, b, c to be adjusted via a regression and allows with the respective set of these parameters the representation

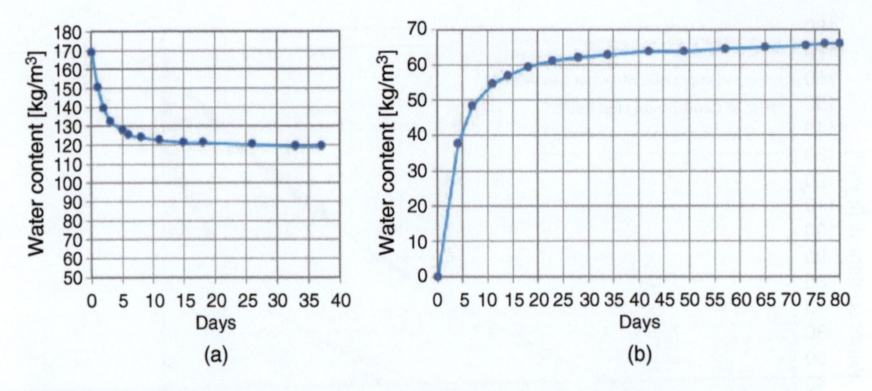

Figure 4.20 REF mortar slices 40·40·7.5 [mm³] , 84d water prestorage, measuring of the two water content values corresponding to 81% RH of the isotherms **Figure 4.19**. (a) Measured desorption water content curve at 81% RH starting from prestorage water content. (b) The adsorption curve at 81% RH starts from standard drying water content.

of the desorption as well as adsorption curve. For the function (4.3), the inverse function (1.52) as well as the functions $\theta(p_{cap})$, $\theta(p_V)$, their inverse functions as well as partial derivatives can be represented in a closed way with the same set of the coefficients mentioned below .

$$\theta(\varphi) = a \cdot \left[\left(\frac{A}{\varphi}\right)^{c} - 1\right]^{-1/b} \quad [\mathrm{m^3/m^3}] \tag{4.3}$$

The coefficients ($Aabc$) for the desorption and adsorption curves of the reference mortar REF are, in this order :

Des : 2.47655 3.01954$D-7$ 0.61668 3.17587$D-4$
Ads : 1.04043 0.03989 2.13534 1.17173

For curves showing the water content of test specimens, for example, as a function of relative air humidity or time, [kg/m³] instead of [m³/m³] was predominantly used as the unit for the water content by multiplying the water content by 1000 for example (4.3). Where dry bulk density is required in certain conversions, $\varrho = 2.10$ [g/cm³] was generally used for the mortar REF.

As a regression function for the measurement points of the sorption isotherms, other formulations can also be used. If one decides, as it is done here, to carry out all following computational calculations with closed functions of the sorption isotherms and the functions derived from them, it is to be checked first whether these function transformations and derivations are possible without further effort and are representable with the same parameter set.

4.4.2 Causes of Differences Between Adsorption and Desorption Isotherms

Several correlations can be considered as causes of hysteretic moisture storage behavior.

With cylindrical pores open at both ends, for example, the phenomenon that no meniscus initially forms when pores are filled but only a condensate from the water vapor on the inner surface, the thickness of which increases with increasing relative vapor pressure until the sorbate layers collapse. Desorption after pore filling then occurs via meniscus formation, compare also **Figure 1.28a,b**. One can show, see for example Espinosa [6], Espinosa and Franke [15], or Cohan [16], that at least theoretically the amount of sorbate taken up at an air humidity φ_{Ads} is released again in this case at a desorption air-humidity ${\varphi_{Ads}}^2$. Further discussion of this mechanism can be found in Section 1.4.6.2. The mechanism can be expected at pore radii larger than about 1 [nm]. For the validity of the Kelvin equation, see also Section 1.3.2.

The effect can be demonstrated using the synthetic material MCM-41 made of mesoporous silicates, which have regularly arranged parallel cylindrical pores open on both sides, compare the TEM image (**Figure 2.1a**) in Section 2.1.

By Min Wu et al. [17] comparative sorption experiments were performed on MCM-41 in water-vapor atmosphere as a function of relative air humidity. **Figure 4.21** shows the measured adsorption and desorption isotherms. The curves on this synthetic material confirm the previously described mechanism of water vapor sorption and desorption for this pore structure.

In Figure 2.1b these are nitrogen sorption isotherms. It is evident that the sorption isotherms of the variants with constant radii of 2.2 [nm] show only a small hysteresis in nitrogen sorption, compare **Figure 1.30b**.

In the practice of porous construction materials, this cause of hysteresis has little significance because of the nonconstant pore radii of the pore networks and the action of closed pore roots. Nevertheless, it can be demonstrated that water-vapor sorption at the internal surfaces contributes to hysteretic moisture storage behavior even in cement-bonded products.

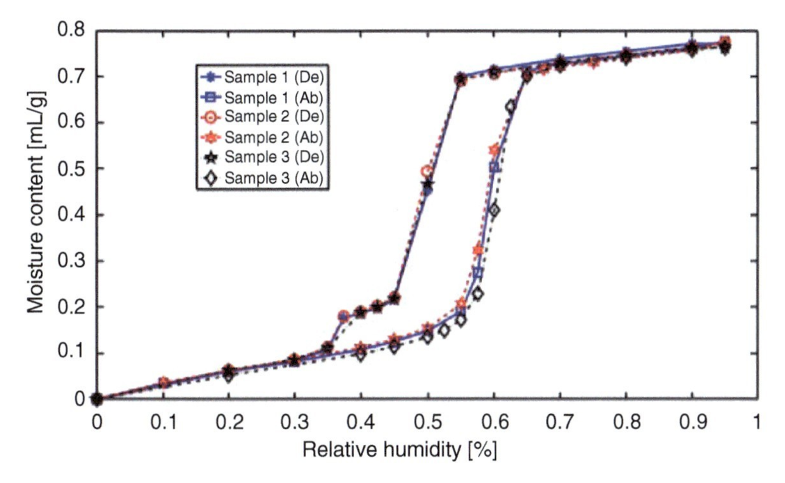

Figure 4.21 Water-vapor adsorption and desorption curves measured on synthetic material MCM-41 with regularly parallel cylindrical pores with sample radius of about 2 [nm]. Source: Wu et al. [17]/with permission of Elsevier.

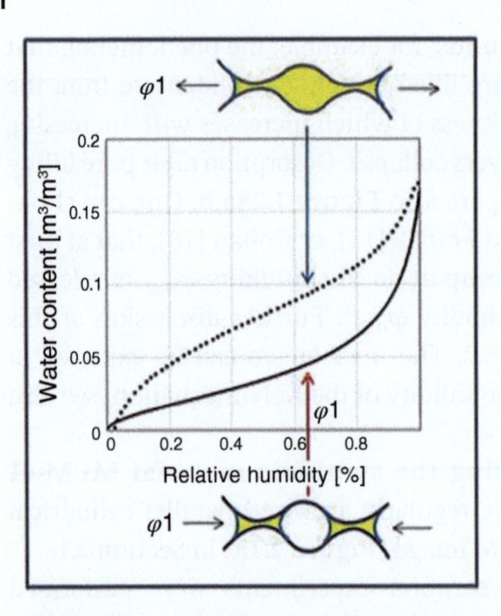

Figure 4.22 Schematic illustration of the cause of the difference between water adsorption- and desorption curve in the case of cement bound materials and materials with similar pore networks.

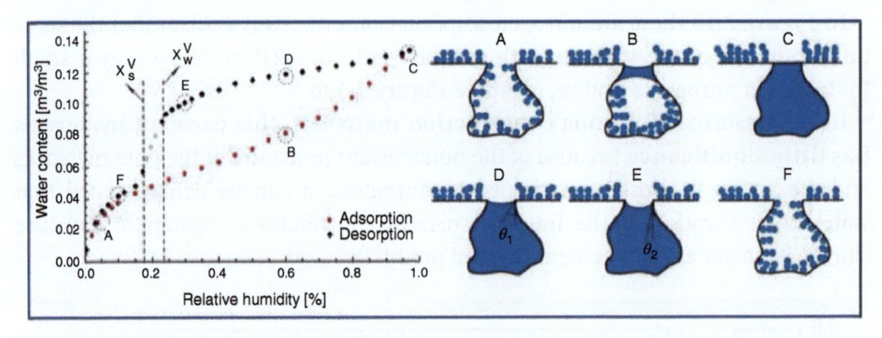

Figure 4.23 Schematic illustration from Zandavi and Ward [18] and Zandavi [19]/American Chemical Society of water adsorption- and desorption curves of a pore in shale rock.

Main cause the presence of pronounced hysteresis between liquid adsorption and desorption are the pore constrictions present in pore networks, commonly referred to as bottle-neck pores. Their effect is shown schematically in **Figures 4.22 and 4.23**: The liquid trapped in larger pores after adsorption can only escape during desorption when reaching lower vapor partial pressures resp. relative air humidity where the correspondingly higher capillary pressures corresponding to the smaller radii of the access pores. **Figure 4.22** shows this effect for pores with narrower accesses and exits, **Figure 4.23** from Zandavi [19] shows the behavior for assumed bag pores.

4.4.3 Questions with Respect to Modeling of Storage and Transport Processes

A number of questions arise regarding storage and transport behavior, the answers to which are necessary for at least satisfactory modeling of the moisture budget. These include the following relationships:

- Existence and course of the moisture storage functions for drying and adsorption for the material under consideration (sorption isotherms)
- Moisture storage via scanning curves and their shapes.
- Differences in saturation moisture for hygroscopic and capillary water uptake and over-hygroscopic behavior
- Influence of degree of drying and material age on storage behavior and the transport velocity during capillary water uptake and vapor transport
- Influence of pore size distribution, air void shape, and test specimen shape on storage and transport.

These relationships are discussed below in this book based on the results of experimental investigations and theoretical considerations.

4.5 Water Storage Behavior Under Changing Moisture Boundary Conditions with Consideration of the Air-Pore Content

Presented first are investigation results on the storage behavior of the cement-bound material, in particular the selected reference material REF, which are required for the description and modeling of the moisture content and moisture transport. These results are needed because in the relevant literature there is no sufficiently suitable information on the considered resp. interesting behavior, neither of cement mortars nor of concrete. Results are shown, which were obtained on slices on the one hand and on cubes and prisms on the other. It is shown that the water absorption capacity differs significantly and systematically for slices on the one hand and cubes and prisms on the other hand.

4.5.1 Illustration of the Structure-Related Pore Volume Fractions in Relation to the Total Storage Capacity

For a better understanding of the moisture storage behavior and with regard to a subsequent computational modeling, the tool from Section 2.2 provides excellent services. A useful function is thus fulfilled, e.g. **Figure 4.24a** in the representation of measured time-dependent changes in moisture content under changing

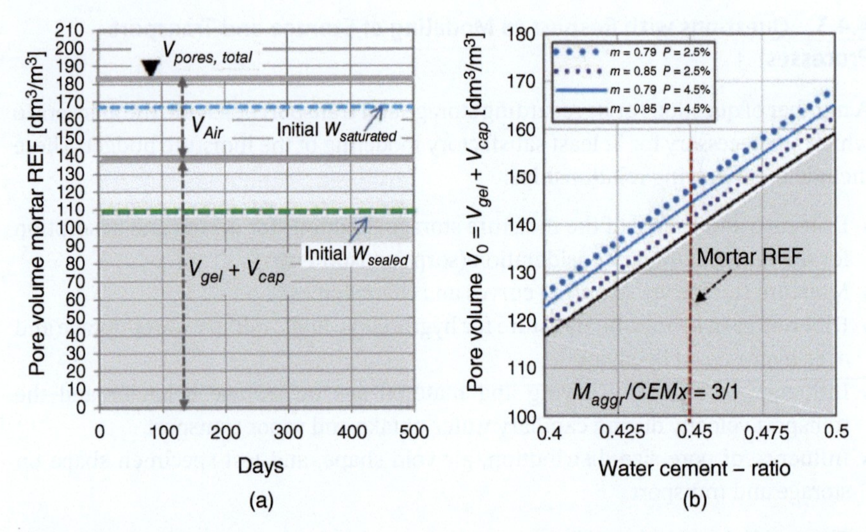

Figure 4.24 Illustration of the volumes of the pore types of mortar REF, calculated with the tools of Section 2.2, (Mix Portland cement CEMx, $M_{Aggr}/CEMx = 3/1$, $V_{Air} = V_0 \cdot 4.5\%$, hydratation degree $m = 0.82$) relative to the water content W; (a) The light-gray zone and the two horizontal dark-gray lines show the limits of $V_{gel} + V_{cap}$ and the total porosity, respectively, including the initial air pores V_{Air}. The dashed green line indicates the calculated water content in the case of sealed storage. The dashed blue line illustrates the often measured water content after longer water storage. (b) $V_{gel} + V_{cap}$ - dependence on W/C, V_{Air} and on hydration degree m. Water content relative to dry mass: $m_W = V_0/(\varrho_{dry,REF} \cdot 1000)$; $\varrho_{dry,REF} \approx 2.10$.

storage conditions in relation to the calculated volumes of the pore types resp. the structure-related pore volume fractions.

In the following reproduction and explanations of test results on storage behavior, use is therefore made of the supplementary elucidation given by **Figure 4.24a**. **Figure 4.24b** points out the dependence of the pore volume $V_{gel} + V_{cap}$ per [m^3] on **the degree of hydration m** according to Eqs. (2.11) and (2.13) and on the air-pore volume P (besides the dependence on the W/C value).

4.5.2 Long-Term Moisture Storage Behavior of Slice-Shaped Test Specimens

First of all, test results are reported, which were obtained on slices of dimensions $4 \cdot 4 \cdot 0.7$ [cm^3], which were worked out from prisms initially water-stored after fabrication. The objective was to verify the moisture storage behavior of the REF mortar under different storage conditions. The aim was to check whether the water contents specified by the sorption isotherms, resp. the initial moisture contents, were regained during alternating storage in different humidities resp. dryings and/or capillary water absorption.

The dimensions of the test specimens were the same as those used to determine the sorption isotherms. The slices were stored under water after fabrication (1 day

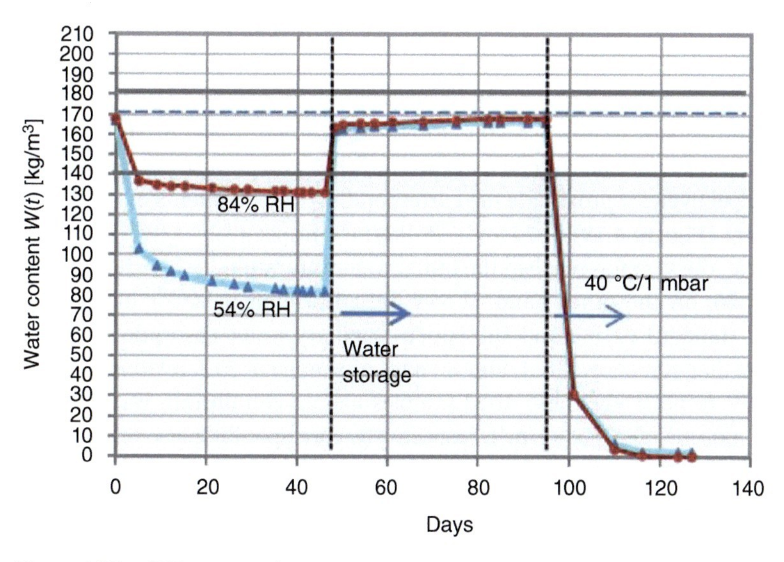

Figure 4.25 REF mortar-slices 7.5 [mm] (series C1 and C3): 240d water prestorage. Water content $W(t)$ curves measured, corresponding to the exposure and time data specified in the picture. Water content relative to dry mass: $m_{W(t)} = W(t)/(\varrho_{dry,REF} \cdot 1000)$; $\varrho_{dry,REF} \approx 2.10$.

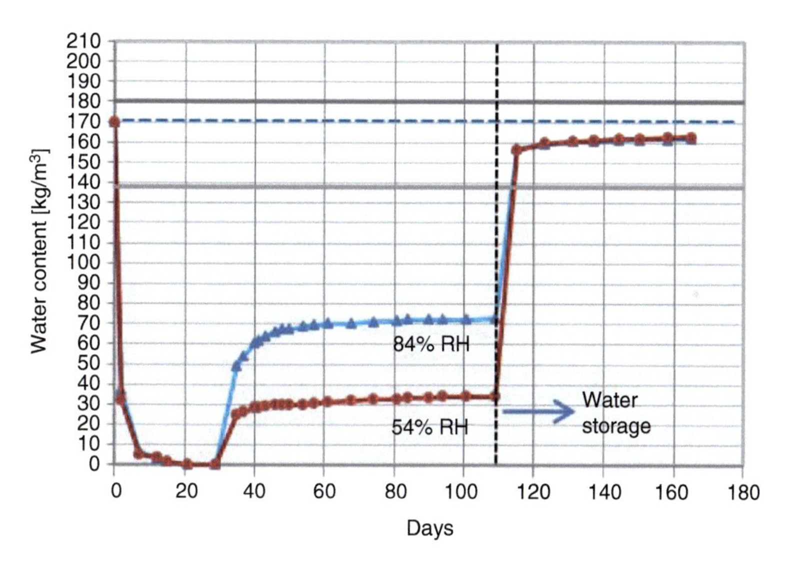

Figure 4.26 Water content curves measured on REF mortar-slices 7.5 [mm] (series C2 and C4): 240d water prestorage, 29d standard-drying, 82d at RH 84%/20 °C (on C2) and RH 54%/20 °C (on C4), respectively, finally 56d water storage.

of wet storage) until the actual start of the experiment, for varying lengths of time, starting from 28d up to a maximum of 420d.

Figures 4.25 and 4.26 now show selected results of these slice tests. The measuring points in the diagrams are in each case the mean values from three samples. The sample age was of secondary importance.

Figure 4.25 shows the water content changes of two water-stored sample series by desorption at 54% as well as at 84% RH until sample equilibrium after 46 days, subsequent capillary water absorption until equilibrium by immersion over 49 days as well as subsequent standard drying. From the diagram, it can be seen that the equilibrium water contents after 54% RH and 84% RH storage of 80 [kg/m^3] and 131 [kg/m^3], respectively, approximate the corresponding water contents of the desorption isotherm (**Figure 4.19**) of 74 and 128 [kg/m^3], respectively. After subsequent water storage, the samples reached the initial water saturation.

Figure 4.26, on the other hand, shows the adsorption behavior of the slices after initial standard drying and subsequent capillary water absorption by water storage. The diagram shows that after 80 days of storage at 54% relative humidity and 84% RH, respectively, these dried samples had an equilibrium water content of about 34 [kg/m^3] and 72 [kg/m^3], respectively, corresponding also to the values of the adsorption isotherms (**Figure 4.19**). During subsequent water uptake by immersion for about 50 days, the two series of slices reached the same equilibrium water content, but not the initial value.

In addition, the results of the two **Figures 4.25 and 4.26** make clear the effect of **the hysteretic storage behavior on resulting equilibrium moisture contents.**

Figure 4.27 shows an additional experimental series which, starting from water storage, was first subjected to desorption at 75% RH, then at 84% RH to equilibrium, then immersed for about 70 days to the initial saturation value, then dried again by desorption at 84% RH to equilibrium, and then stored in water.

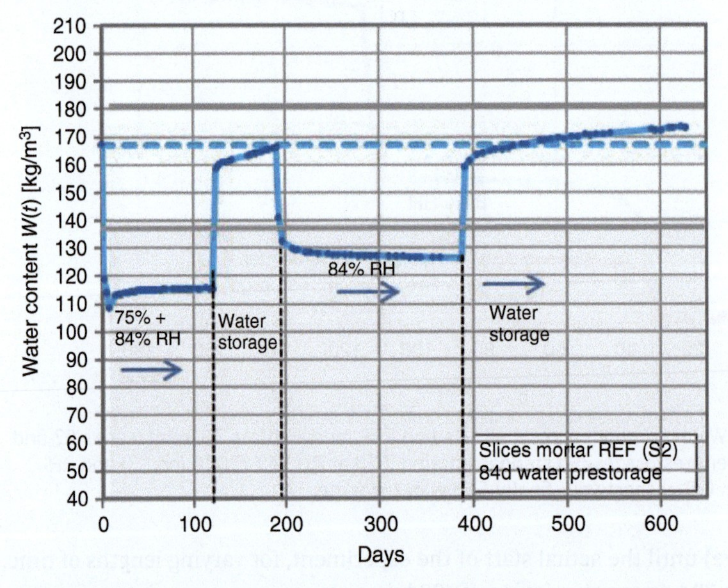

Figure 4.27 Water content $W(t)$ curves measured on REF mortar-slices 7.5 [mm] : 84d water prestorage, 120d at RH 75%/85% at 20 °C, 70d water storage, 197d at RH 85%/20 °C and 236d water storage following. Water content relative to dry mass: $m_{W(t)} = W(t)/(\varrho_{dry,REF} \cdot 1000)$; $\varrho_{dry,REF} \approx 2.10$.

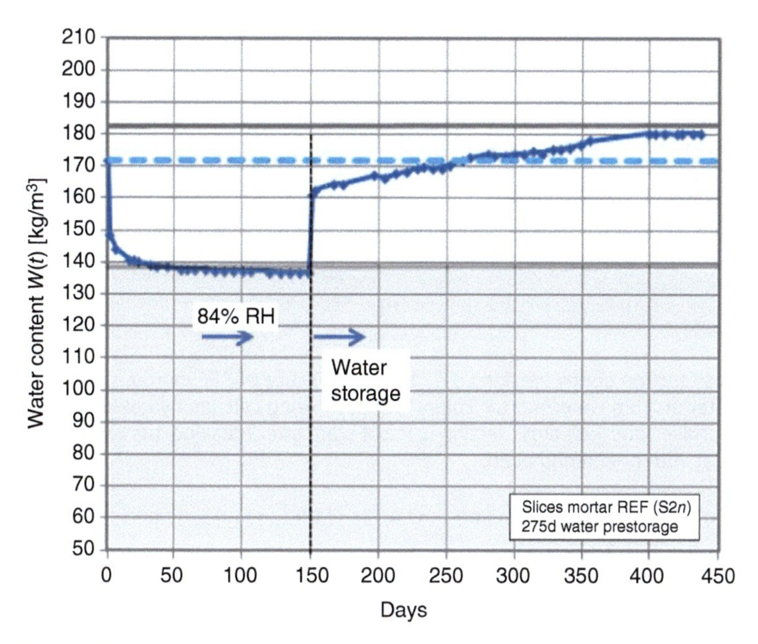

Figure 4.28 Water content curves $W(t)$ measured on REF mortar-slices 7.5 [mm] : 275d water prestorage, 140d desorption at 84% RH/20 °C and 294d water storage following.

After the 1st and 2nd drying phases at 84% RH, the equilibrium moisture contents at 84% RH did not reach the same value. This is due to the hysteresis effect of the shorter initial drying portion at 75% RH before the first 84% RH drying phase. A more detailed explanation of this can be obtained from Section 4.6.2. During the final long immersion period, the slices exceeded the initial saturation state in contrast to the much shorter post-water-stored slice series in **Figure 4.26.**

In the experiments of Figure 4.28, after about 400 days of water storage of the (previously not dried) slices, a complete filling of all pores ascertainable by standard drying, i.e. also of all air pores, took place. **This process of air-pore filling will be discussed further** in Section 4.5.4.

The result of these investigations is that after initial water storage of the slices until equilibrium saturation, an appreciable proportion of the air pores are water-filled. During long-term water storage, this pore fraction can increase further to complete pore filling, at least for slice-shaped test specimens, as shown by the results in **Figures 4.27 and 4.28.** After intensive drying from the initial saturation state, e.g. by standard drying, the filling of the air pores is obviously at least slowed down during renewed water storage.

4.5.3 On the Question of Dissolution of Portlandite from Slice-Shaped Test Specimens During Water Storage

Figure 4.29 shows the cross-section (fracture surfaces) of a slice of mortar REF stored in a larger volume of water for at least five years. Significant carbonation and

Figure 4.29 Cross section got by breaking a $4 \cdot 16 \cdot 1\,[\mathrm{cm}^3]$ slice of REF mortar, stored 5 years under water at a big water/mortar-volume ratio, showing calcium hydroxide leaching in the surface zone. Left cross section is in the state after breaking, the counterpart on the right treatet with phenolphthalein.

possibly portlandite loss have occurred on the surfaces. The question must therefore be answered as to whether falsification by dissolution of portlandite may have occurred in the slice tests previously shown. This is the case with the mortar slice shown in **Figure 4.29**. During the sometimes long storage periods after prism manufacture up to the start of the test, the test specimens were either sealed or stored under water. The water storage took place in covered plastic boxes using tap water. Water loss during prolonged storage was compensated for by the addition of tap water.

SchulteHolthausen, Raupach, Merkel and Breit have **experimentally investigated the dissolving influence of deionized water and tap water on different types of concrete** in [20].

When exposed to soft resp. deironized water, they measured a total removal resp. mass loss of approx. 200 [g] per [m^2] concrete surface within half a year for concretes with $W/C = 0.50$ with regular water bath renewal and brushing weekly of the exposed concrete surfaces.

With thin material slices of approx. 7 [mm] and the assumption of exposure on both sides and that the mass loss leads to a corresponding increase in material porosity, this would result in a porosity increase of approx. 2%. Exposure to tap water, 6 bath changes per day, and weekly brushing resulted in **no damage to surfaces in tap water** even after 280 days of exposure. Microscopic analysis of the surfaces showed thin calcium carbonate deposits, to which a protective effect of the surfaces is attributed.

Extensive research on the subject was done by M. Schwotzer [22] and Schwotzer et al. [21]. Some of the results obtained are reproduced in **Figures 4.30 and 4.31**. For this purpose, thin slices of CEM I 42.5 R 1 [mm] thick were prepared from the center of HCP prisms prestored in water for more than 100 days and stored under the boundary conditions mentioned in **Figure 4.30** on the one hand in deionized water and on the other hand in tap water. Despite the small thickness of the disks, there was **no appreciable loss of material on exposure to tap water**.

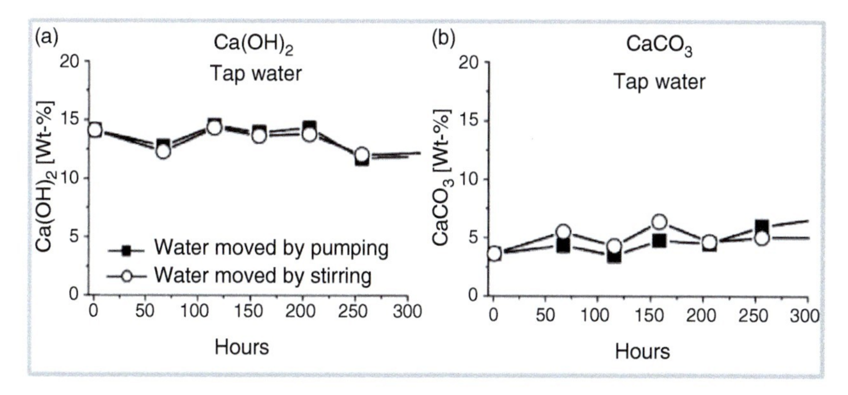

Figure 4.30 TGA tests of thin HCP-slices (size 1 [mm]) on possible dilution of portlandite content after storage in moved tap water at 11 °C. Source: From Schwotzer et al. [21]/with permission of Elsevier. (a) Portlandite content dependence on exposure time. (b) Calcium carbonate content dependence on exposure time.

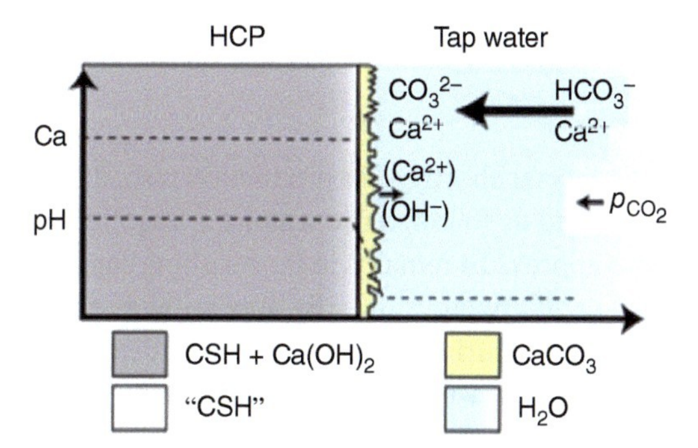

Figure 4.31 Schematic figure of the precipitation of a thin calcium carbonate layer on the HCP or concrete surface in the presence of tap water. Source: From Schwotzer [22]/Schwotzer, Matthiasr.

In addition, if the water in the water storage boxes is not exchanged during storage, as in the experimental studies for this book, therefore is an Ca-ion enrichment in the storage water and thus an additional fundamental reduction in solubility can be observed.

In addition, Schwarzer's investigations showed **a precipitate of calcium carbonate** on the surfaces of the water-stored samples, indicated in **Figure 4.32**. The consequences of such a precipitate on transport processes are discussed in more detail in Section 5.3.7.

4.5.4 Behavior and Durability of Air-Pores in Cement-Bound Materials

The durability of air voids in cement-bonded materials will be considered, especially when the material is stored in water. The focus is on the **air pores introduced during mixing of the concrete** according to Section 4.3.3. However, the results of the following considerations can also be broadly applied to the behavior of air pores artificially created by air pore agents to improve frost resistance.

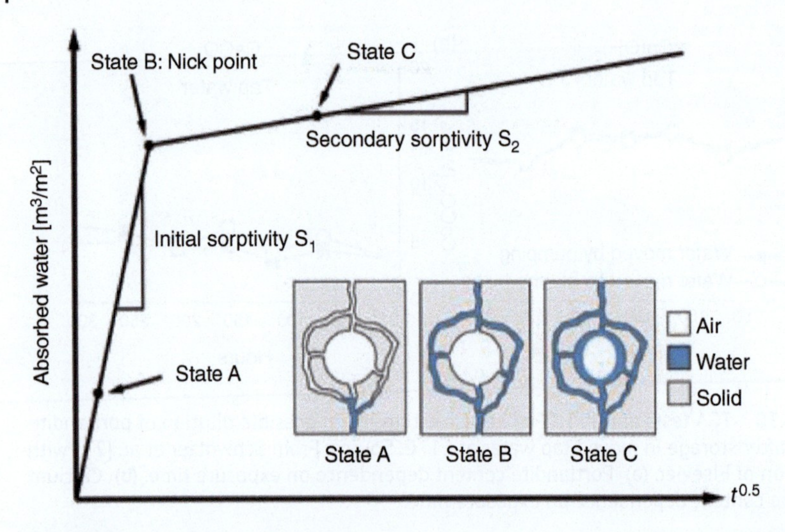

Figure 4.32 Schematic illustration of the water suction into capillary pores including an air void and schematic water absorption curve versus time. Source: Picture from Eriksson et al. [23]/Springer Nature/CC BY 4.0.

The pore situation in the HCP of the mortar shortly after production resp. during the first days of hydration can be described as follows. The water not required for hydration at water-cement ratios above approx. 0.4 remains in the resulting capillary pores, with a proportion of additional capillary pores initially appearing as empty space due to the internal shrinkage process during hydration, which, depending on the external storage conditions, can be filled by suction and redistribution by additional air resp. additional water. Compared with these capillary pores, the production-related large air pores initially remain unaffected by this filling process resp. continue to be filled with air. Apart from the positive effect of artificial air pores on frost resistance, the **importance of the air pores created during processing for the material behavior is usually underestimated**. For this purpose, the material REF should be considered first. The measured air-pores volume is 4.5% related to the total volume of 1000 [dm^3]. The HCP volume is given by Eq. (2.9) including the air-pore volume to

$$V_{HCP} = 1000 - M_{Aggr}/\varrho_{Aggr} = 430 \text{ [dm}^3\text{]}.$$

The volume fraction of air pores contained is 45/430 ≈ 10%.

However, if the air pores are related to the total porosity V_{pores} from Eq. (2.10) after sufficient hydration, the air-pores fraction is 45/185 = 24%

For the concrete C35/45 also used with $W/C = 0.45$ and 2.0% air-pores content, a proportion of 17% results, for the concrete C25/30 ($W/C = 0.60$) with 2.5% air-pores a part of 14%, each related to the total pore volume V_{pores}.

Reference is first made to the pictorial representation of the pore groups present in the sample material used, mortar REF, and their volume fractions, as well as the expected initial water contents, on **Figure 4.24a**, and further to the interrelationships of Section 4.5.1.

Figure 4.38 shows the results of water absorption tests on laterally sealed prisms made of mortar REF. The results make it clear that only the gel and capillary pores are filled by the capillary suction test, but not volume fractions of the air pores present.

This corresponds to the behavior shown schematically in **Figure 4.32**, there the pore filling up to the "Nick point" after reaching stage B, where only the capillary pores and not also the air pores are filled [23].

This is due to the fact that the capillary pressure (tension) present at the air-pore surface in the course of capillary pore filling is interrupted, compare **Figure 1.11** in Section 1.3.1.3. Although the cement paste inner surface in the area of the air-pore envelope is also covered with a water film for surface-energy reasons, since **the air-pore radii are 50–1000 times larger than the largest adjacent capillary pores** the capillary suction of the pore water drops sharply there to nearly 0 during the transfer from capillary pores to air pores. Therefore, the air cannot be initially displaced from the air pores in contrast to the capillary pores where this occurs almost simultaneously due to the high capillary pressures.

The water storage tests show, however, that at least part of the air pores also fills with water. In the present case of the REF mortar, an air-pore volume of about 45 [dm/m^3] was measured on the hardened material. Tests conducted for 500 days showed that in the case of larger test specimens, approx. 50% of the air-pore volume can fill with water, in the case of thin test specimens, complete saturation can take place.

It follows that the air content in the pore system can have a significant influence on storage and transport properties. Therefore, the mechanisms leading to water storage in air pores will be investigated in more detail below.

The storage capacity of water for inert gases (nitrogen, oxygen, air) depends on the external partial pressure according to Henry's law and depends on the temperature (Section 1.3.6). According to this law, the soluble gas quantity increases with the partial pressure of the gas and is zero at zero partial pressure.

In water uptake experiments into porous materials, the air-saturated water under atmospheric pressure p_{atm} is absorbed. In the case of only partial water saturation of the pore system and subsequent capillary condensation, the air component is dissolved in the condensed water. Excess air is discharged to the outside via pores that are not yet filled.

Figure 4.33 shows the absolute capillary pressure in pore systems with pore radii in the size of the area shown in the figure. Contrasted therein is the positive plot of the capillary pressure (draft) and the constant atmospheric pressure of 10^5 [Pa]. The difference between the two pressures gives the absolute pressure in the water-filled part of the pore system as a function of the corresponding maximum pore radius filled. For a maximum pore radius of, for example, 10 [μm], the capillary pressure in the pore system is $\approx -10^4$ [Pa] according to **Figure 4.33**, and thus the absolute pressure is $\approx (10^5 - 10^4) = 9 \cdot 10^4$ [Pa].

For a pore radius of 1.45[μm], the capillary suction is equal to the atmospheric pressure, so that there the absolute pore pressure is 0 (**Figure 4.33**). For water-filled

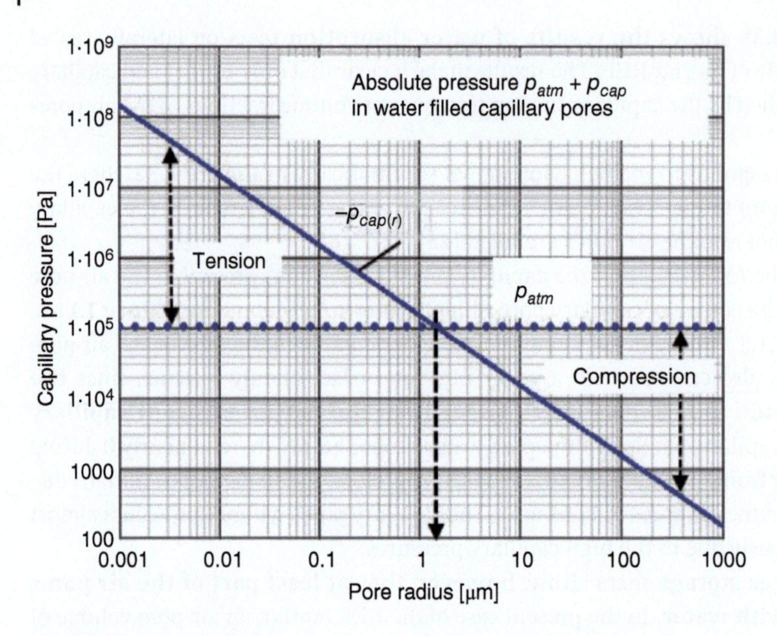

Figure 4.33 Absolute pressure $p_{cap} + 10^5$ [Pa] in capillary pores dependent on capillary radius and atmosphere pressure.

pores with even smaller maximum pore radius, the absolute capillary pressure is always negative (tension), **Figure 4.33.**

It is known from the numerous results of experiments with thin cylindrical single capillaries that the air dissolved in the aspirated water under atmospheric pressure does not escape, despite the stipulations of the classical Henry's law. Therefore, it is spoken of a metastable state with supersaturation.

However, in the remarkable paper by Lionel Mercury et al. [24] it is found on the basis of extensive also thermodynamic calculations that at an external (atmospheric) pressure of 10^5 [Pa] water under negative absolute pressure very well stores gases like nitrogen, oxygen (air) in dissolved form, even in higher concentration than under positive pressure. At a pressure of −100 bar the solubility for the mentioned gases is about 10% higher than under a pressure of 1 bar. This provides the explanation for the fact that no escape of dissolved air is observed either in experiments with single capillaries or in water uptake experiments with capillary-porous materials.

Further, the question to be answered is why the air porosity introduced by processing or artificially introduced, which remains stable in capillary suction tests and has also generally been considered to be largely stable up to now, escapes during water storage, depending on the test specimen shape, with simultaneous water filling of the pores concerned.

For this purpose, the behavior of spherical air pores in the pore-water of a capillary-porous material under atmospheric pressure is considered. Miller [25] makes it clear that the relationships already derived in connection with cavitation in water can also be applied to the behavior of air resp. of air pores in porous materials. If one compares the absolute pressure in a spherical air pore with the absolute pressure of the surrounding pore-water, the following relations apply:

$$p_{atm} + \frac{2 \cdot \gamma_L}{r^\star} = p_{r_{pore}} + p_{atm} \Rightarrow p_{r_{pore}} = \frac{2 \cdot \gamma_L}{r^\star} \Rightarrow r^\star = \frac{2 \cdot \gamma_L}{p_{r_{pore}}} \tag{4.4}$$

Where $p_{r_{pore}}$ is the capillary pressure prevailing in the vicinity of the air pore. $r^\star$ is also called the critical radius. For air-pore radii $r < r^\star$, the gas pressure (air) inside the bubble is larger than the ambient pressure, so according to Henry's law, these **air pores with $r < r^\star$ disappear due to solution in the surrounding water**, corresponding also to the energetic consideration reaching the same result. Bubbles with larger diameters are unstable and can continue to grow unless they are prevented from doing so by the surrounding solid material. For air pores with the radius of the capillary pore in which they are located, the magnitude of the internal pressure is equal to the surrounding capillary traction. **It follows that air pores with smaller radius than the surrounding capillary pore will always be dissolved.**

Air pores introduced during concrete mixing are surrounded or enveloped by water or cement paste and therefore have a largely spherical shape. During this process, the developing air pores are not under capillary pressure, but under atmospheric pressure. The relations of Eq. (4.4) also apply to this state of formation of the air pores. This means that for them in part 3 of Eq. (4.4) $p_{r_{pore}} = p_{atm} = 10^5$ [Pa] has to be inserted. From this, the associated boundary radius follows to $r^\star = 1.45\ 10^{-6}$ [m] ≈ 1.5 [μm].

Therefore air pores introduced via material processing are always greater than 1.5 [μm] and have correspondingly large radii also in the hardened material. This also corresponds to the measurements of the air pore sizes, compare Section 4.3.3.

In addition to the physical phenomenon that leads to the definition of the critical radius $r^\star$, there exists a second physical process that can limit the lifetime of air bubbles. For example, if the escape of bubbles with $r > r^\star$ is prevented from escaping due to impact from an aqueous environment, a reduction in the size of the gas or air bubble can occur through the mechanism of concentration equalization by diffusion across the air-pore/water interface. The precondition is that the gas concentration in the water is smaller than the saturation value of the gas under consideration and the concentration of this gas in the bubble exceeds the concentration in the water. This condition is fulfilled using Henry's law if

$$p_{bubble} \cdot \frac{1}{K_H} > p_{water} \cdot \frac{1}{K_H} \tag{4.5}$$

If a material sample, on which, for example, a capillary water uptake test was first performed, is subsequently immersed or stored in water, the menisci of the entrance pores at the material surface and from there in the capillary pores change to the capillary pressure ≈ 0.

Let the pressure p_{water} be the pressure in the capillary pores, whose absolute pressure now increasingly corresponds to the atmospheric pressure $p_{atm} = 10^5$ [Pa], the pressure in the gas resp. air bubble, on the other hand, is

$$p_{bubble} + p_{atm} = 2 \cdot \gamma_w / r_{bubble} + p_{atm}.$$

Thus the bubble has an overpressure of at least $\Delta p_{(t)} = 2 \cdot \gamma_w / r_{bubble(t)}$

compared to the adjacent water under atmospheric pressure, see Eq. (1.48) in Section 1.3.7.

Let the air pores in the pore system with radius r_0 now be considered corresponding to the previously mentioned gas bubbles to $r_{bubble(t)} = r_{(t)}$.

The diffusion of air $\Delta m_{(t)}$ per time step Δt through the boundary surface of the air pore $A_{(t)}$ to the water is now expressed by the following equation, where the overpressure is expressed in bar (instead of [Pa]) by dividing by 10^5:

$$\Delta m_{(t)} = \beta \cdot \frac{1}{K_H} \cdot \frac{\Delta p_{(t)}}{10^5} \cdot A_{(t)} \cdot \Delta t \tag{4.6}$$

Where β is the gas/water transfer coefficient. As air diffusion from air pores or air bubbles increases, the air content in the bubbles decreases, so their diameter also decreases and the bubble pressure increases at the same time. Corresponding calculations must therefore be performed in appropriate time steps and the boundary conditions must be adjusted before each new time step. The volume reduction under its own pressure plays a significant role only from air-pore radii smaller than approx. 3[μm] and can therefore be neglected.

$$\Delta V_{(t)} = V_0 - \frac{\sum_{t=0}^{t} \Delta m_{(t)}}{\varrho_{Air}} \qquad \Delta V_{(t)} = \frac{4\pi}{3} \cdot \left(r_0{}^3 - r_{(t)}{}^3\right) \tag{4.7}$$

$$r_{(t)} = \left(r_0{}^3 - \Delta V_{(t)} \cdot \frac{3}{4\pi}\right)^{\frac{1}{3}} \qquad A_{(t)} = 4\pi \cdot r_{(t)}{}^2 \tag{4.8}$$

The calculation of the behavior of air pores by this method with initial radius r_0 of 50, 200, 400, and 600 [μm] was carried out with the following parameters:

Transition coefficient = $0.9 \cdot 10^{-4}$ [m/s] for nitrogen according to [26], Henry coefficient at 20 °C : $1/K_H$= 0.024 [g/(dm^3· bar)] determined from Henry coefficients for N_2 and O_2 corresponding to their partial pressure fractions, air density ϱ_{Air}= 1.20 [g/dm^3].

The calculation results are shown in **Figure 4.34**. Displayed is the radius reduction of air pores with the given initial pore radii as a function of time in s.

According to measurements on the solid, the air void content of the REF mortar is about 4.5 *vol.%*. These air resp. compaction pores obviously have a continuous size distribution with **pore diameters** from 5 [μm] to circa 2–3 [mm], see the measurement results on comparable mortars produced in the same laboratory, **Figure 4.10**. According to this, about 50% of the processing-caused air-pores of the mortar have REF **pore-radii** ≤ 250 [μm], given for comparison in **Figure 4.34**.

The results shown indicate that by the mechanism of dissolving, air-pores can indeed vanish into the surrounding pore-water by the prevailing pressure differences, and the smaller the initial size of the air-pores, the more readily.

Because the air pores are not surrounded by free water but by water-containing capillary pores, the air transfer per unit time and hence the **dissolution time is greatly prolonged** compared with the results from **Figure 4.34**. In addition, the air concentration going into solution from the air pores must be removed

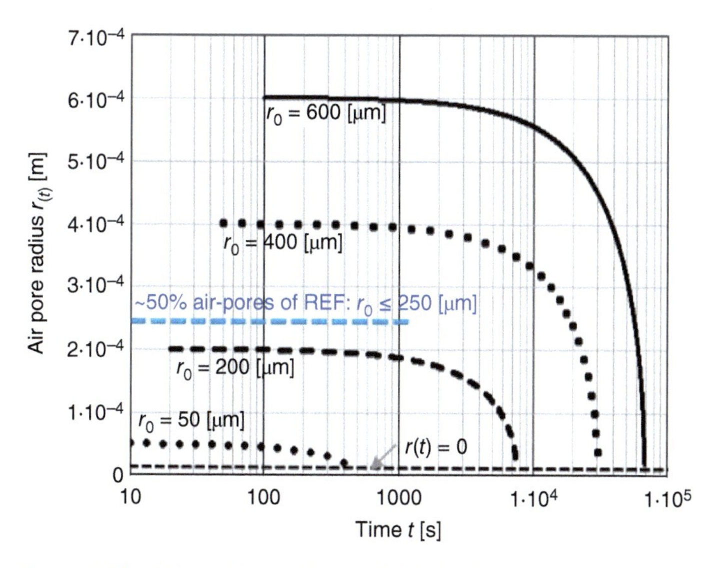

Figure 4.34 Dissolution of air-pores (air bubbles) with initial size r_0 in surrounding pore-water with gas concentration (air) corresponding to atmosphere pressure.

by diffusion, otherwise concentration equilibrium between pores and water will occur and air dissolution will stop. Because of the influence of the capillary pore volume and the size of the capillary pores, the W/C value plays a role in this, furthermore the mixing ratio m_{CEM}/m_{Aggr}.

From Figure 4.13a,b it is evident that the exchange from the inner surface of the air pores into the adjacent capillary pores can take place via a corresponding rough inner surface. Experimental investigations show, moreover, that ettringite, portlandite or ASR-gel dissolved in the pore solution can penetrate into the air pores and crystallize there and can influence the air dissolving.

Because of the path dependence of the dissolved air removal, the shape and size of the cement-bound material bodies also play a major role in the timing of the process. The hygric boundary conditions are also crucial.

Because of the high effort involved, further computational modeling of the dissolved air removal process is omitted here. However, in the discussion of the different water uptake experiments, the air-pore influence and its plausibility will be at least qualitatively checked.

4.5.5 Influence of Test Specimen Shape on Water Uptake into Air Voids

In the following, the behavior of cube-shaped and slice-shaped samples of the REF mortar series is illustrated by some selected experimental results.

Figure 4.35a reflects the results of measurements on a cube series (hardened 84d by initial water storage) about the time-dependent water absorption after standard drying (40 °C/1 mbar). **Figure 4.35a** shows the possible water saturation when vacuum and pressure of 150 bar are applied. It becomes clear that after the 84 days initial water storage, already about 40% of the total air-pore volume was water-filled

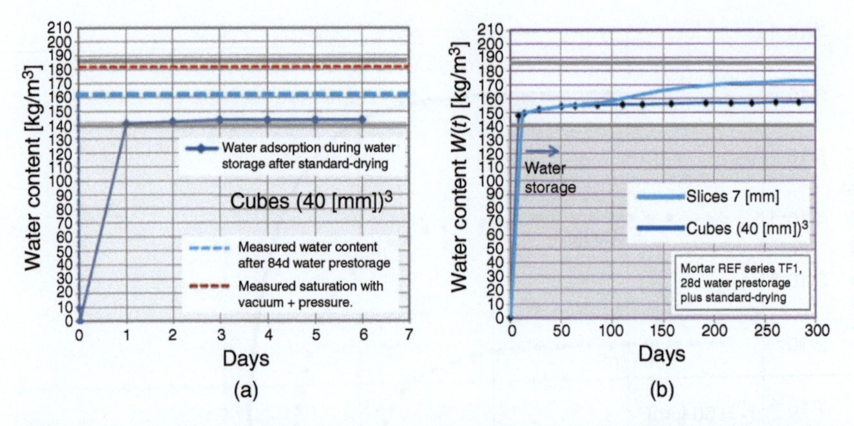

Figure 4.35 Comparing tests of water absorption $W(t)$ on mortar-REF-samples. (a) Water adsorption of cubes, size (40 [mm])^3, after 84d water prestorage plus standard-drying. (b) Measured influence of water storage time on absorption on slices size 7 [mm] thick and cubes (40 [mm])^3 , respectively, after 28d water prestorage plus standard drying. Water content relative to dry mass: $m_{W(t)} = W(t)/(\varrho_{dry,REF} \cdot 1000)$; $\varrho_{dry,REF} \approx 2.10$.

by dissolution of the smaller air pores. Furthermore, in the case of the cubes with an edge length of 4 [cm], all gel and capillary pores are already filled after one day when the cubes are again stored in water after drying, but after one week, no air pores have yet been dissolved.

If the vacuum pump is switched on upon the waterstored samples for subsequent measurement of the total pore content, in the first step of the process, air bubbles escape from the test specimen over a period of several hours. In the 2nd step, a water pressure of 150 bar is then applied (and then the water saturation value or total pore content is determined by standard drying).

Even partial drying to equilibrium at defined relative humidities does not initially restore the original initial water content of the cube-shaped test specimens during subsequent water storage to equilibrium. This behavior of the cubes is shown to be systematic and reproducible.

Figure 4.35b now shows for comparison the test results of the water absorption as a function of time of cubes and slices, cut after only 28 days of water storage from REF mortar prisms of the same production batch. Standard drying of the material samples was carried out before the start of the test. **The behavior of the cubes** corresponded in the 1st week of water exposure to the water uptake of the cube series of **Figure 4.35a** with a slightly larger initial value (influence of the degree of hydration), then increased within approx. 60 days to a water content higher by approx. 10 [dm^3/m^3], which remained almost unchanged until the end of the experiment of 300 days or increased only very slightly further.

The water content of the parallel water-stored 7 [mm]-thick slices behaved identically to the cubes up to approx. 60 days, and then increased continuously until the end of the experiment after 300 days with a further increasing tendency, whereby approx. 25 [dm^3/m^3] of the contained air-pore volume had already been filled with water by this time point. This result clearly shows that over the time period considered, the dissolved increased air concentration was removed by diffusion from both surfaces of the disks via the maximum 3.5 [mm] long paths, but not to a sufficient extent from the outer surfaces of the cubes. These results prove that the dimensions of the material bodies or components exert an influence on the possible water saturation.

4.5.6 Capillary Water Uptake After Drying

In the following, the behavior of post-fabrication water-stored prisms of REF mortar series is presented, showing experimental results on a remarkable rewetting behavior after drying from saturation state.

In Figure 4.36, the results are reproduced from a series of prisms of length about 150 [mm]. After about 290 days of water storage, the prisms were dried to 128 [kg/m^3] from the initial water content (which also included 30 [dm^3/m^3] of original air-pore volume) to equilibrium by all-sided vapor desorption over 136 days at 20 °C and 84% RH. This value corresponds to the corresponding value of the desorption isotherm. After sealing the longitudinal sides, a capillary water uptake test was performed. The resulting equilibrium saturation of the prism series corresponded to the saturation value of all gel and capillary pores without filling the air pore volume.

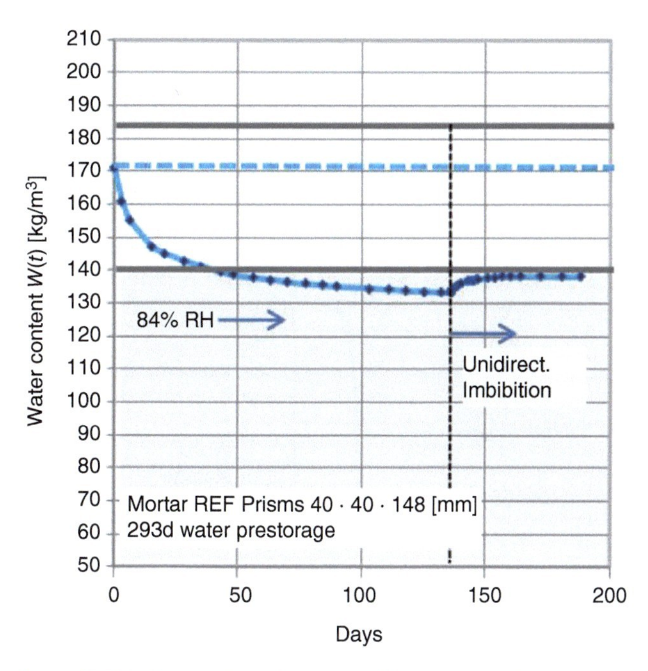

Figure 4.36 Water adsorption curves $W(t)$ measured on mortar REF prisms 40·40·148 [mm^3] after 293d water prestorage: 136d desorption at 84% RH (20 °C) and 50d unidirectional capillary water imbibition (after sides coating). Water content relative to dry mass: $m_{W(t)} = W(t)/(\varrho_{dry,REF} \cdot 1000)$; $\varrho_{dry,REF} \approx 2.10$.

In the series Figure 4.37, 80 [mm]-long prisms were dried to the equilibrium water content corresponding to 65% RH for 320 days after 28 days of initial saturation, subjected to a capillary water uptake test for 45 days after sealing the longitudinal sides, and then further stored in water for about 130 days.

The results show in **Figure 4.37** that the water content obtained after desorption at 65% RH at equilibrium corresponds to the value of the desorption isotherm. The subsequent capillary water uptake filled only the still free capillary pores. During the subsequent water storage for about 130 days, the initial water content was reached again after 28 days by filling about 50% of the air pores.

For the experimental series of Figure 4.38 80-[mm]-long prisms (of the same manufacturing batch as those in **Figure 4.37**) were standard dried from 28 days of initial saturation, after sealing the longitudinal sides, a capillary water uptake test was performed for 45 days, followed by water storage for about 100 days.

The results obtained show that also in this case, only the gel and capillary pores were filled in equilibrium due to the unidirectional capillary water uptake. During the subsequent water storage, water was again absorbed into the air-pore volume, but not yet up to the initial water content during the test period.

By further prism series **Figures 4.39a and 4.40a**, it was also verified with regard to the modeling of the transport behavior, whether the rewetting behavior observed in **Figures 4.36 and 4.37 after long-term desorption** in air humidities of 84% and 65% also occurs on prisms with capillary water absorption after standard drying (**Figure 4.38**) and, for comparison, after standard drying and partial saturation by

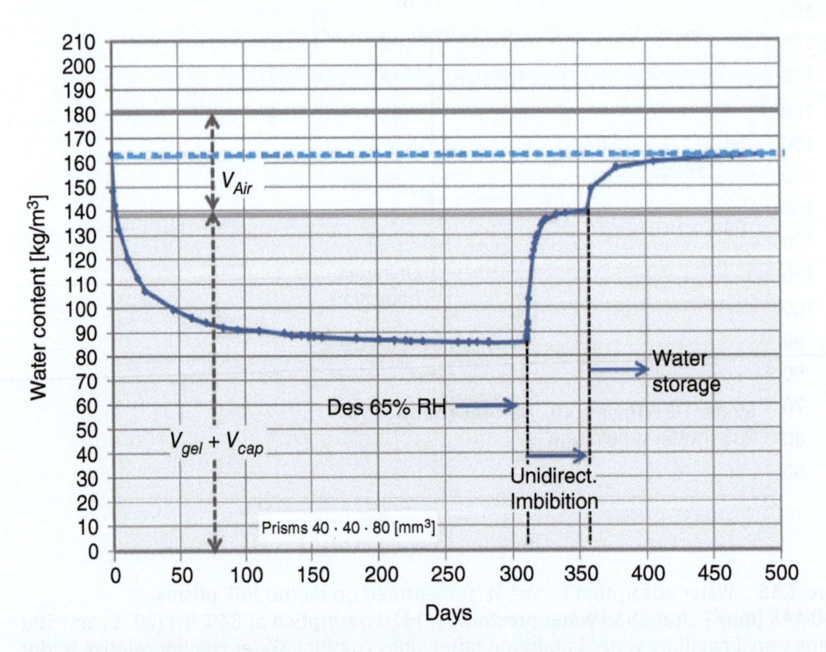

Figure 4.37 Water adsorption curves measured on mortar REF prisms 40 · 40 · 80 [mm³] after 28d water prestorage: 320d desorption at 65% RH (20 °C), 45d unidirectional capillary water imbibition (after sides coating), and 127d water storage with lateral coating left.

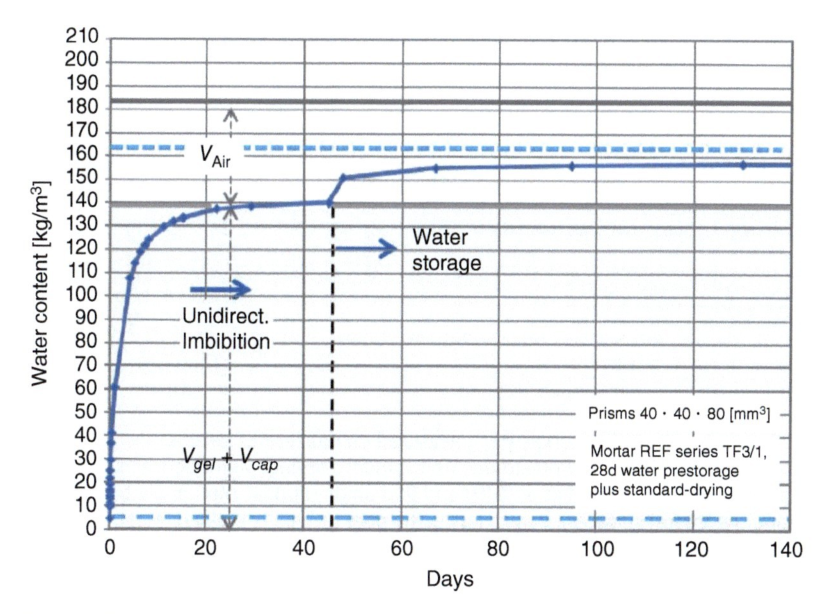

Figure 4.38 Water adsorption curves $W(t)$ measured on mortar REF prisms 40·40·80 [mm³] after 28d water prestorage plus standard drying: 45d unidirectional capillary water imbibition (after sides coating) and 98d water storage with lateral sides coating left.

long-term vapor adsorption up to a selected air humidity equilibrium and subsequent 1D capillary water absorption.

The results in **Figures 4.39a and 4.40a** show that even after the long vapor absorption over 840 days at 84% RH from standard drying and the adsorption over 870 days at 54% RH, the reached water contents correspond quite exactly to the water content values of the adsorption isotherm for the relative air humidities set.

Another important result of the subsequent capillary water uptake tests (with sealed prism sides) is that the results presented before are confirmed: Only the gel and the capillary pores were completely filled, while no filling of the contained air void volume occurred during the 84 days of capillary water uptake experiments.

Figures 4.39b and 4.40b show the recordings of the water uptake curves of each of the three tests in both series. The curves are largely superimposed and correspond to the expected curve progression in relation to the initial saturation, see also Section 5.2.1.

Figure 4.41 shows a schematic overview of the water uptake behavior and saturation states during water storage depending on the storage time and material drying.

4.5.7 Over-hygroscopic Range in Cement Mortar and Concrete

The transition from the hygroscopic to the over-hygroscopic moisture range is usually assumed to be between 95% RH and 98% RH [5], [27]. According to Scheffler [27], the starting value of the over-hygroscopic range as the transition point between the two moisture ranges should be where the sorption isotherm begins to show a sharp rise.

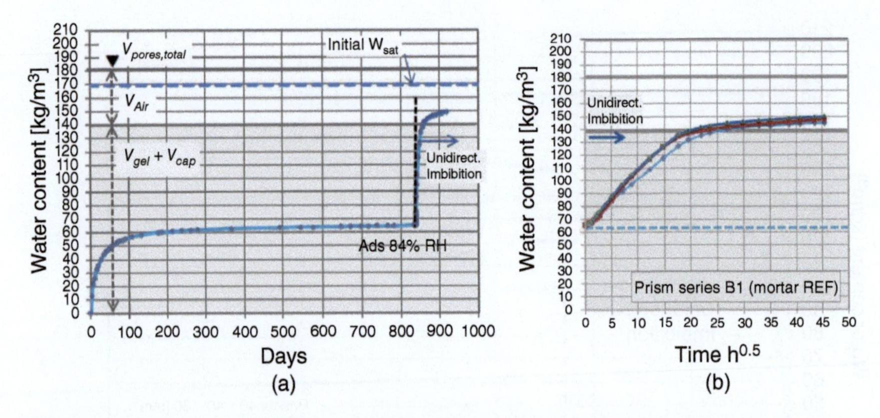

Figure 4.39 Water adsorption curves $W(t)$ measured on mortar REF prisms 40·40·150 [mm³] after 84d water prestorage and standard drying. (a) Influence of storage on water adsorption during 840d at 84% RH and 91d unidirectional capillary imbibition (after coating the lateral surfaces), in relation to the pore groups contained (mean values of three samples). (b) The individual capillary imbibition curves of the three prisms of figure (a) after the storage at 84% RH (20 °C).

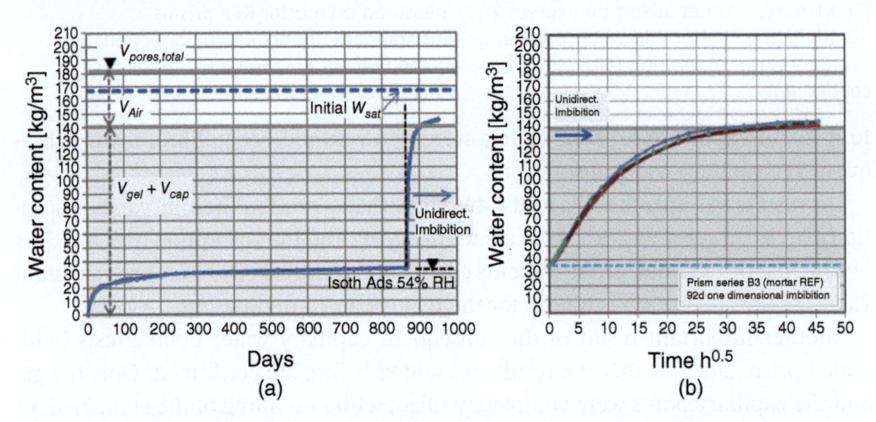

Figure 4.40 Water adsorption curves measured on mortar REF prisms 40·40·150 [mm³] after 84d water prestorage and standard drying. (a) Influence of storage on water adsorption during 870d at 54% RH and 92d unidirectional capillary imbibition (after coating the lateral surfaces), in relation to the pore groups contained (mean values of three samples). (b) The individual capillary imbibition curves of the three prisms in figure (a) after storage at 54% RH (20 °C).

By this definition, it becomes clear that the transition between the moisture ranges is chosen arbitrarily, because also above the "transition point" in the so-defined over-hygroscopic range a hygroscopic water absorption by vapor uptake from the air takes place. The reason for this definition is the increased effort to set a relative air humidity of about 100% in adsorption experiments. The physically correct **boundary between hygroscopic and over-hygroscopic range is a relative air humidity close to** 100%, because up to this value, water content can be stored in the material without liquid-water action. Water absorption in the over-hygroscopic

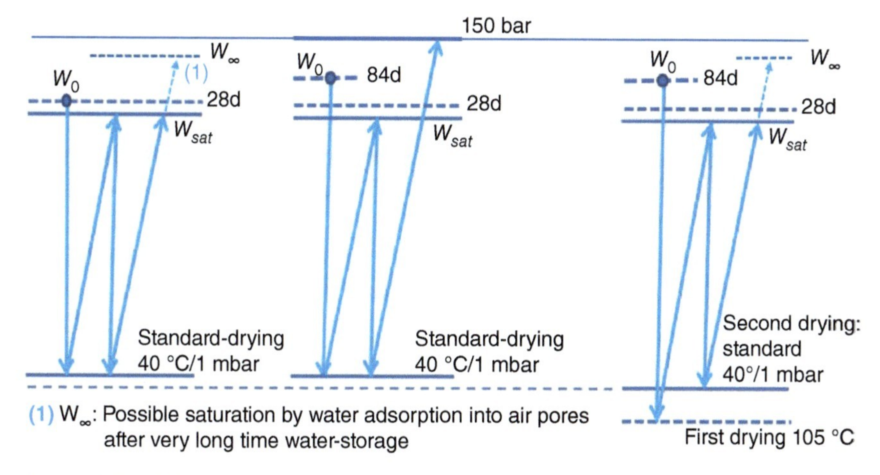

Figure 4.41 Schematic diagram about the dependence of possible water content at different saturation conditions due to the material age and the drying procedure.

range is then determined by submerging the material samples. The test results presented below show that the setting of approx. 100% RH in the determination of the values for the adsorption and desorption isotherms is possible without significant test difficulties.

Classical mortars or concretes, which have no or only a low air content, do not possess a super-hygroscopic range. With regard to the hysteretic storage behavior, the over-hygroscopic water storage caused by air-pore filling is to be treated differently from adsorption and desorption in the governing hygroscopic storage range, compare the Section 4.6.2.

In this respect, the remark in [27] that the hysteresis model developed by Espinosa and Franke [15] for HCP and mortar unfortunately does not apply to the super-hygroscopic range is also correct in principle but not of practical importance for concrete material.

On the other hand, cement-bound products such as the mortar REF used as a reference (with over 4% processing air-pore content) absorb a water fraction when stored in water by filling an additional part of the air pores, which could possibly be called a-hygroscopic fraction if it cannot be filled by vapor sorption, compare **Figure 4.19**. This would be justified if the magnitude of the air content introduced by processing is not known.

To clarify the relationships, desorption tests were performed on 430 days of hydrated water-stored slices, which by now had an initial water content of 171 [kg/m^3], at 100% RH, 97.5% RH, and 92.5% RH on three slices each. **The desorption level of 100% RH was introduced as part of the project** in this series of experiments. For this purpose, tests were first carried out with a view to realizing the 100% RH environment in order to obtain correct water absorption. Slice-shaped test specimens were stored in a closed box above an approximately 2 [cm]-deep water surface. By placing an inclined plate above the test specimens (without contact to them), it was ensured that water drops formed on the walls of the box could not drip onto the test specimens. The magnetic stirrers used in previous tests could be dispensed with. Hygrometer measurements confirmed that the expected approx. 100% RH was established inside the box at 20 °C, not least for thermodynamic reasons.

To set defined relative humidities in the desorption experiments over salt solutions, the samples were stored in closed boxes without water contact above the surface of the bath.

Figure 4.42b shows the results of the measurements at the above three relative humidities at 20 °C. One can see from the initial saturation a significant decrease in water content at 100% RH storage and, as expected, a higher decrease in discs at 97.5% RH and 92.5% RH. It is clear that some of the air-pore volume leads to over-hygroscopic water uptake.

Figure 4.42a further shows that the comparison samples at 100% RH storage for about 180 days (after standard drying) by vapor adsorption have a water content nearly equal to the desorption value and therefore **confirm it formal as the limit to the over-hygroscopic range**.

If the slices (**Figure 4.42b**) are subsequently water-stored after 80 days of desorption, the water content increases rapidly to a saturation level between the original initial water content and the 100% RH value, **Figure 4.43**. Water uptake remains at this level for about five days, then slowly but steadily increases beyond the initial water content toward long-term complete saturation of the slices.

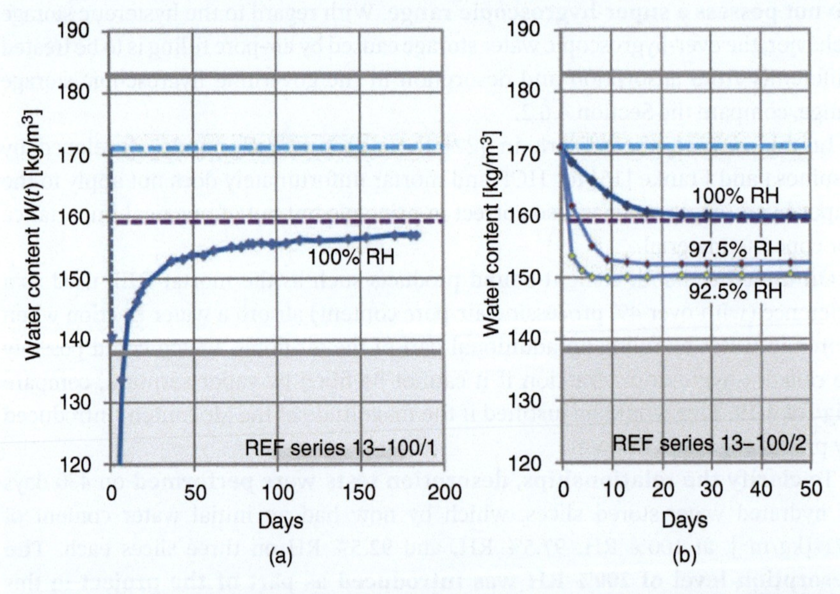

Figure 4.42 Results of adsorption/desorption measurements $W(t)$ **on mortar REF slices** (≈7 [mm] thick) after 430d water storage. Comparison of adsorption and desorption and the volume of $V_{gel} + V_{cap}$. Water content relative to dry mass: $m_{W(t)} = W(t)/(\varrho_{dry,REF} \cdot 1000)$; $\varrho_{dry,REF} \approx 2.10$. **(a)** Dashed blue line = initial water content per [m³] after prestorage; then exposure to ≈ 100% RH after standard-drying. **(b)** Desorption of three series (three samples each) at the RH indicated.

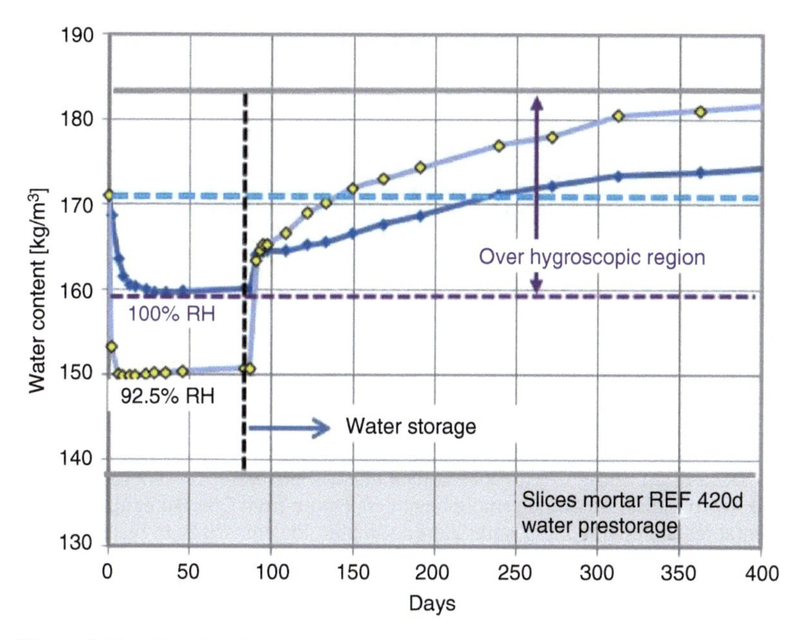

Figure 4.43 Results of adsorption/desorption measurements $W(t)$ on mortar REF slices (7 [mm] thick) after 430d water prestorage, showing the **over-hygroscopic region under the present conditions.**

This saturation behavior beyond the initial saturation could be observed despite the longer investigation periods **only on the thin slice-shaped test specimens**, but not on the comparatively large prisms. This confirms the mechanism described in Section 4.5.4 that **an escape of the elevated dissolved gas or air concentrations from the pore solution to the surrounding atmosphere** will only be detectable at thin test bodies in the time periods investigated here. If, on the other hand, 100% RH tests are carried out on larger test specimens, the behavior shown in **Figure 4.50** is evident, namely that in the case of the cement-bonded material at hand, an over-hygroscopic region is present only if a partial filling of (manufacturing-related) air pores takes place.

In the case of porous concrete and, for example, ceramic bricks, as well as products with a high proportion of large pores, a **pressure plate test to record the super-hygroscopic range** has so far been carried out to determine the moisture storage behavior during desorption, in addition to a sorption isotherm usually limited to 95% RH. Results on this are presented in Section 4.6.1. However, the combination of pressure plate test and desorption isotherm can also be useful for conventional cement-bound products. In Section 4.5.8, such investigation results are presented.

4.5.8 Air-Pore Influence on Sorption Isotherms

Figure 4.44 shows in comparison results of Qier Wu et al. [28] and other authors on the experimentally observed steep desorption behavior near the 100% value of

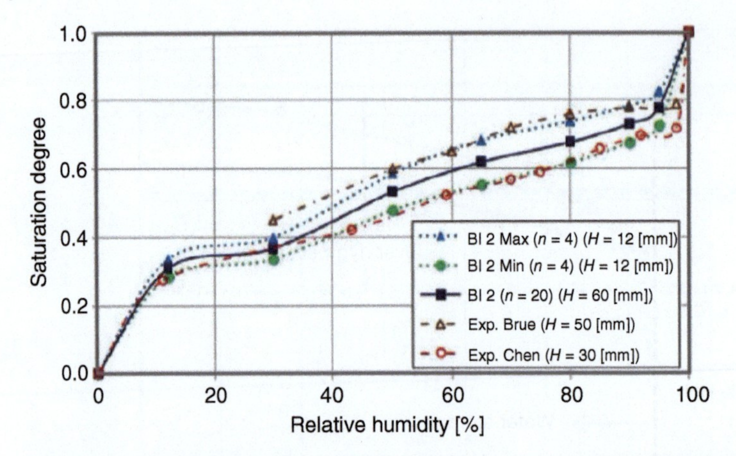

Figure 4.44 Desorption isotherm with special look at the sharp water content reduction near 100% RH (with indication of the sample height H). Figure from Qier Wu et al. [28] with his experimental results on 8 months water stored samples, in comparison to results adapted from Brue et al. [29] and Chen et al. [30] (all samples 6 months water-stored and supposed saturated).

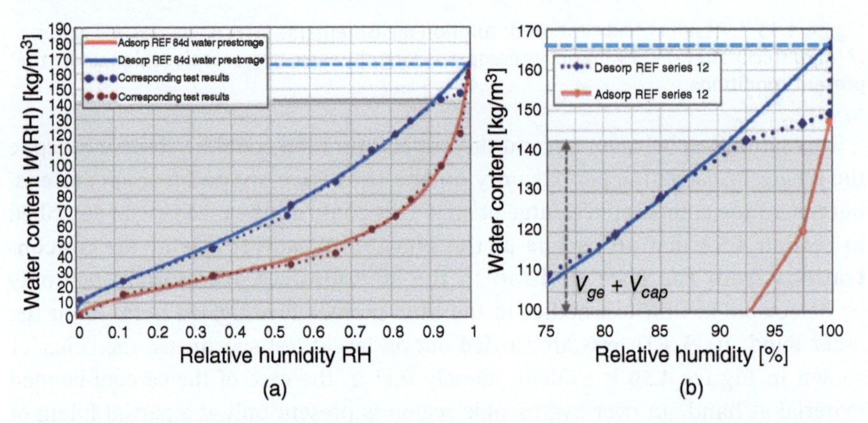

Figure 4.45 Experimental control of the desorption behavior of water-stored mortar REF close to 100% RH. (a) Isotherms got from the initial experiments on REF series 12 after 84d water prestorage. (b) Accurate desorption behavior near 100% RH according to complementary tests on mortar REF. Water content relative to dry mass: $m_{W(RH)} = W(RH)/(\varrho_{dry,REF} \cdot 1000)$; $\varrho_{dry,REF} \approx 2.10$.

RH in desorption isotherms on water-saturated test specimens. Similar behavior was observed in the sorption isotherms of REF mortar after prior water storage. Since this behavior can have considerable influence on transport processes, the mechanism is discussed in more detail below.

First, the measured sorption isotherms (desorption and adsorption isotherms) for REF mortar series are shown in **Figure 4.45a**. These measurement results are used as a basis and reference for evaluations of water absorption measurement results. Also noteworthy here is the "erratic" behavior of the desorption isotherms above approx. 90% RH, which was repeated in desorption isotherms of later test series. Since the highest desorption air humidity had been here, supplementary experiments on adsorption/desorption at 100% RH were made up. From this, the behavior shown in **Figure 4.45b** can be deduced with initial desorption to the 100% value.

To verify this behavior, complementary experiments were performed on long-term water-stored samples. The behavior of the samples is again characterized by the fact that when the samples are desorbed or dried, even when stored at 97.5% RH, there is initially a steep drop in moisture content, and subsequent storage at 92.5% RH leads to a only moderately lower moisture content at equilibrium, compare **Figure 4.42**, whereby the behavior shown in **Figure 4.45a,b** was confirmed in principle.

It should be noted again that during this desorption, a fraction of the previously water-filled air-pore volume is preferentially emptied. This is caused by the fact that water-filled air pores act like a water reservoir from which the adjacent, much smaller capillary pores preferentially remove water during drying. However, this process is also time-dependent and depends on the permeability of the pore system.

Supplementary tests were carried out on two selected concretes, whose concrete engineering data can be taken from Section 4.2. These concretes were C25/30 ($W/C = 0.60$) and C35/45 ($W/C = 0.45$) with a maximum grain size of 8 [mm]. The desorption behavior in the moisture range above 90% RH was investigated using pressure plate measurements, and vapor desorption measurements up to 95% RH. The **results of these measurements can be seen from Figures 4.46a and 4.47a**, averaged from at least 3 individual measurements and plotted as usual as a function of capillary pressure and the junction curve intercept between 4 and ca.5 [hPa].

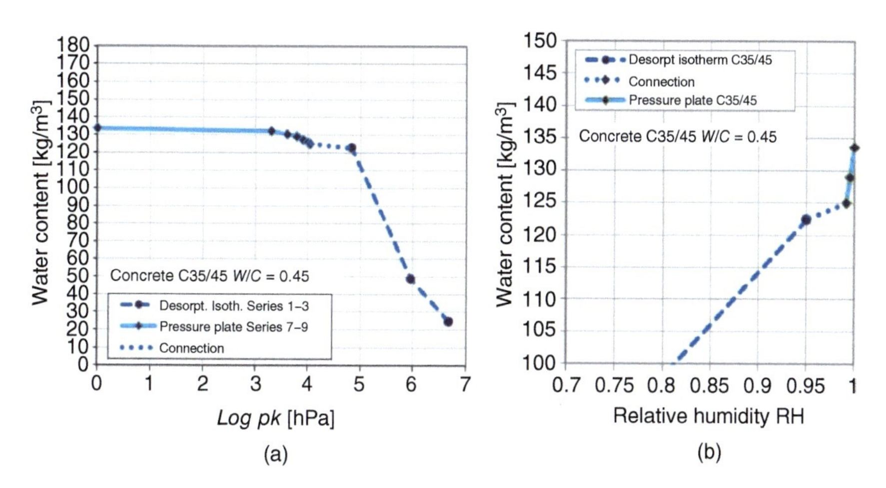

Figure 4.46 Measured water storage function $W(pk)$ by combining the results of pressure plate tests and desorption-isotherm measurements (RH $\leq$ 95%) on concrete C35/45 ($W/C = 0.45$, age 24 months, standard-drying). (a) Water storage function $W(pk)$ via capillary pressure. (b) Water storage function from figure (a) converted into RH-dependence W(RH).

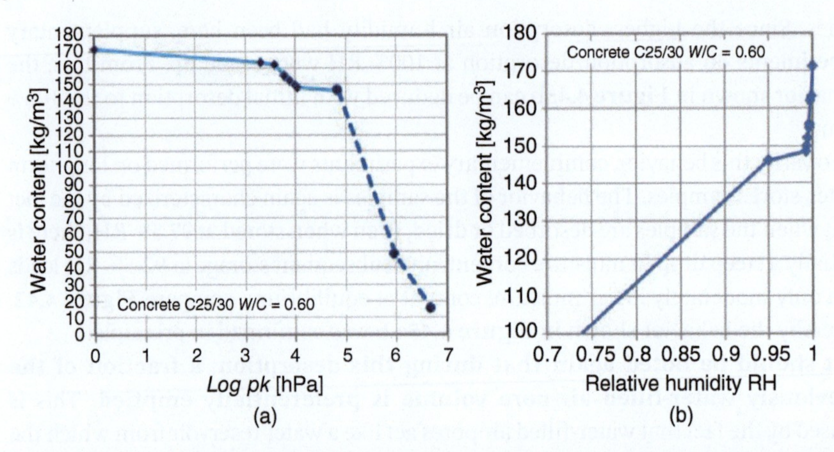

Figure 4.47 Measured water storage function by combining the results of pressure plate tests and desorption-isotherm measurements (RH $\leq$ 95%) on concrete C25/30 ($W/C = 0.60$, age 24 months, standard drying). (a) Water storage function via capillary pressure. (b) Water storage function from figure (a) converted into RH-dependence.

First of all, the well-known "steps" between the pressure plate results and the vapor desorption results between approx. 2 and 5 [hPa] are striking. If one now plots these desorption curves against the relative vapor pressure, one obtains the curves of **Figures 4.46b and 4.47b.**

These additional investigations on concrete therefore showed similar step-like behavior completely independently of the investigations on the REF mortars. Further investigations on the question of the influence of the air pores were not carried out here.

4.5.9 Water Storage Behavior of Initially Sealed Hardened Test Specimens

Concretes hydrated in air at medium relative humidities generally exhibit a lower degree of hydration after equal times than concretes stored at high humidity or, in particular, water-stored concrete, compare Section 2.2.8.

However, water-stored concrete also have a state inside, at a distance from the surface depending on their dimensions and permeability, which corresponds to the state of sealed-cured concrete.

Due to the abovementioned different possible storage conditions, mortar or concrete components or test specimens of the same production batch (possibly different dimensions) may have significantly different water contents.

In Figure 4.48, the water contents measured on prisms made of reference mortar REF after 28d initial storage under the different conditions mentioned are now plotted for comparison. The light blue dashed lines indicate the water content after 28d, the dark blue dashed lines indicate the resulting water content after additional water storage for 10 days and the red dashed lines indicate the saturation water content after exposure to water under vacuum and subsequent pressure. The gray areas indicate the calculated water content of the gel and capillary pores, the

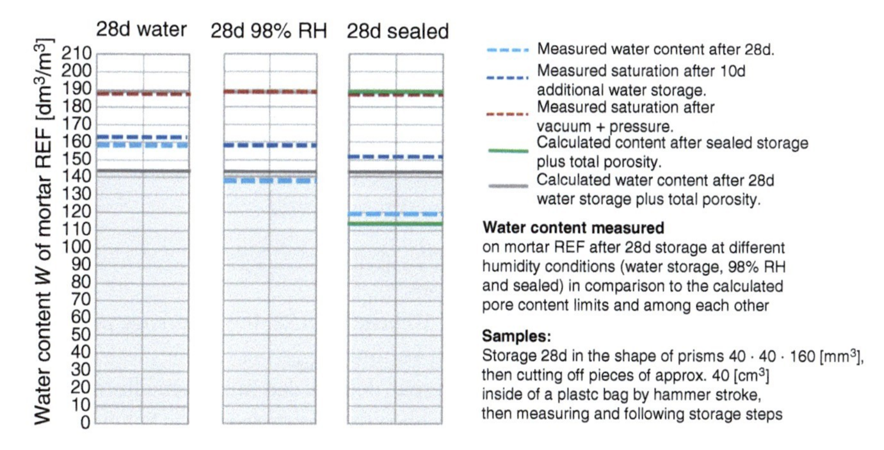

Figure 4.48 Water content measured on a series of prisms and cubes of mortar REF after 28d storage at different humidity conditions. Comparison with calculated corresponding pore content values, using tools of Section 2.2, assumed hydration degree $m = 0.79$; see also **Figure 4.24a**. Water content relative to dry mass: $m_W = W/(\varrho_{dry,REF} \cdot 1000)$; $\varrho_{dry,REF} \approx 2.10$.

calculated total pore volume (volume from gel, capillary, and air pores) agrees with the measured pore volume from vacuum and pressure. For comparison, the green line on the 28d-sealed series indicates the calculated water content of this series (calculated from Section 2.2).

It is particularly striking that the total pore volume after 28d hydration (measured under vacuum and pressure) is identical despite the different storage conditions, so that the degree of hydration of the 3 prism series of the same production batch can hardly be different.

The pore formation during hydration is explained as follows:

Immediately after production of the cement-bound product, a fresh mortar or concrete has only water-filled space between the solid particles (cement plus aggregate), which constantly decreases with increasing hydration as the water is incorporated into the cement phases to form the hydrate. In its initial state, therefore, a fresh mortar has only air pores introduced by processing as pores. By determining the water content by drying the material in the hydration phase, a constantly decreasing pore space can be documented. As a result of the somewhat lower volume of the hydration products compared to the initial volume of cement plus water, the so-called shrinkage pores (referred to as chemical shrinkage) arise as additional volume (in addition to the unused water still present) as additional gel and capillary pore volume.

If a concrete or mortar product is not stored under water in the course of hydration (and thus a water supply from outside is prevented), hydration comes to a standstill when the water used for concrete mixing as planned has been used up, cf. Section 2.2.5. Only at $W/C \gtrsim 0.45$ can hydration be largely complete. In the case of prevented water supply during hydration, a pore volume corresponding to the resulting shrinkage pores always remains water-free or air-filled. The processing-related air pores are water-free in this state anyway. **The pore volume created by chemical shrinkage** is worth noting. It can be related to the total pore volume V_{pores} Eq. (2.10) using the Eq. (2.14) (neglecting the processing-caused air-pores P) as

$$\overline{V}_{Shrink} = V_{Shrink} \; / \; V_{pores}$$

$$\overline{V}_{Shrink} = \frac{m \cdot Wchem0 \cdot (1 - k_{chem})}{W/C - m \cdot Wchem0 \cdot k_{chem}} \; [-] \tag{4.9}$$

With a hydration degree of $m = 0.80$ and $Wchem0 = 0.28$, the shrinkage pore volume fraction results in 22% of the total pore volume V_{pores}.

Due to chemical shrinkage, deviating microstructures can form at medium to low W/C values and sealed storage, which influence the moisture storage behavior and transport processes in the system, which is confirmed in the results reported below.

Figure 4.49a,b give the corresponding results of the MIP analyses. **Figure 4.49a** shows the pore-sum curves of the mortar REF after 28d resp. 10 and 12 weeks of water storage and sealed storage, respectively, for comparison. It is striking that, independent of storage and storage duration, identical behavior is present in the extended gel pore region ≤ approx. 30 [nm]. Furthermore, it is striking that the maximum size of the capillary pores, which is about 100 [nm] to about 300 [nm] in water storage (compare Section 2.1.3), expands to about 1 [μm] after sealed storage. This behavior is also evident in the corresponding pore distribution curves (b) of the MIP measurements. It must be concluded **that this porosity formation is characteristic of the material interior at an appropriate distance from the material surface.**

On test specimens of the same material batch (slice thickness 20 [mm]) of the mortar REF, water absorption tests were carried out after standard drying, 128d at approx. 100% RH and subsequent 380d water storage. The results of these test series with the three differently prestored specimens can be taken from **Figure 4.50**. It

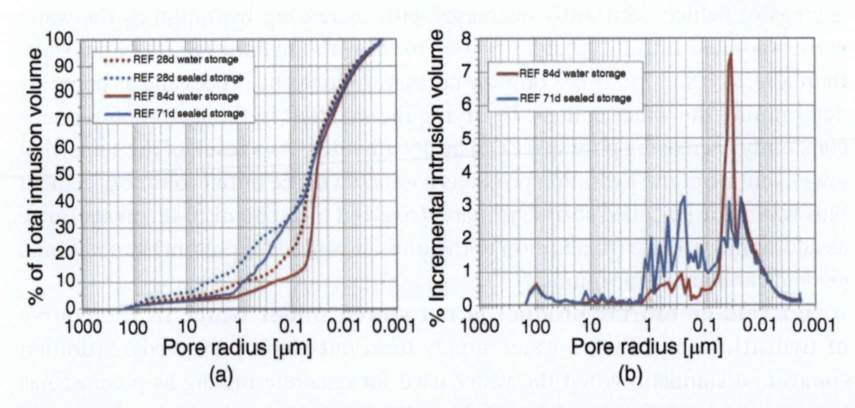

Figure 4.49 Comparison of cumulative pore volume curves (a) and corresponding pore size distribution curves (b) of 84d water-stored mortar and of 71d sealed-stored mortar REF ($W/C = 0.45$).

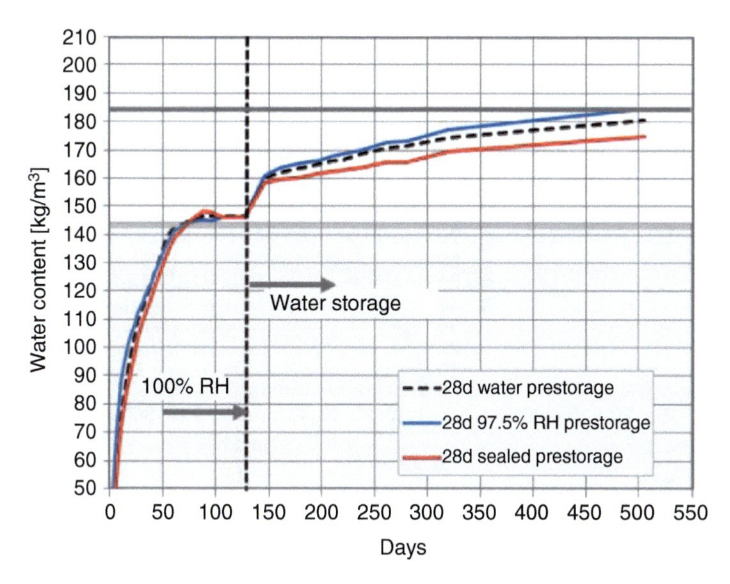

Figure 4.50 Comparison of measured water storage curves $W(t)$ of 3 REF mortar-series (slices 20 [mm] thick) after different prestorage methods (28d sealed, 28d at 97.5 RH, 28d water storage). After standard-drying slices exposed 128d to 100% RH (20 °C) and following 380d water storage.

can be seen that from approx. 100d storage at 100% RH, a moisture equilibrium is established in the material that corresponds to the filling of the gel and capillary pores. At a relative air humidity of RH = 99.9%, equivalent pore radii of 1.0 [μm] are filled, **which corresponds to the filling of the capillary pores in the sealed stored material**. Processing-induced air pores were not filled during this storage. Therefore, this water content would be as a limit to the over-hygroscopic range of the present material. Here a contradiction arises with the results reported in **Figure 4.43**. The visible difference is due to the partial filling of the air pores, which has already taken place on the thinner samples, there resp. the sample thickness of only approx. 7 [mm], where a partial filling of the air pores takes place after previous standard drying and storage at 100% RH, compare explanations on this in Section 4.5.5.

Furthermore, it was checked whether and, if so, how the different initial storage conditions affect the **unidirectional capillary water uptake on prisms after standard drying**. Corresponding series of tests were carried out on laterally sealed prisms of length 150 [mm] for a maximum of 12 000 hours. The results can be seen from **Figure 4.51a,b**. There is a clear difference in the water uptake velocity between the initially sealed hydrated prisms and the initially water-stored prisms. The lower water-uptake velocity of the sealed prestored test specimens is attributed to the partially coarser capillary pores, which develop a lower capillary pressure at the same water-filling degree.

The question of the extent to which the different initial storages affect the adsorption isotherms is reviewed in Section 4.6.

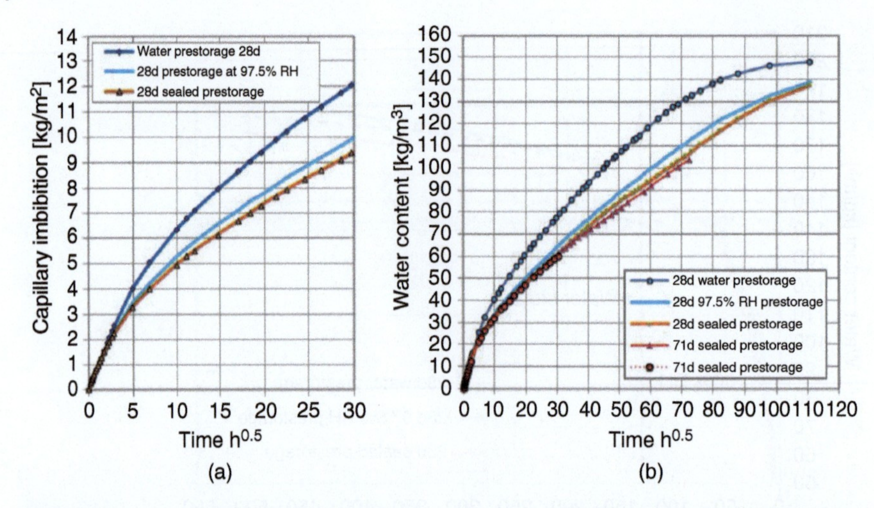

Figure 4.51 Mortar-REF, Tests on influence of prestorage methods on 1D capillary water imbibition, prisms $4 \cdot 4 \cdot 15\,[cm^3]$, mean values of three prisms per method. (a) Capillary imbibition over 900 after 28d sealed storage and 97.5% RH and water storage, respectively, and standard drying. (b) Capillary imbibition curves after longtime imbibition over 12 000 hours maximum after standard drying.

4.6 Adsorption and Desorption Isotherms

4.6.1 Overview of Storage Functions for Different Building Materials

Figures 4.52, 4.53b, 4.54b and 4.55b from Funk [5] present investigation results from the University of Dresden/Germany Plagge et al. [31], [32], [33]). **Figures 4.53b and 4.54b** demonstrate that, contrary to popular belief, ceramic bricks also exhibit distinct desorption isotherms that deviate from the adsorption curves. It can be clearly seen from the measurement curves for the brick material Wienerberger **Figure 4.53b**, for example, that in the case of partial drying from moisture saturation down to, e.g. 80% RH and renewed moisture absorption, a strong difference results in the course of cure between drying and moisture absorption.

Desorption isotherms were compiled from vapor desorption tests and pressure plate test results. These results suggest that these brick materials have distinct superhygroscopic storage behavior even when compared to 100% RH adsorption tests. **The results of various building materials presented in Figure 4.52** show only a slightly pronounced hysteretic behavior. This is due to the fact that the sorption experiments are limited to an upper relative air humidity of 92%. This is particularly evident in the curves for the ceramic brick products. Their hysteretic behavior is especially evident during drying after capillary water absorption.

Figure 4.55a,b proves that also lime-sandstone material shows a pronounced hysteretic storage behavior. Extensive measurements of transient moisture absorption and moisture distribution on specimens of calcium silicate material, aerated concrete, brick material, and sand-lime brick are reported by Scheffler [27]. Adsorption and desorption curves in the hygroscopic moisture range are given for one

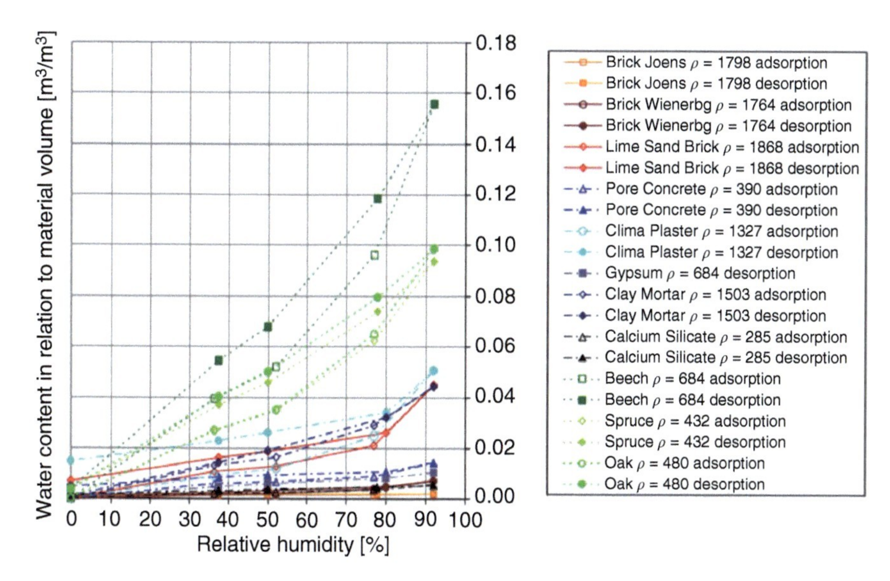

Figure 4.52 View of adsorption and desorption curves of selected building materials (except normal concrete and cement mortars). Source: Data and figures from Funk [5]/Max Funk.

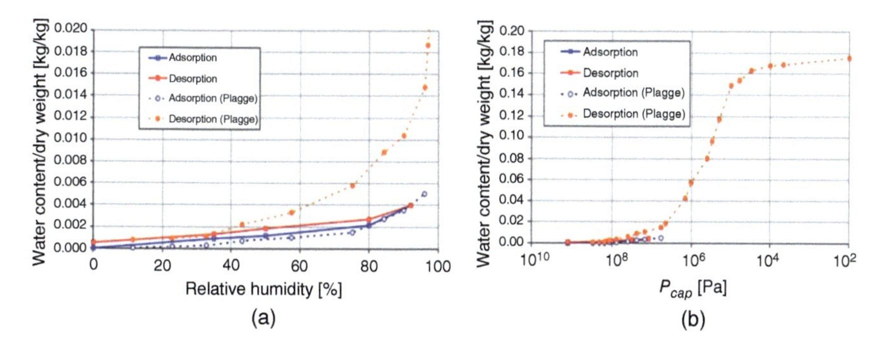

Figure 4.53 Vapor sorption and over-hygroscopic storage behavior of brick Wienerberger (Germany), test results from Funk [5] and data from Plagge and Teutsch [33]. **(a)** Hygroscopic adsorption and desorption behavior. **(b)** over-hygroscopic desorption behavior from pressure plate tests.

representative from each of the addressed material groups. In the superhygroscopic range, the capillary pressure-dependent adsorption curves were assumed to be desorption curves from the point of view of similarity. To what extent these curve sections correspond to reality is not known due to the lack of measured values. The subject of scanning curves was not dealt with. However, the transient moisture distribution results obtained are very valuable and can be used to evaluate simulation programs.

In the soil mechanics literature, numerous results on moisture storage functions and hysteretic behavior have been documented. These results are usually

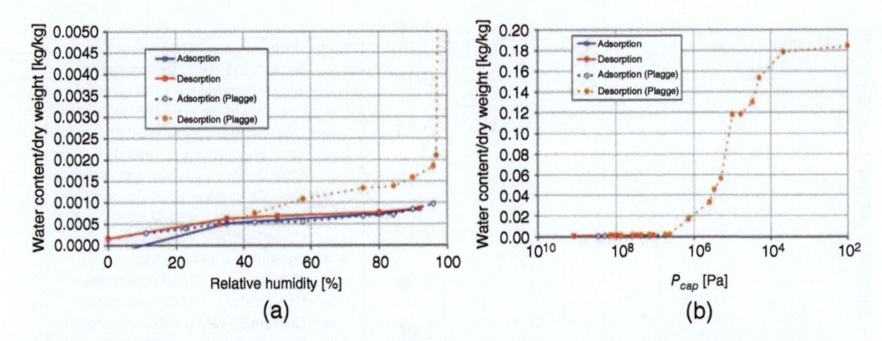

Figure 4.54 Vapor sorption and storage behavior of brick Joens (Germany), test results from Funk [5] and data from Plagge and Teutsch [33]. **(a)** Hygroscopic adsorption and desorption behavior. **(b)** Over-hygroscopic desorption behavior from pressure plate tests.

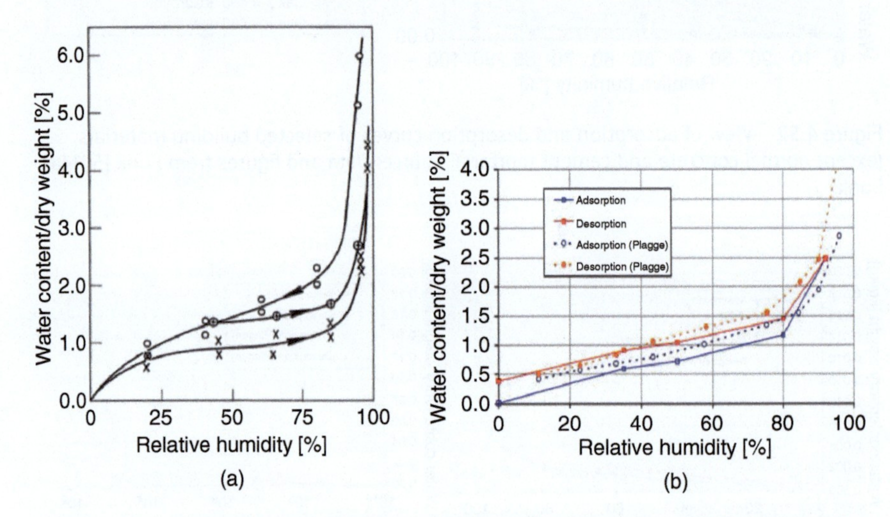

Figure 4.55 Adsorption and desorption behavior of sand-lime bricks. **(a)** Test results from Ahlgren [3]; (dry density = 1830 [kg/m^3], Porosity = 31%). **(b)** Test results from Funk [5] and with data from Plagge and Teutsch [33].

used to validate simulation models for predicting the main sorption curves and hysteresis curves.

Experimental and modeling approaches for clay soils are contained, for example, in Mualem 1974 [4]. The essays, especially those of Mualem [4], propose computational models for predicting one of the principal curves resp. the hysteresis curves on the basis of the two principal curves.

To measure the desorption curves on soil samples in the laboratory, either the pressure plate method or suction stress measurements with a suspended water column are used. The gas pressure in the pressure plate test resp. the water underpressure in the case of hanging water column are adjusted stepwise until the respective equilibrium is reached. When the hanging water column method is used, the adsorption curve can also be measured by stepping down the negative pressure after a desorption experiment.

Prediction models based on domain theory are obviously **widely used in soil mechanics for modeling moisture storage functions**. Thereby, starting, e.g. from measured data for the main desorption curve, the missing adsorption curve and the scanning curves are modeled on the basis of an assumed domain. According to this, a domain is a collection of pore structures with a base structure recurring according to a given regularity under fixed physical laws resp. processes in the course of humidification and dehumidification. Extensive explanations of the domain models are also contained in Funk [5]. A number of other publications, for example, by Jaynes 1984/1985 [34] or Pham et al. 2005 [35], compare the predictions of numerous theoretical models of hysteresis of moisture storage in soil mechanics with experimental results and evaluate the goodness of the models.

From this concise explanation, it is clear that domain theory, in particular, can be of assistance in investigating the effects of certain pore structure assumptions on the moisture storage function. However, assigning such modeling results to specific materials without checking the agreement between experimental results of moisture storage and the modeled curves is unreliable. **It should also be noted that no hygroscopic range is considered for moisture storage in soil mechanics tests and modeling.**

It is therefore state of the art that even non-cementitious materials can exhibit hysteretic wetting and drying behavior.

4.6.2 Sorption Isotherms and Scanning Isotherms of Hardened Cement Paste and Concrete

As a basis for clarifying further relationships, adsorption isotherms and desorption isotherms were also performed on pure **HCP** (without aggregate) with different W/C values and different curing times. More details on the composition of the mixtures can be obtained from **Table 4.1**.

Silica gel drying (referred to as 3% **drying)** was used as the drying method, compare the figure (**Figure 4.41a,b**) for the effect.

Figure 4.56 contains the results for HCPII, nine months prestored at 98% RH for water-cement ratios of 0.40 and 0.60. The curves are plotted as water content $m_{W,HCP}$ related to dry bulk density, in %. One can see a pronounced difference between adsorption and desorption and the clearly different water contents of the HCPs with $W/C = 0.40$ and $W/C = 0.60$.

Figure 4.57a,b shows a similar behavior of the investigated HCPIII for W/C values 0.40 and 0.60 with pre-storage times of 12 weeks and 15 months at 98% RH.

Based on the CEMIII/B 42.5 R cement used for HCPIII, a mortar MIII was also investigated. The resulting sorption isotherms for initial storage durations of 12 weeks and 15 months at 98% RH and W/C values of 0.40 and 0.60 are shown in **Figure 4.58a,b**. These values are also related to the dry weight after 3% drying.

In the publications of R. Espinosa [6], Espinosa and Franke [15], and [36], the measurement results on scanning isotherms obtained within these projects were reported in detail. There it is stated how, on the basis of thermodynamic derivations, an estimation of the course of scanning isotherms becomes approximately possible,

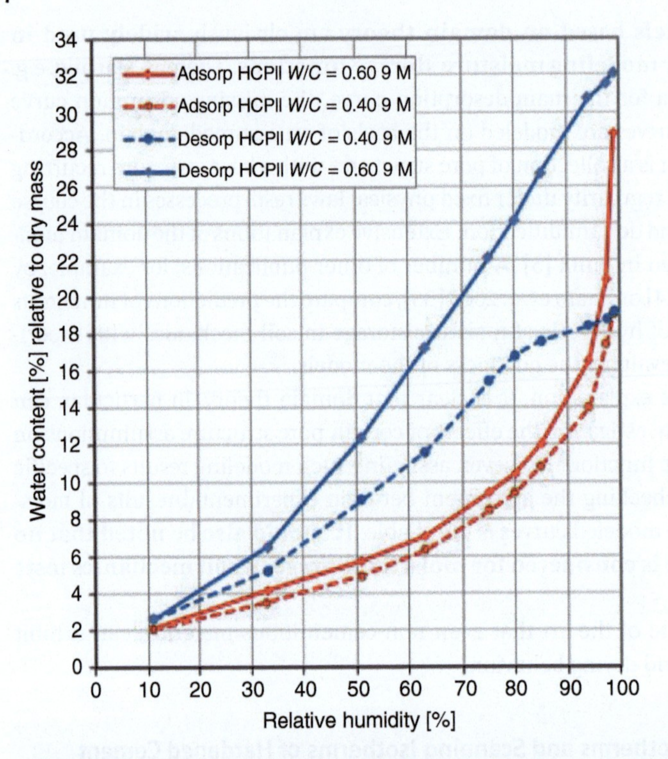

Figure 4.56 Measured adsorption and desorption isotherms on **Hardened Cement Paste HCPII** ($W/C = 0.40$ and 0.60, mixture compare **Table 4.1**), prestored for 9 months at 98% RH (20 °C), desiccator-method above salt solutions.

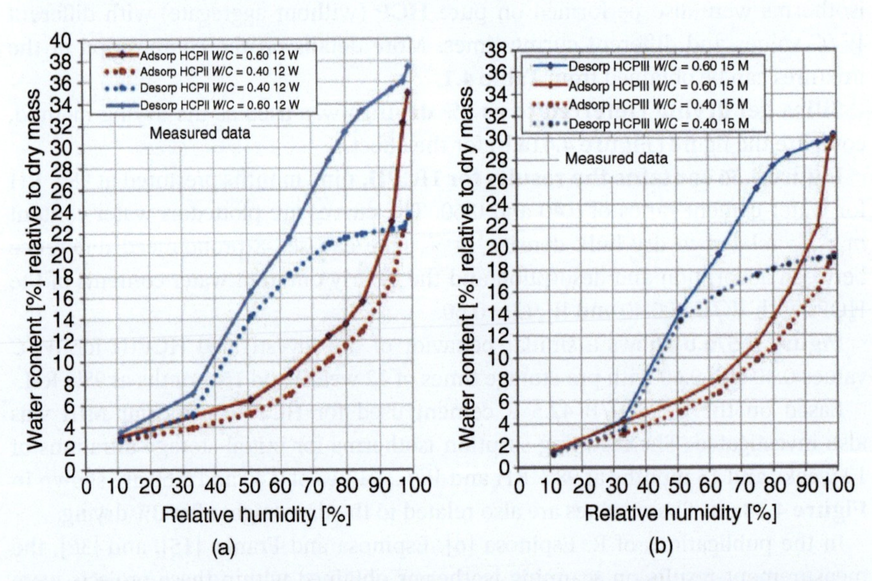

Figure 4.57 Measured adsorption and desorption isotherms on **Hardened Cement Paste HCPIII** ($W/C = 0.40$ and 0.60, mixture compare **Table 4.1**), desiccator-method above salt solutions, prestored at 98% RH (20 °C). (a) after 12 weeks prestorage (b) after 15 months prestorage.

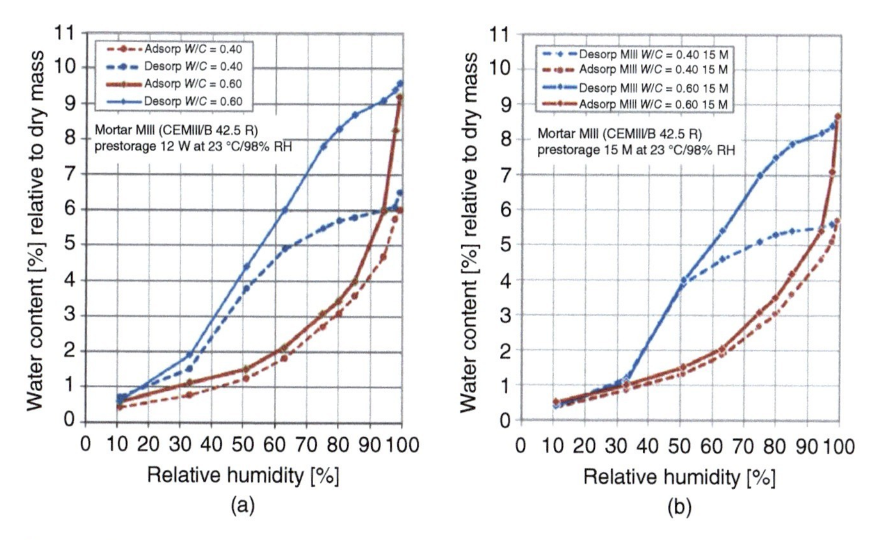

Figure 4.58 Measured adsorption and desorption isotherms on **mortar MIII** (W/C = 0.40 and 0.60, mixture compare **Table 4.1**), desiccator-method above salt solutions, prestored at 98% RH (20 °C). (a) After 12 weeks prestorage; (b) After 15 months prestorage.

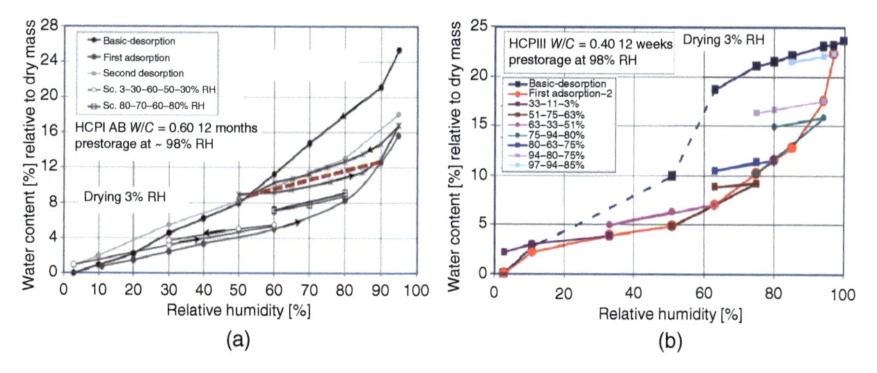

Figure 4.59 Scannings measured on **HCPI and HCPIII/TUHH. (a)** Main Isotherms and scannings measured by desiccator-method, Second Desorption and scannings near red line by DVS (dynamic vapor sorption analyzer), 3% RH-silica-gel drying. HCPI (CEMI-AB, W/C = 0.60, 12 months prestorage at 98% RH) . Source: Espinosa and Franke [15]/with permission of Elsevier. **(b)** Isotherms and scannings measured on HCPIII (mixture see **Table 4.1**) by desiccator method above salt solutions, 3% RH-silica-gel drying, W/C = 0.40, 12 weeks of prestorage at 98% RH.

provided that the necessary parameters are available [36]. A procedure deviating from this is presented in Section 4.6.4.

Figures 4.59a for HCPI (CEMI AB) and Figure 4.59b for CEMIII show measurement results for the course of the scanning isotherms also from these projects; in each case, 98% RH was used as initial storage instead of water storage.

Both figures give an impression of the expected course of scanning isotherms for cement-bound products made of different cements with different water-cement ratios. The scannings move approximately linearly between the desorption and adsorption isotherms.

When evaluating the slopes of scanning isotherms, it should be noted that at the transition of a scanning into the target isotherm, an exaggerated transition curvature can be suspected, caused by the fact that a measuring point still on the scanning is connected to a measuring point on the target isotherm, which belongs to an advanced state on this isotherm and is not the actual "contact point." This can be seen in model form, for example, in **Figure 4.60a** for the increasing scanning starting from 20%. Also, the mean slope of the scannings cannot always be assumed to be constant for the entire air-moisture range.

Based on the dashed red line in **Figure 4.59a**, a mean slope of the scannings can be indicated for the HCP from CEMI. It results in a first approximation from a connection between the absorption curve at 90% RH and the governing desorption curve at 50% RH. This also applies to other cement-bound products. A more detailed consideration is given below, in particular in Section 4.6.4.

In **Figure 4.59a** and the parallel **Figure 4.68**, it is also clear that the adsorption and desorption scannings can take a slightly concave shape, visible in the neighborhood of the red dashed approximation line.

The basic desorption isotherm and 1st adsorption isotherm (basic adsorption) were carried out in the TUHH projects with the direct desiccator method, i.e. each measuring point shown, for example of the desorption line, was obtained by desorption from the initial saturation (of usually at least three separate samples) at the associated air humidity. Exceptionally, the use of stepwise approach or DVS is always indicated, compare Section 4.6.3. Secondary desorption isotherms and scannings were obtained stepwise using the desiccator method, and some curves and scannings were also obtained using the DVS method.

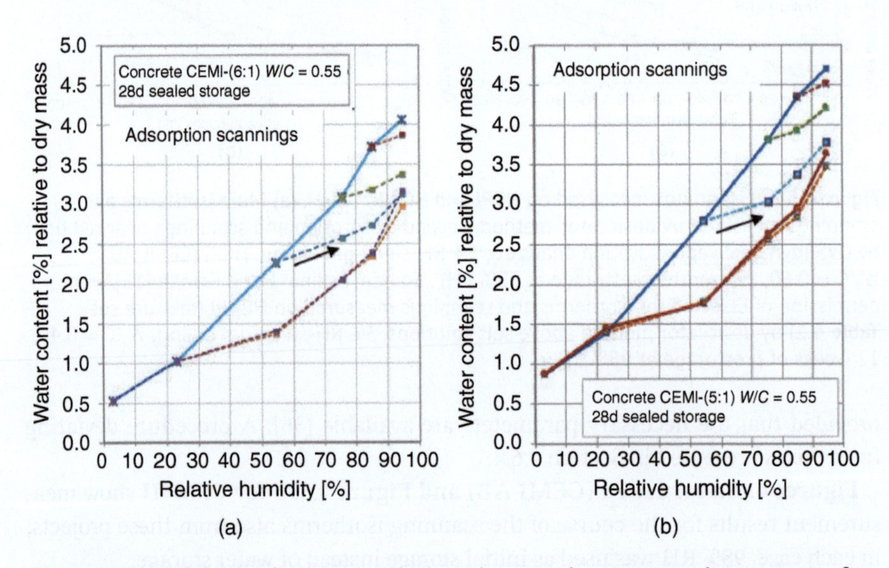

Figure 4.60 Adsorption isotherms and adsorption scannings, **measured on concrete after sealed storage**, referring to 105 °C dry mass. **(a)** Concrete CEMI 32.5 R, $W/C = 0.55$, $Aggr/CEM = 6/1$. **(b)** Concrete CEMI 32.5 R, $W/C = 0.55$, $Aggr/CEM = 5/1$ results from Deckelmann and Schmidt-Döhl/TUHH [37].

The adsorption isotherm in Figure 4.59a starts at 3.0% **RH**. Thus, the residual water content still present in the samples at this point, which may well be up to 10 of the initial water content, is not included in the curves. In addition to drying at 3% RH, freeze drying was also used as a drying method in [6], [15], [36]. In this process, there is no ice formation, but drying via sublimation. Nevertheless, this process is more intensive compared to 3% drying and microstructural changes cannot be ruled out.

The isotherms measured on HCPIII, compare **Figure 4.59b** make it clear how differently the sorption isotherms can run depending on the pore structure, and the slope of the scannings also varies accordingly, compare also Section 4.6.4.

At this point, it should already be pointed out that the samples stored in water or hydrated at high RH **show significantly different desorption curves compared to material sealed or hydrated in the core of concrete components**. This is discussed in more detail in Section 4.6.3.

Figure 4.60a,b give the sorption curves and scanning isotherms measured on different sealed hardened concretes. Desorption measurements were made using the desiccator method immediately after unsealing without further wetting. Corresponding to the lower water content of the material, the desorption and adsorption isotherms are closer together above about 50% RH with steeper increases in the scanning isotherms.

Figure 4.61a,b of Veronique Baroghel-Bouny show results of mixtures with low W/C values, HCP from CEMI with $W/C = 0.35$ and UHPC concrete with $W/C = 0.27$ and 10% SF with isotherms determined by the desiccator method, from Baroghel-Bouny [38]. The very extensive results in this publication by Veronique Baroghel-Bouny are the basis for evaluations, conclusions, and the basis for modeling by other authors.

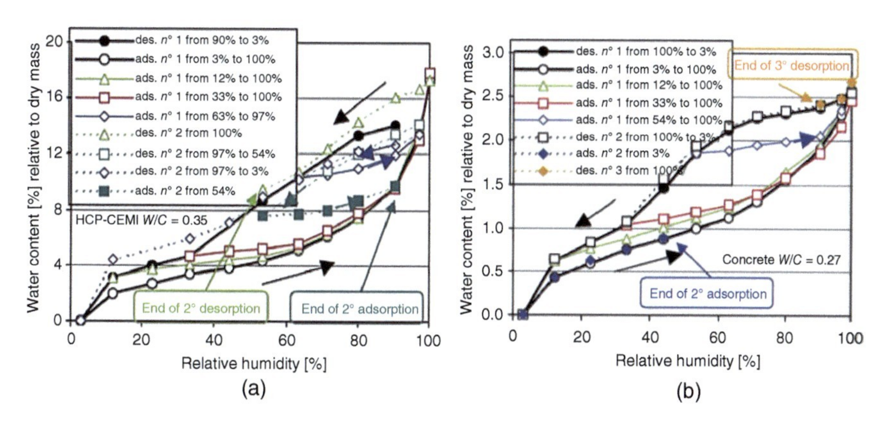

Figure 4.61 Measured adsorption and desorption isotherms and scannings. Source: Results from Baroghel-Bouny [38]/with permission of Elsevier. **(a) Hardened Cement Paste HCP** CEMI 52.5, $W/C = 0.35$, 2 years sealed curing, then vacuum water saturation and 100% RH storage, crushed specimens. **(b) Concrete** (420 kg CEMI 52.5 + 10% SF, $W/Binder = 0.24$, superplasticizer, max. aggregate size 20 [mm], ≥ **6 months sealed curing, then vacuum water saturation and** 100% **RH storage**, age > 1 year at start of sorption experiments, desiccator-method above salt solutions, and 3% RH-silica-gel drying).

The measurement results reproduced in Figure 4.61a confirm that caution is required in the evaluation and interpretation of the results also as a basis for physical conclusions and for modeling, if measurement results from several series of the same material are presented in the same diagram without sufficient labeling.

For the concrete with $W/C = 0.27$ of Figure 4.61b it is particularly remarkable that the three measured desorption isotherms starting from 100% RH show the same shape, with significantly lower slope above about 60% RH, i.e. a more "bulbous" behavior compared to the HCPI, resulting in an influence on the shape and slope of the scannings, compare Section 4.6.4. The measured scannings show a largely linear course.

It should be explained in [38] how the 100% air humidities were realized in comparison to the 97% RH starting values, because at 100% RH there is a risk of condensation in the desiccators, see Section 4.5.7.

4.6.3 Primary and Secondary Desorption Isotherms, Reversibility of Structural Changes

In projects to determine sorption isotherms, the desorption isotherm is usually first measured from the moisture saturation state (**designated as primary or basic desorption curve**), and the adsorption isotherm is usually then determined from a defined drying state (designated as **primary adsorption curve**) by measuring the equilibrium moisture contents at now increasing relative humidities of the air.

In research projects, a 2nd desorption curve and, if necessary, also a 2nd subsequent adsorption curve are often measured on the same material samples after a high moisture content or moisture saturation has been reached, as a check.

In cases where the water saturation at the beginning of the 2nd desorption curve is below the initial water saturation (of the 1st desorption), the 2nd desorption curve will inevitably run below the 1st curve. Of interest here is only at what relative humidity the 2nd curve merges into the 1st (primary) desorption curve or the 2nd desorption can be considered as a scanning isotherm.

It should be noted that in some projects 2nd desorption curves have exceeded the course of the 1st desorption isotherm at relative humidities below 50% RH, see for example, **Figure 4.61a** or also Espinosa [6] and Espinosa and Franke [36], **Figure 4.62**.

However, the difference of the curves in **Figure 4.62** below 50% RH would be significantly lower in the presence of a measured value (for the 1st curve) also below 34% RH. Furthermore, the drying method used (freeze-drying) may have an influence. The Baroghel-Bouny [38] experiments were carried out on concrete and cement mortar of CEMI, $W/C \geq 0.35$ by desiccator-method and 3%-silica-gel drying.

Jennings et al. [14] that the main cause of this behavior is irreversible water removal from the "interlayer space," which is not compensated even after prolonged water storage.

However, in the experimental studies of Jennings et al. [14] **no overlap was found** between primary and subsequent desorption curves. Tests were performed

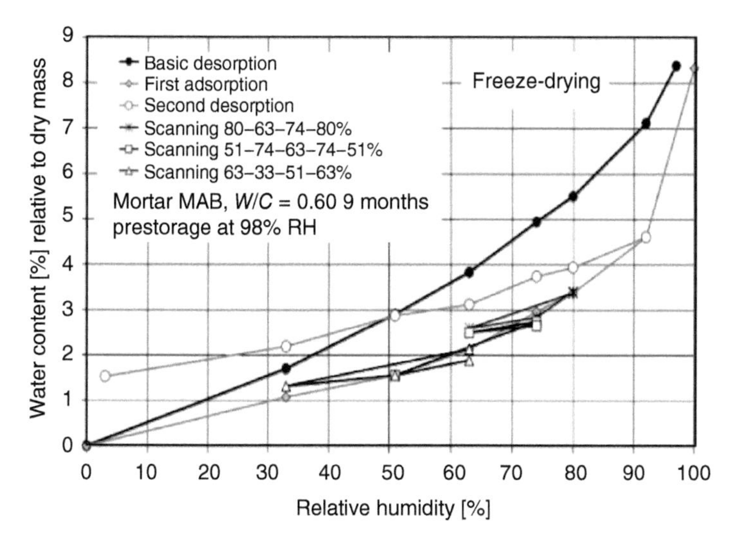

Figure 4.62 Phenomenon of crossing of primary desorption isotherms by subsequent desorption isotherms under certain conditions: Isotherms and scannings measured by Dynamic vapor sorption analyzer on **mortar MAB**, $W/C = 0.60$ (**9 months of prestorage at** 98% **RH, freeze-drying**)

on six-month **sealed stored samples** HCP CEMI, $W/C = 0.45$ and 0.65 in a DVS apparatus. Samples were ground before the start of the experiment and adjusted to 98% in the apparatus at the beginning. Wu et al. [17] also found no overlap in their DVS experiments, which are mathematically corrected for sorption equilibria. Jennings et al. assume that irreversible changes can be excluded if the desorption does not fall below 25% RH. Pore modification by drying has also been dealt with by other authors.

Saeidpour and Wadsö [39] **derive adsorption and desorption curves from DVS tests** on HCP samples of CEMI and CEMIII, compare **Figure 4.74a** to show hat **no structural changes** are expected when drying is limited to 30% RH. In [39], the experimental results were mathematically corrected to the "true" moisture equilibrium at the RH steps (unlike the experiments in [14]). Desorption/adsorption experiments with desorption to 50% using the desiccator method were also found by Espinosa and Franke [36] to have no change influences, compare **Figure 4.68.**

In the DVS experiments of Wu et al. [17] were also conducted using ground HCP samples (from CEMI and CEMIII, $W/C = 0.40$) and MCM-41 samples. The HCP samples were initially stored in the form of small cylinders in lime-hydrate water for about six months and were also completely saturated with lime water by vacuum before comminution. After crushing, the material was stored in a post-saturated state with lime water in closed containers for about seven days until the start of DVS tests.

Results of these DVS measurements are given in **Figure 4.76a**. After measurement of the primary desorption curve and the subsequent first adsorption curve, the samples were "restored" by adding several drops of distilled water to the sample holder of the DVS instrument (on the respective sample), and a rest period of three days followed until the start of the measurement of the secondary desorption curve. In **Figure 4.76a**, it can be seen that the secondary desorption curve is analogous or identical to the primary desorption curve with a slightly lower water content. It is assumed that this lower water content is caused by an unfortunately not completely successful post-saturation by the restoration. The adsorption curves immediately following the 1st and 2nd desorption curves were identical.

The authors conclude that either no structural changes occurred during the drying in the DVS apparatus or at least a complete restoration took place or possible structural changes due to intensive drying are reversible in the subsequent adsorption phase resp. water saturation. Even after the measurement of the presented scanning curve and subsequent desorption (directly on the 2nd adsorption curve), no indications of nonreversible structural changes can be read from the results [17].

Baroghel-Bouny [38] also concludes that the differences between the base desorption isotherm and the secondary desorption isotherm observed by the desiccator method are not permanent resp. are reversible and are not caused by either progressive hydration or chemical aging.

The test results presented in this section show that, despite the considerable advantages of the DVS method in terms of test duration and sample preparation, there is still a need for research into the causes of the differences in results compared with the desiccator method. As an interim solution for the application of the DVS method, it seems reasonable to use the tightened boundary conditions of Saeidpour and Wadsö [39] for the DVS method, i.e. to set the limit value of 0.0001%/minute within a maximum of 2000 minutes as a criterion for the transition from a completed air humidity stage to the subsequent air humidity setting in the DVS device.

4.6.4 Modeling the Course of Scanning Isotherms

Recent scientific publications [40–42], and [43] address the state of the art of modeling sorption isotherms or scannings for cement-bound products to date. In each case, the authors conclude that the modeling methods that have become known are only conditionally suitable for cement-bound products. This is obviously also due to the fact that these methods come from the field of soil mechanics and consider the fluid partial pressure in the pores as the driving force.

In the modeling of sorption scannings described below, the local capillary pressure is not assumed, but the local vapor pressure.

For predicting the moisture behavior or modeling the scanning curves, it is important to note that they remain at least approximately unchanged between initial and final points on the same curve (or straight line in the middle region of the scanning) also on repetitions.

An important property with respect to modeling is the slope of the scanning curves. This is presented in [15] based on sorption characteristics and a shape factor to characterize the pore structure resp. bottleneck porosity. Since this factor has to be estimated, the slope of the scanning curves calculated in this way does not have a higher accuracy than the proposal initially made below. Indeed, if one evaluates the numerous results contained in [6] and [15] on the course of the scanning curves for hardened cement paste and mortar, respectively, but also in the older results, for example, of Ahlgren [3] for concrete as well as Feldmann and Sereda [2], **the slope** of the connecting straight line in the $\theta(\varphi)$ diagram **between the value of the desorption isotherms at 50%** RH and the value of the adsorption isotherms at **90%** RH results as a very good approximation for both HCP, mortar, or concrete. In **Figure 4.59a**, this slope is shown as a red dashed line.

In the following, a method is proposed on the basis of which a more precise course of the scannings can be determined for different cement-bound materials, starting from their adsorption and desorption isotherms. A decisive influence is exerted by the respective adsorption isotherm. In view of modeling a scanning isotherm, with reference to **Figure 4.63**, a reversal point θ_{rev} or θ_0 on θ_{Des} is chosen as the starting point for an adsorption scanning.

The initial state belonging to θ_0 in the pore system can be characterized as follows:

A material section with homogeneously distributed water content θ_0 at equilibrium is considered. At equilibrium, there is a homogeneous vapor-pressure distribution in the pore system over the pore portion filled with θ_0 and the pore portion not filled with water.

The filled pores facing the empty pore space or their current water surface have meniscus radii corresponding to the current internal vapor pressure φ_0.

However, this state can also result from adsorption from the dry state to the same vapor pressure φ_0. Both states differ essentially by the associated water content of

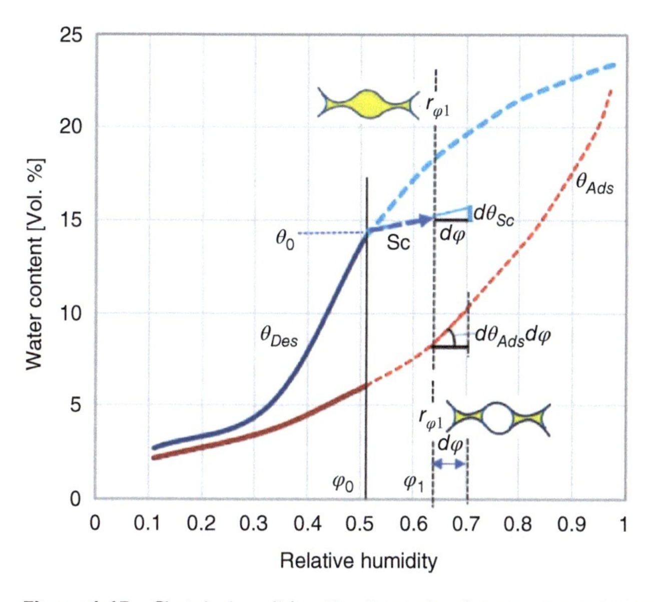

Figure 4.63 Sketch describing the dependencies or parameters used for modeling the course of scanning isotherms. Let θ_{Des} and θ_{Ads} be the present sorption isotherms, let θ_0 be a (relative) water content on θ_{Des} and the starting point of an adsorption scanning isotherm such as **Sc**. φ_0 corresponds to the vapor pressure prevailing in the pore system at the pore filling θ_0.

the material. The effective pore radii $r_{\varphi 0}$ must be the same in both cases for a given homogeneous internal vapor pressure. (Here, the sorptive moisture fraction on the inner surface of the unfilled pore fraction is naturally neglected.) Corresponding to the highly simplified pore sketches in **Figure 4.63**, the filling of θ_{Des} to θ_{Ads} differs mainly in that "behind the pore radii" $r_{\varphi 1}$ of the desorption isotherms water fractions are trapped via bottle-neck pores, compare also **Figure 4.22**. When the vapor pressure is increased by $d\varphi$, the slope increase of the adsorption scans is lower compared to the slope of the adsorption isotherms at the corresponding φ value due to the on average larger pore radii in the adsorption scanning region.

The real difficulty in modeling is to give a formulation as general as possible resp. usable for different pore configurations. Here, one can imagine complex analyses based on the vapor pressures present at specific pore distributions and the associated proportionally stored water contents.

On the other hand, the course resp. the shape of the adsorption and desorption isotherms is already a consequence of these relationships. Therefore, it is obvious to use these isotherms resp. characteristic values derived from them directly as the basis for the modeling.

In further explanations, reference is first made to Figure 4.63.

The adsorption scanning Sc had already progressed to φ_1. The water content increase $d\theta_{Sc}$, when the vapor pressure is increased by $d\varphi$, is initially unknown and is therefore searched for.

Starting from given isotherms θ_{Des} and θ_{Ads}, the characteristic values $Ai_{(\theta 0,\varphi 0)}$ and k_a and the respective slope $d\theta_{Ads}/d\varphi$ of the adsorption isotherms, **the evolution of the water content $\theta_{Sc(\varphi)}$ of an adsorption scanning** as a result of a stepwise increase of the vapor pressure by $\Delta\varphi$ can be described as follows:

$$\theta_{Sc(\varphi+\Delta\varphi)} = \theta_{Sc(\varphi)} + Ai \cdot \left(k_a \cdot \frac{d\theta_{Ads}}{d\varphi} \right)_{(\varphi)} \cdot \Delta\varphi \tag{4.10}$$

Ai in (4.10) is introduced as a measure of the internal surface area at the starting point of adsorption scanning, expressed by the relative water-free pore volume

$$Ai = Ai_{(\theta_0)} = \frac{\theta_{sat} - \theta_0}{\theta_{sat}} \tag{4.11}$$

θ_{sat} **corresponds to the water saturation**, which is present at approx. 100% RH resp. the maximum value of the measured desorption isotherms. θ_{sat} is therefore usually lower than the maximum water content immediately after water storage, since this also contains a partial filling of the processing-related air pores. For the REF material, θ_{sat} is about 140 [kg/m^3], compare for example **Figures 4.36–4.38**. Since there is usually no superhygroscopic region in cement-bound materials, θ_{sat} corresponds to the sum of gel and capillary pores, if necessary to be determined according to Section 2.2.4.

$$k_a = k_a(\varphi) = \frac{\theta_{sat} - \theta_{Sc(\varphi)}}{\theta_{sat} - \theta_{Ads}(\varphi)} \tag{4.12}$$

$k_a(\varphi)$ **in Eq.** (4.10) is, according to Eq. (4.12), the ratio of the free internal pore surface areas at the current water content of the adsorption scan $\theta_{Sc(\varphi)}$ to the free internal surface area at the water content of the adsorption isotherm at the same vapor pressure φ, expressed overall in terms of the ratio of the corresponding water-free pore volumes.

In Eq. (4.10), the value $\theta_{Sc(\varphi+\Delta\varphi)}$ after completion of $\Delta\varphi$ is the initial value $\theta_{Sc(\varphi 0)}$ for the next step for $\varphi_1 = \varphi_0 + \Delta\varphi$. Before each step, $k_a(\varphi)$ and the slope $\Delta\theta_{Ads}/\Delta\varphi$ of the adsorption isotherms at φ are recalculated. The step size $\Delta\varphi$ was usually chosen to be 1% or 2% RH. A change in this value has only a minor influence on the slope of the scannings.

The slope of the adsorption isotherms for the current φ is determined either from the derivative of the function of the adsorption isotherms or by interpolation directly from the measured values $\theta_{ads}(\varphi)$ as

$$d\theta_{ads} d\varphi(\varphi 0) = \frac{\theta_{ads}(\varphi 0 + \Delta\varphi) - \theta_{ads}(\varphi 0)}{\Delta\varphi} \tag{4.13}$$

Figure 4.64a shows, for the water-aged mortar REF, the results calculated with these correlations according to Eq. (4.10), labeled as "modeling-0" in **Figure 4.64a**. The calculated adsorption scannings start from the desorption isotherm at selected starting points θ_0 as blue dashed lines upward in an initial slight concave curvature. As the adsorption isotherm is approached, the scannings take on a steeper curvature until they have the same slope as the adsorption isotherm.

The adsorption scanning starting at $\varphi = 50\%$ RH is compared with measurement results of Goedeke/TUHH [44] from an independent project. The same water-stored material REF was used. Measured values and calculations agree. Also, in comparison to the other adsorption scannings measured on other materials, this modeling-0 (**Figure 4.64**) obviously yields correct progressions.

However, the physically correct behavior of the transition of the adsorption scannings into the course of the associated adsorption isotherm raises the question of whether and at which value the scanning curves should reach the adsorption

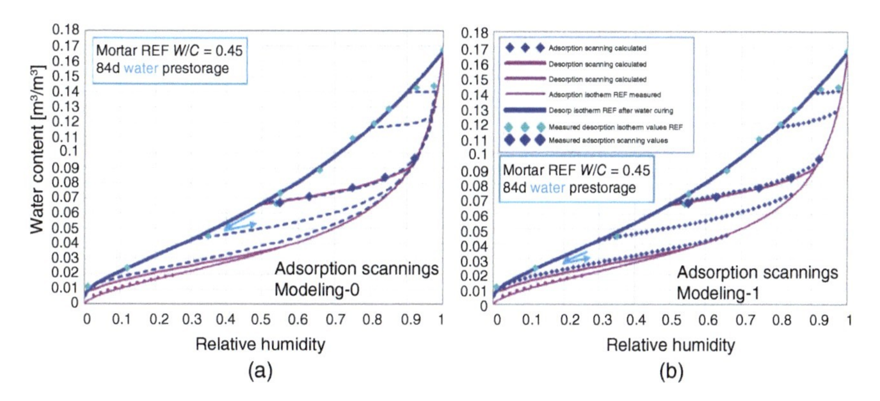

Figure 4.64 Course of adsorption and desorption scannings for the water-saturated mortar REF, calculated according to the explained concept on the basis of the measured adsorption and desorption isotherms. The points on the adsorption scanning from the starting point of 50% RH are measured points from Goedeke/TUHH [44]/Technische Universität Hamburg-Harburg. The effect of high initial water content on desorption scannings above 90% RH (2nd desorption isotherm) is not yet included in these diagrams.

isotherm resp. the further course is described by the latter. With regard to modeling of the moisture transport in larger material units with non-constant moisture content under consideration of the hysteretic storage behavior, it is **sensible to define clearly defined contact resp. transfer points between scanning and isotherms.**

For this purpose, a slightly **modified modeling is proposed below (modeling-1) Figure 4.64b**, from which clear touch points now result without causing significant sacrifices in prediction accuracy.

The calculations revealed that the best fit of the scanning isotherms to the modeling-0 trajectory shown in **Figure 4.64a** occurs when the effective internal storage area is defined by the associated still-water-free relative pore space φ-depending on $Ai_{(\theta_{Sc},\varphi)}$.

Figure 4.65 resp. Eq. (4.14) **show this relation**:

$$Ai_{(\varphi)} = \frac{\theta_{sat} - \theta_{Sc}(\varphi)}{\theta_{sat}} \tag{4.14}$$

The calculation algorithm is then given by the following Eq. (4.15)

$$\theta_{Sc(\varphi+\Delta\varphi)} = \theta_{Sc(\varphi)} + Ai_{(\varphi)} \cdot \left(\frac{d\theta_{Ads}}{d\varphi}\right)_{(\varphi)} \cdot \Delta\varphi \tag{4.15}$$

The resulting adsorption scanning curve using the same starting points is shown in **Figure 4.64b**. One can see the desired slightly flatter curve as the scanning approaches the adsorption isotherm. **Figure 4.66** contains the measurement results from [37] for a concrete material identified in more detail in the figure. The trajectories of the plotted adsorption scannings calculated with Eq. (4.15) agree with

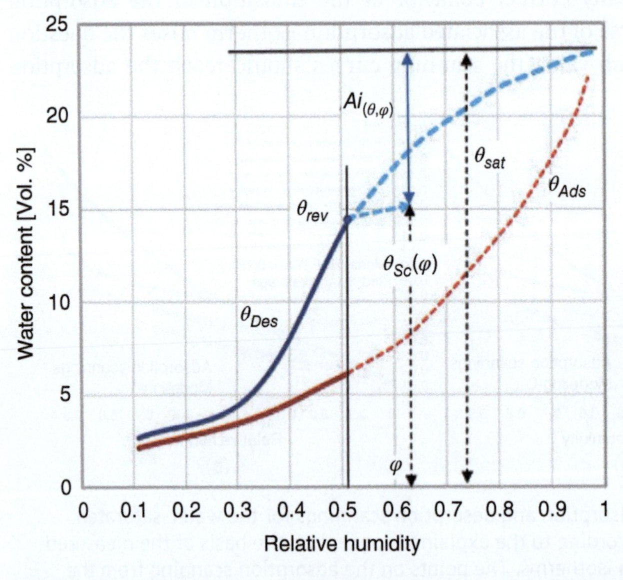

Figure 4.65 Sketch describing the effective inner water surface Ai_{φ} at the vapor pressure φ of a scanning isotherm Sc, started at the reversal point θ_{rev}. θ_{Des} and θ_{Ads} are the present sorption isotherms.

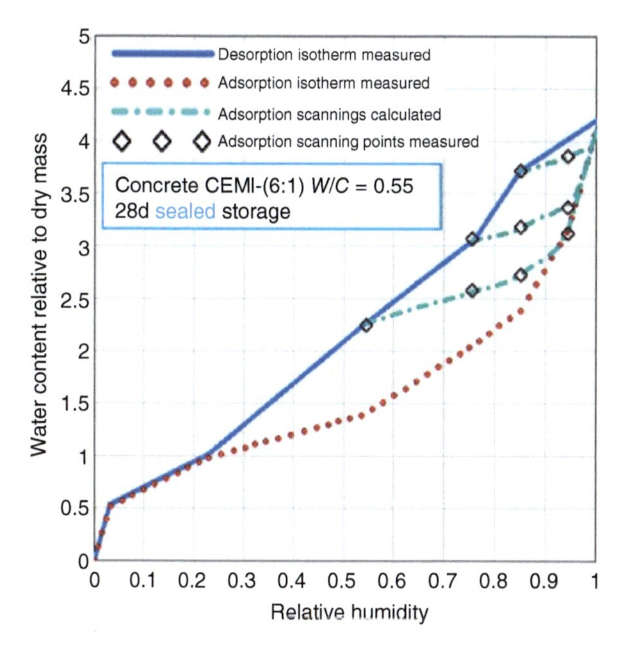

Figure 4.66 Comparison of measured and calculated adsorption scannings of a CEMI-concrete ($W/C = 0.55$) **after sealed curing only**. Experimental results from Deckelmann and Schmidt-Döhl/TUHH [37].

the measured values. The value 4.4 was used as θ_{sat} in (4.14), the steps $\Delta\varphi = 0.020$. The slopes were determined according to Eq. (4.13).

In the following, the calculation of desorption scanning is considered. According to the presented modeling based on the pore structure and vapor pressure-dependent internal surfaces, desorption scannings are to be calculated accordingly, **assuming that the pore structure, i.e. the distribution of bottle-neck pores in the material, is homogeneous.**

In the case of desorption scannings, a point $\theta_{(\varphi 0)}$ of the adsorption isotherms is assumed to be equal to $\theta_{Sc(\varphi 0)}$. The continuation of the calculation is then done with

$$\theta_{Sc(\varphi-\Delta\varphi)} = \theta_{Sc(\varphi)} - Ai_{(\varphi)} \cdot \left(\frac{d\theta_{Ads}}{d\varphi}\right)_{(\varphi)} \cdot \Delta\varphi \tag{4.16}$$

The difference to the adsorption scans is only that now the sorption loss is included for each reduction of φ by $\Delta\varphi$. The governing slope is still $d\theta_{Ads}/d\varphi$ of the adsorption isotherms.

Concerning the course of adsorption-scannings and desorption-scannings Diagrams with measurement results from de Burgh and Foster [42] are shown for comparison in **Figure 4.67**. The material used there is HCPI from Portland cement containing 7.5% limestone powder (a standard Australian cement). The W/C value is 0.45. The 3–4 [mm]- thick disc-shaped specimen bodies were cut from sealed stored cylinders and then stored in lime water until testing, for a total of at least 170 days. The desorption curve was determined directly from this water storage. The sample bodies for the adsorption isotherm were previously dried with

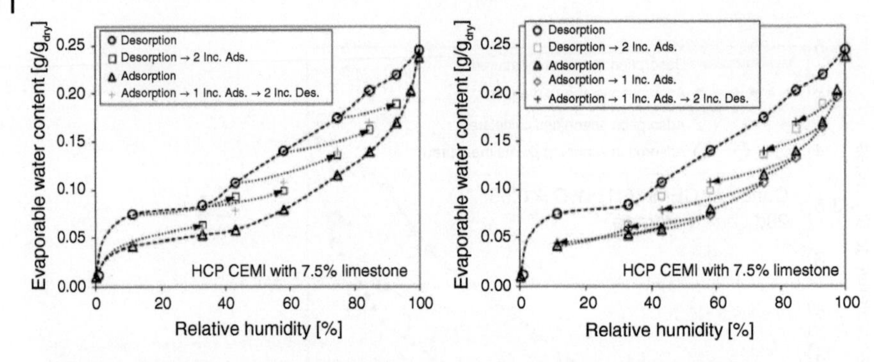

Figure 4.67 Observed course of adsorption scannings and desorption scannings on HCPI of Portland cement containing 7.5% limestone powder (a standard Australian cement), W/C value is 0.45. Source: Diagrams with measurement results from de Burgh and Foster [42]/with permission of Elsevier.

a molecular sieve material. The storage was carried out in special climate boxes, and the weight changes were measured by a precision balance that could be coupled in each case. The values shown in **Figure 4.67** were finally related to the weight after oven-drying.

de Burgh and Foster also conclude from their results the concave shapeshown for both the adsorption and desorption scannings, or the similarity of these curves corresponding to the calculated curves in **Figure 4.64**. The desorption-scanning curve starting from the adsorption isotherm above 90% RH, however, shows a much higher slope in **Figure 4.67**.

This phenomenon of a significantly different 2nd desorption curve, which can be observed in particular with water storage and higher W/C values, is not yet reflected in the calculations for **Figure 4.64**. How desorption scanning in this water saturation region can be done, at least approximately, is presented in the following Section 4.6.5, along with other results.

The value of the θ_{sat} parameter in Eq. (4.14) depends on the W/C value.

If isotherms, especially desorption isotherms, are present, which were measured after water storage, θ_{sat} corresponds to the water saturation. As a rule, θ_{sat} is lower than the maximum water content immediately after water storage, since this also includes partial filling of the processing-related air pores.

For the material REF, θ_{sat} is about 140 [kg/m^3], compare, for example, **Figures 4.36 to 4.38**, i.e.θ_{sat} **corresponds to the sum of gel and capillary pores.**

4.6.5 Dependence of Desorption Isotherms on Initial Storage Conditions

Section 4.5.9 already discussed the higher degree of water saturation that occurs by design in initial water storage compared to externally closed or sealed storage, i.e. inside concrete components. In the case of so-called moisture storage above water, the specimens have a water content corresponding to approx. 98% RH. This water content is usually between that after sealed storage and that after water storage. In sealed hydrated material, the pore fraction from chemical shrinkage as well as the air pore fraction are anhydrous and dependent on the W/C value. The water content after this storage corresponds to the adsorption moisture at 93–97% RH, compare **Figure 4.81a**.

The effect of the different initial water content on the desorption and adsorption isotherms, among others, were measured on the reference mortar REF by the desiccator method. Significant effects on the adsorption isotherms are not measurable, although the MIP analyses above pore radii of about 10–25 [nm] show clear differences, both in the pore-radius distribution and in the cumulative curves, compare **Figure 4.49a,b.** In contrast, there are clear effects on the desorption curves, see for example the measured results for the mortar REF in **Figure 4.69.**

This clear effect of initial storage on the course of the primary desorption isotherms is therefore not only due to the different initial water saturation degree of the material. The W/C value and the degree of hydration through their influence on the pore structure of the HCP and the cement type also affect the course of the sorption isotherms. Independent of the question of a possible overlap discussed in Section 4.6.3, the following 2nd desorption curve (starting from the primary adsorption isotherm) shows a further clear deviation of the desorption behavior from the behavior after water storage in the sense of a flatter curve up to approx. 50% RH, compare Section 4.6.4.

The different course of the particularly affected desorption scannings in the pore region with higher water saturation is attributed here to a changed distribution of the bottle-neck pores compared to the gel pore region. Based on the relationships worked out in Section 2.2.7.3, the **gel pore region was defined as the region below φ = 91.5%** RH.

In Figures 4.68, 4.69, 4.74a, 4.76a, and 4.77, the results of various authors illustrate this. Å*hs* [45] determines sorption isotherms on concrete samples taken from

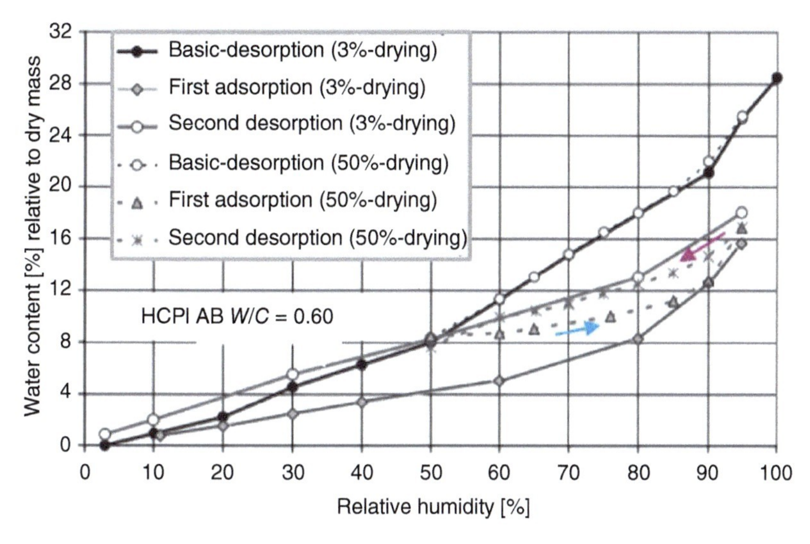

Figure 4.68 Sorption isotherms and scannings determined on HCPI. Comparison of the behavior after standard desorption (down to 3%) and desorption down to 50% RH only and 2nd desorption/scanning, starting from 95% RH. Prestorage at 98% RH. Source: From Espinosa and Franke [15]/with permission of Elsevier.

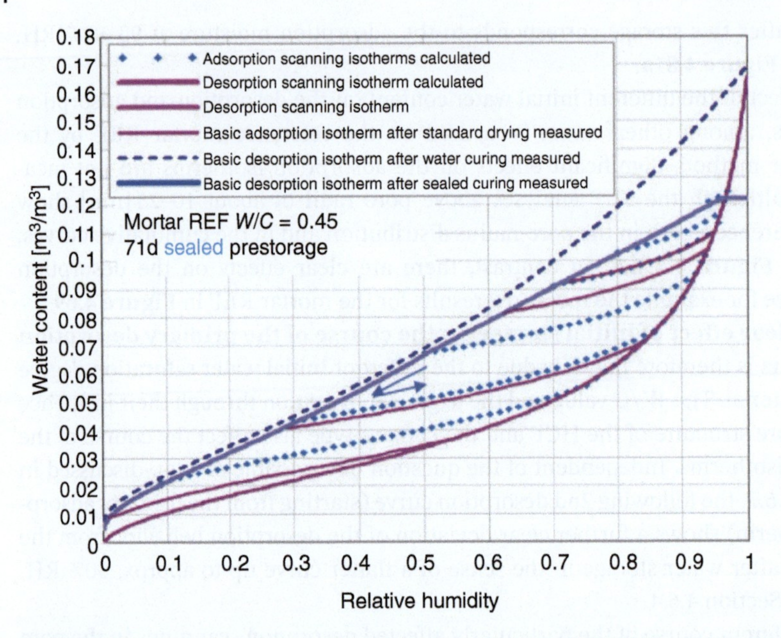

Figure 4.69 Comparison of sorption isotherms of mortar REF ($W/C = 0.45$) **after sealed curing** and after water storage, respectively, with corresponding calculated adsorption and desorption scannings. The blue desorption curve corresponds to the moisture content at the end of curing (desiccator-method above salt solutions, standard-drying at 40 °C/1 mbar, Franke/TUHH.

a component hardened at about 60% RH and finds significant differences between sealed storage and water storage.

Figure 4.68 [15] shows results on HCPI stored above water at about 98% RH after desorption to 3% and to 50% RH for comparison. After subsequent determination of the adsorption curves to 95% RH, the 2nd desorption curve was measured starting from this moisture condition. The results show that **the adsorption scanning starting at 50%** RH from the 2nd desorption curve meets the primary adsorption curve at about **90%** RH and then rises with it to 95% RH. It can be further seen that the desorption curves/scannings of both series, starting at 95% RH, follow the same course and together meet the coincident primary desorption curves of both sample series at about 50% RH.

Figure 4.69 gives the base desorption isotherm measured by the desiccator method on mortar REF **after sealed storage**, and the isotherm measured on the basis of the explanations in Section 4.6.4 according to the Eqs. (4.15) and (4.16) calculated adsorption and desorption scannings. Here, the adsorption scanning starting at about 50% RH corresponds to the measured course, compare **Figure 4.64**. For the adsorption scannings starting at 70% and 80% and calculated according to Eq. (4.15), the plotted desorption isotherm measured from sealed storage was taken as a basis. This approach is justified for modeling the adsorption scannings, as shown by the comparison between the curves calculated according to Eq. (4.15) and the measurement results of Deckelmann and Schmidt-Döhl/TUHH [37] in **Figure 4.66**.

The **question arises, in which way the desorption scannings or the 2nd desorption isotherm in the range of high water saturation levels can be predicted** or modeled with reasonable effort.

For modeling the progression of desorption scannings in the high water saturation degree range, it is crucial to note that there is significantly less crosslinking or concentration of bottle-neck pores in the capillary pore region compared to the gel pore region. As noted again previously, the gel pore region θ_{gel} **was defined as the integral pore volume between** $\varphi = 0\%$ RH and $\varphi = 91.5\%$ RH. In this range, the present modeling assumes a homogeneous distribution of bottle-neck pores in contrast to the pore volume θ_{cap} of capillary pores.

Therefore, deviating from the procedure explained in 4.6.4, the following empirical procedure is proposed for modeling the desorption scannings for the water saturation region above θ_{gel} based on the experimental results available here for 2nd desorption curves.

This method uses both main desorption isotherms and the desorption scanning calculated according to 4.6.4, starting from the starting point θ_{gel}, hereafter referred to as $\theta_{Sc,basic}(\varphi)$.

The dashed line in **Figure 4.70** is a modeled desorption scanning curve $\theta_{Sc}(\varphi, \varphi_0)$ started from the point $\theta_{Ads}(\varphi_0)$ of the adsorption isotherm. The other points of the modeled curve are obtained by dividing the perpendicular distances between the θ-values of the base scanning $\theta_{Sc,basic}(\varphi)$ and the main desorption isotherm (blue curve) for given phi values. As a result, the calculated scanning curves, when approaching these reference curves, take their respective shapes according to the existing experimental results.

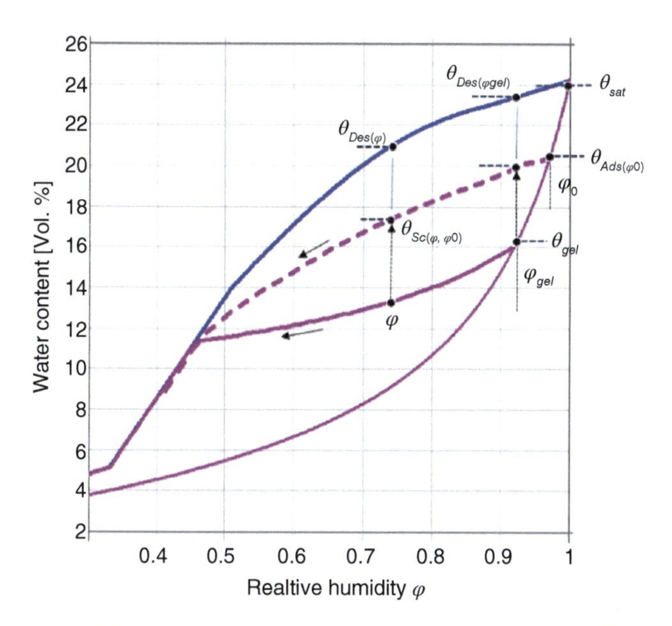

Figure 4.70 Schematic diagram with labels for the Eqs. (4.17) to (4.20) for **modeling secondary desorption isotherms** or for desorption scannings starting in the capillary saturation region above 92% RH.

In this regard, reference is also made to **Figure 4.75 of Baroghel-Bouny** [38]. It shows the strong dependence of the shape of the desorption isotherms on the W/C value, or the steeper shape of the curves with an increasing W/C value resulting from the increasing permeability or the decreasing fraction of bottle-neck pores in the region of the capillary pores in question. This behavior is accounted for by the procedure shown below.

Referring to the labels of Figure 4.70, the following equations are used for this purpose.

$k\varphi(\varphi_0)$ describes the relative position of the starting point on the adsorption isotherms θ_{Ads}

$$k\varphi(\varphi_0) = \frac{\theta_{Ads(\varphi_0)} - \theta_{gel}}{\theta_{Des(0.99)} - \theta_{gel}} \tag{4.17}$$

The course of the sought desorption scanning is then obtained for $\varphi \leq \varphi_{gel}$ to:

$$\theta_{Sc(\varphi)} = \theta_{Sc,basic(\varphi)} + (\theta_{Des(\varphi)} - \theta_{Sc,basic(\varphi)}) \cdot k\varphi(\varphi_0) \tag{4.18}$$

Equation (4.19) provides the point at φ_{gel} also needed for modeling the scanning process in the range $\varphi > \varphi_{gel}$.

$$\theta_{Sc(\varphi_{gel})} = \theta_{gel} + (\theta_{Des(\varphi_{gel})} - \theta_{gel}) \cdot k\varphi(\varphi_0) \tag{4.19}$$

The scanning history in the range $\varphi > \varphi_{gel}$ is then, under the (usually) satisfied condition $\theta_{Ads(\varphi_0)} \geq \theta_{Sc(\varphi_{gel})}$:

$$\theta_{Sc(\varphi>\varphi_{gel})} = \theta_{Sc(\varphi_{gel})} + \frac{\theta_{Ads(\varphi_0)} - \theta_{Sc(\varphi_{gel})}}{\varphi_0 - \varphi_{gel}} \cdot (\varphi - \varphi_{gel}) \tag{4.20}$$

According to this modeling, the 2nd desorption curves or scannings reach the desorption isotherms at vapor pressures from $\varphi = 0.40$ to $\varphi = 0.50$, compare also the measurement results in **Figure 4.68.**

Figure 4.71a shows several 2nd desorption curves for the water-stored material REF, which were determined for two different starting points using the method shown previously. In **Figure 4.71b**, it is shown that with the starting point corresponding to the measured water content after sealed storage, a desorption curve is obtained which matches the measured desorption curve.

The desorption scannings obtained using the modeling for the HCPIII, $W/C = 0.40$ (12 weeks hydrated at 98% RH) are shown in **Figure 4.72a,b**. One can see the fit of the desorption scannings to the primary desorption curve of this cement paste from blast furnace cement. **Figure 4.72a** shows the "loop" that results for an adsorption start of $\varphi \approx 0.46$ when, after traversing the path via adsorption scanning, adsorption isotherm, and back via desorption scanning, it should reach the start point again.

Figure 4.73 makes it clear for the capillary pore region that an adsorption scanning starting from a 2nd desorption curve (or for example, a primary desorption isotherm after sealed storage) toward the main adsorption isotherm is identical to the corresponding adsorption scanning arriving from the primary desorption isotherm.

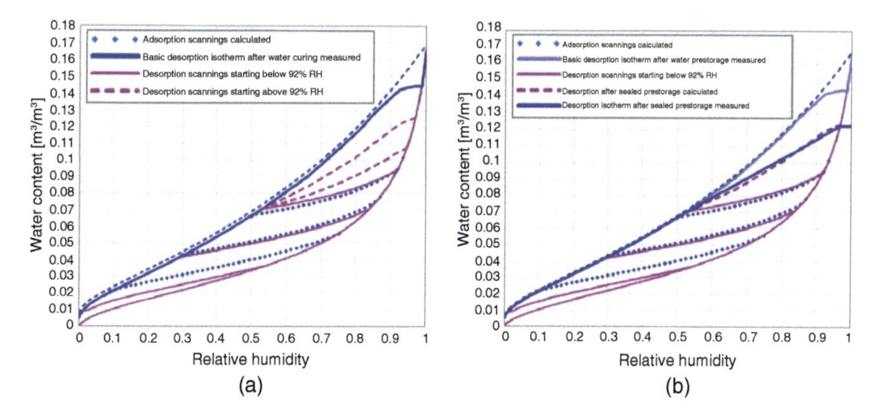

Figure 4.71 Sorption isotherms of mortar REF ($W/C = 0.45$) **after water storage** and after sealed curing, respectively, with corresponding calculated adsorption and desorption scannings with starting point above 92% RH especially. (a) Modeled second desorption isotherms from 94% and 96% RH. (b) Focus on measured desorption after sealed storage and the corresponding modeled curve.

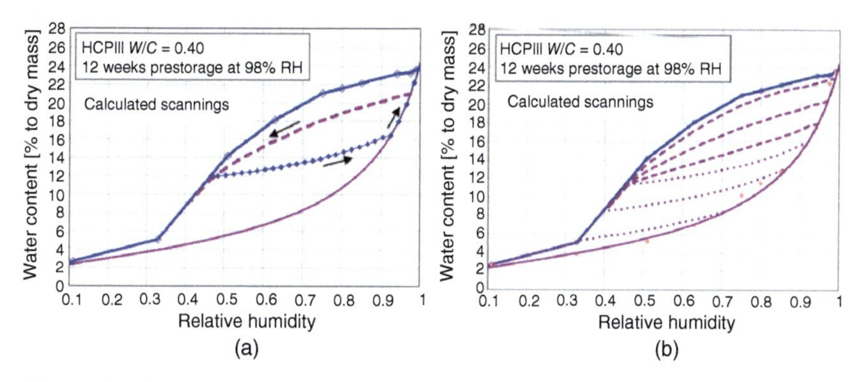

Figure 4.72 Scannings calculated based on the procedures in Sections 4.6.4 and 4.6.5 , for the HCPIII hardened cement paste, $W/C = 0.40$, hydrated for 12 weeks at 98%. (a) Loop of adsorption and desorption curves in the capillary saturation region. (b) Desorption scannings, as dashed lines in the capillary saturation region especially.

Furthermore, the computational checks show that also the adsorption scannings, which run from a 2nd desorption isotherm to the adsorption isotherm, do not cross this 2nd desorption isotherm in any case, but can at most run parallel to it.

In the following, we consider whether the proposed method for modeling scannings can also reproduce measurement results taken from the relevant literature.

Figure 4.74a reproduces adsorption and desorption results of Saeidpour and Wadsö [39] for HCP from CEMIII and W/C values of 0.40, 0.50, and 0.60. Measurements were made with DVS with mathematical correction of the results, compare Section 4.6.3. For the three W/C values investigated, the 2nd desorption curves were correspondingly lower than the base desorption, and there was no crossing of the primary desorption curves.

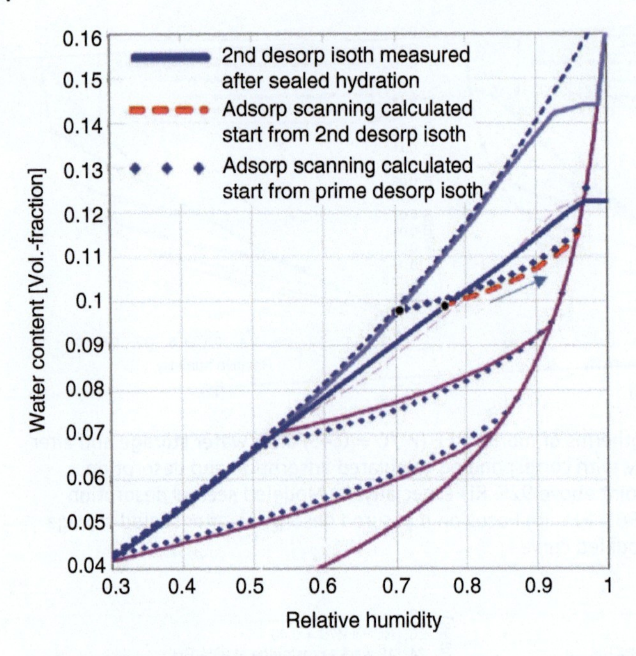

Figure 4.73 Comparison of corresponding calculated adsorption scannings (dashed red line and blue dotted line), starting from the desorption isotherm after sealed curing or from the desorption isotherm after water curing, respectively, (capillary saturation region, mortar REF).

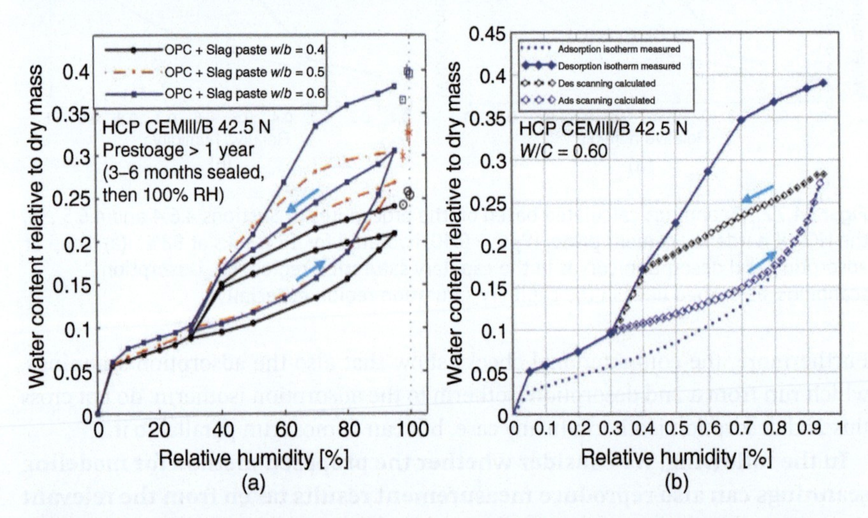

Figure 4.74 DVS measured adsorption and desorption isotherms and scanning isotherms on crushed HCPIII specimens and comparison to scanning calculation proposed. (a) DVS measured desorption isotherms and scannings, prestorage of the samples three to six months sealed + more than six months at 100% RH, results from Saeidpour and Wadsö/Lund [39]/with permission of Elsevier, see also Section 4.6.3. (b) Desorption scanning starting at 95% RH calculated with the desorption scanning modeling for saturations above θ_{gel} presented in Section 4.6.5 using the original desorption isotherm from left figure and the matching adsorption isotherm out of [39]; Calculation of the adsorption scanning by Eq. (4.15).

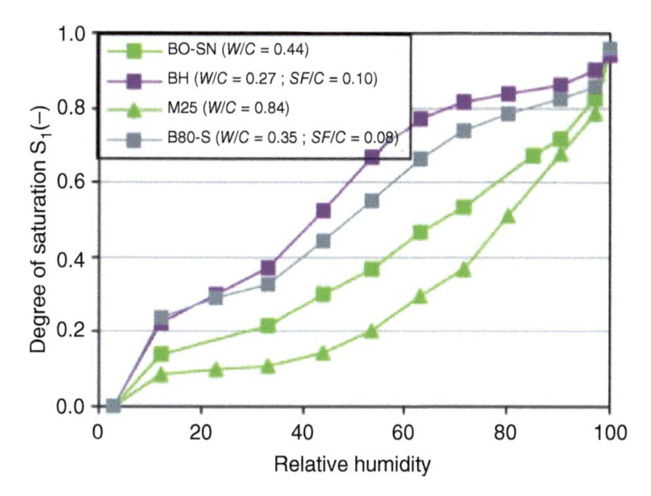

Figure 4.75 Dependence of the shape of desorption isotherms on the W/C value, see remark to modeling possibilities in Section 4.6.5. Source: Baroghel-Bouny [38]/with permission of Elsevier.

The cement paste CEMIII, $W/C = 0.60$, used there was selected to check to what extent the associated measured scannings can also be modeled. In the publication [39], a primary adsorption isotherm is also given for the selected mixture. To verify the procedures, the measured values of the primary adsorption and desorption isotherms were now taken from [39] and using them from the starting point $\varphi_0 = 0.30$ according to Eq. (4.15), the adsorption scanning and with Eqs. (4.18) and (4.20), respectively, the subsequent desorption scanning was calculated. The comparison between calculation and experiment shows an extensive agreement, in particular also with regard to the shape and the course of the desorption scanning.

Figure 4.76a of Wu et al. [17] also shows measurement results for HCP CEMIII $W/C = 0.40$ to the determined secondary desorption curve from starting point of 95% RH. Section 4.6.3 gives more details on the experimental procedure and possible influences on the measured curve.

Nevertheless, it was checked here to what extent the measured 2nd desorption curve with start point $\varphi_0 = 95\%$ RH (**Figure 4.76a**) can be predicted by calculation. For this purpose, the measured data of the adsorption and the desorption isotherms were again taken from **Figure 4.76a**, and the desorption scans were calculated using the Eqs. (4.18) and (4.20), respectively, and reproduced in **Figure 4.76b**. The value θ_{sat} was not determined separately, but $\theta_{sat} = 0.38$ was assumed. In view of the not entirely certain boundary conditions, the agreement between calculation and experiment is considered good.

Zhang et al. [40] examine procedures of different authors with respect to their suitability for modeling sorption isotherms and scannings. According to [40], obviously none of the methods provides a result that is predominantly consistent with the measured values. **Figure 4.77** shows a diagram from [40] (figure right) as an example of satisfactory modeling results, with particular critical discussion of

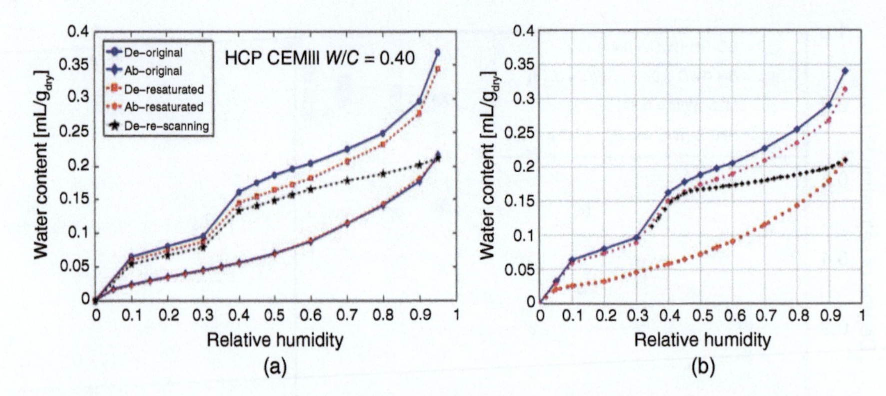

Figure 4.76 DVS measured adsorption and desorption isotherms and scanning isotherms on crushed HCPIII specimens and comparison to scanning calculation proposed. (a) Experimental investigations of Wu et al. [17]/with permission of Elsevier, see details to storage conditions in Section 4.6.3. (b) Calculated desorption scanning starting at 95% RH using the original sorption isotherms from left figure, using desorption scanning modeling presented in Section 4.6.5.

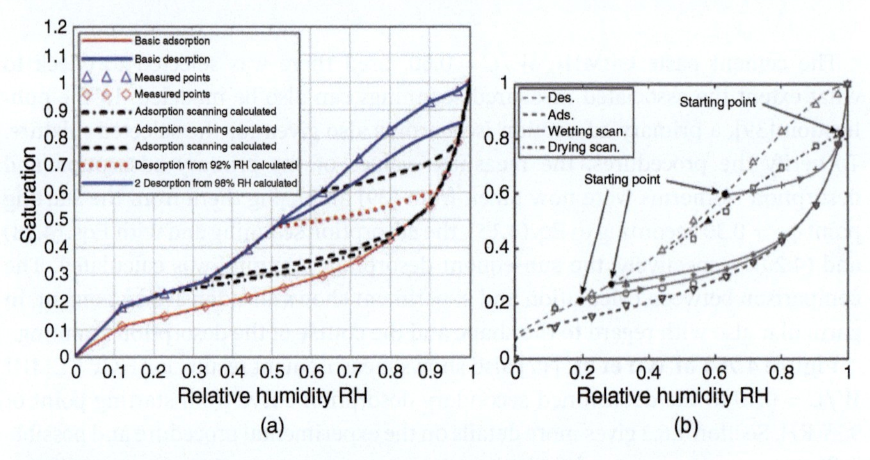

Figure 4.77 Comparison of measured values (**b**) from Zhang et al. [40]/with permission of Elsevier with the modeled curves (a) based on the procedures in Sections 4.6.4 and 4.6.5. (**a**) The original values measured of the desorption and adsorption isotherms out of figure (b) were introduced. On this basis, the scannings were then calculated on the basis of Sections 4.6.4 and 4.6.5.

the uncertainties in modeling the secondary desorption scanning starting from the "starting point."

In view of modeling the curves in **Figure 4.77b**, the measurement points of the primary desorption and adsorption curves were taken from it. On the basis of these, the scannings shown in the figure on the left were calculated, the black-dashed adsorption scannings again according to Eq. (4.15), the secondary desorption curve (blue line) again with Eqs. (4.18) resp. (4.20). In addition, the "base desorption scanning" starting from 92% RH, calculated with Eq. (4.16), is entered in red dotted line in the figure. Further, the value $\theta_{sat} = 1.0$ of the normalized curves was used.

Jiang et al. [43] perform computational pore modeling based on the adsorption isotherms of the cement-bound materials considered. Comparisons with measured values obviously led to satisfactory adjustments to the overall elaborate procedure.

4.6.6 Using Given Isotherms for Other Concrete Compositions

The moisture retention of concrete is significantly influenced by the cement paste proportion and its pore volume and pore size distribution. The latter values are in turn decisively determined by the cement type, the selected W/C value, and the proportion of air or processing pores.

Sorption isotherms are represented as percentage water content m_W related to the dry bulk density ϱ_{dry} or as water content related to the material volume in [kg/m³]. **The experimentally determined water content of a HCP** for a given storage condition is, for example:

$$W_{HCP} = m_{W,HCP} \cdot \varrho_{dry,HCP} \quad [\text{kg/m}^3] \tag{4.21}$$

Now, considering a concrete or mortar with identical cement type and the same W/C value after the same storage conditions as the HCP, the **sorption isotherm of the mortar over its HCP fraction** $V_{HCP,M}$ can be calculated according to Eq. (2.9) as follows.

$$W_{mortar} = W_{HCP} \cdot K_{V,M} \; [\text{kg/m}^3] \quad \text{from} \quad K_{V,M} = \frac{V_{HCP,M}}{1000} \; [-] \tag{4.22}$$

Figure 4.78a,b shows the result of the calculation of the sorption isotherms for the mortar MIII using the HCPIII isotherms in Figure **4.57a** measured after 12 weeks of curing for the W/C-values of 0.40 and 0.60 as a basis. **Figure 4.79a,b** shows the

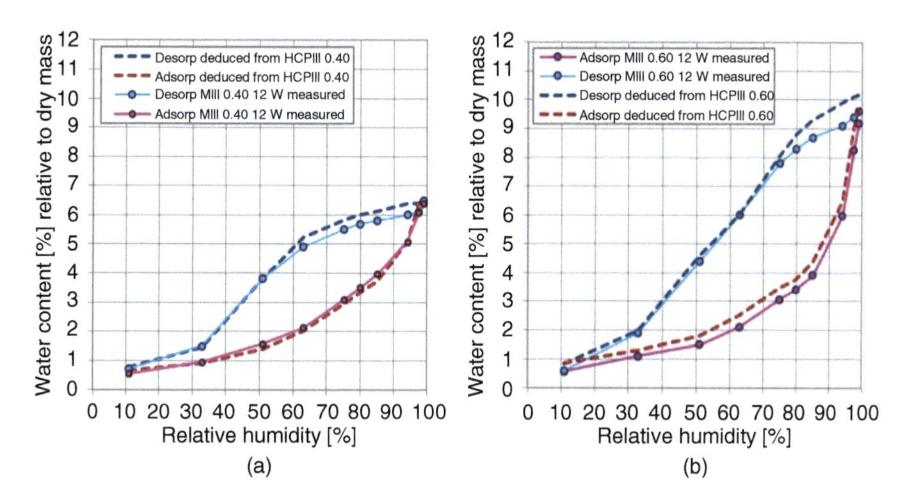

Figure 4.78 Sorption isotherms for mortar MIII deduced by calculation from the measured sorption isotherms of HCPIII (see **Figure 4.57a**) compared to the sorption isotherms measured on mortar MIII (mixtures of HCPIII and MIII: compare **Table 4.1**). (a) MIII with $W/C = 0.40$ **after 12 weeks of curing** at 98% RH/(20 °C). (b) MIII with $W/C = 0.60$ after **12 weeks of curing** at 98% RH/(20 °C).

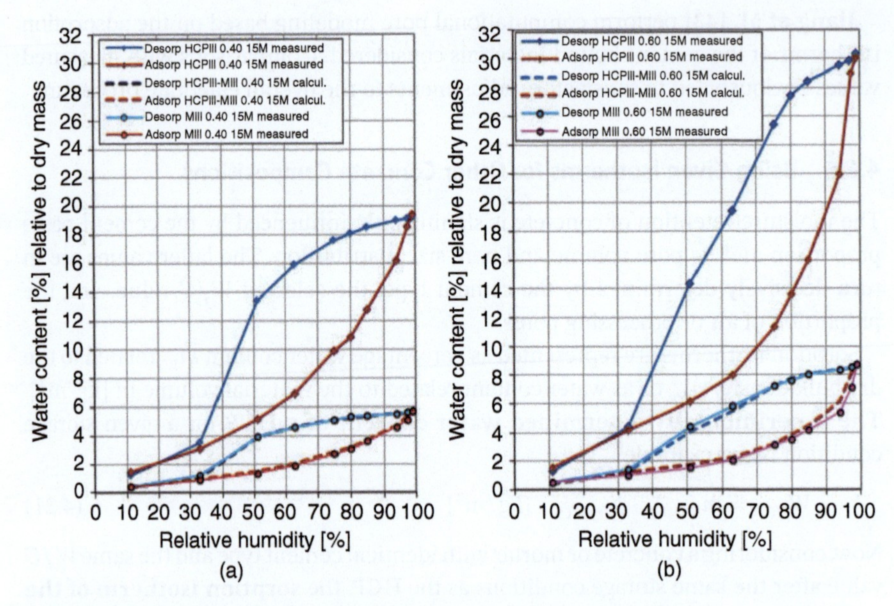

Figure 4.79 Sorption isotherms of HCPIII and mortar MIII measured after **15 months of curing** at 98% RH/(20 °C) **and** comparison of the measured sorption isotherms with the curves derived computationally from the HCPIII isotherms. (a) Results for $WC = 0.40$; (b) Results for $WC = 0.60$.

resulting isotherms for the W/C-values of 0.40 and 0.60 for 15 months of cured HCPIII (**Figure 4.57b**) and MIII (**Figure 4.58b**) as a basis of the comparison. From the calculation based on Eqs. (4.21) to (4.23) follows a very good to perfect agreement between the (from HCP curves) derived resp. calculated mortar isotherms and the measured mortar isotherms.

$$m_{W,M} = \frac{W_{mortar}}{\varrho_{dry,M}} = m_{W,HCP} \cdot K_{V,M} \cdot \frac{\varrho_{dry,HCP}}{\varrho_{dry,M}} \quad [-] \tag{4.23}$$

$$\frac{m_{W,M2}}{m_{W,M1}} = \frac{\varrho_{dry,M1}}{K_{V,M1}} \cdot \frac{K_{V,M2}}{\varrho_{dry,M2}} \tag{4.24}$$

One then obtains from a measured adsorption isotherm (for example, for $W/C =$ 0.50) the curves for other W/C values, for example, corresponding to **Figure 4.80**.

4.6.7 Alternative Experimental Determination of the Slope of Scanning Isotherms

The possibly hysteretic moisture storage behavior of materials is determined in the hygroscopic moisture range by sorption or desorption experiments, usually by exposing dried or moist sample bodies to defined relative air humidity. As Deckelmann and Schmidt-Döhl/TUHH [46] have shown that **this experiment can also be reversed** by measuring the resulting relative humidities RH (in closed containers of adapted size) for given water saturations of a material.

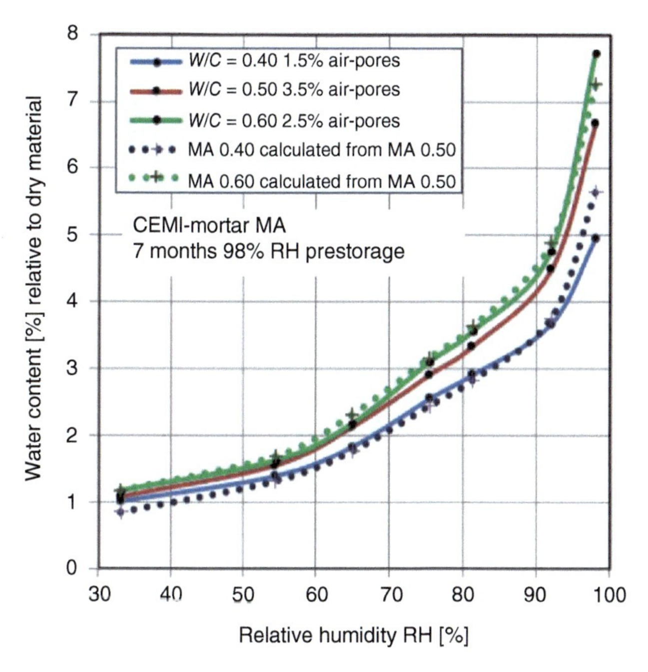

Figure 4.80 Comparison of measured and calculated adsorption isotherms as a function of the W/C ratio (taking into account the air-porosity content): Derivation of the isotherms of mortar MA with W/C ratios of 0.40 and 0.60 from the measured 0.50 isotherm of MA. (MA with $W/C = 0.40$ made with plasticizer, compare **Table 4.1**).

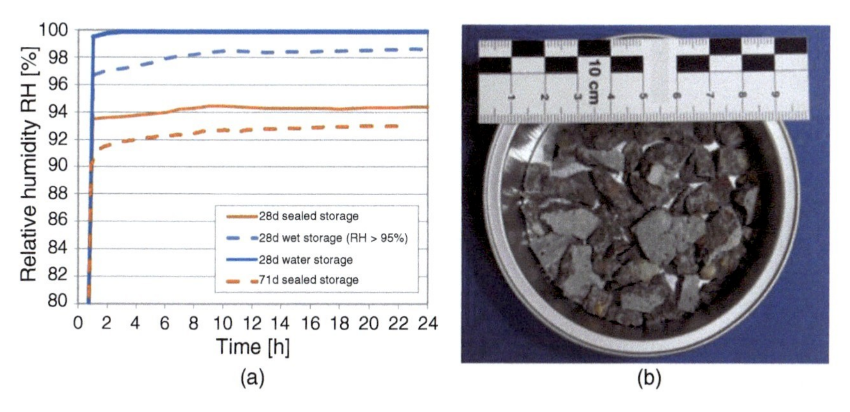

Figure 4.81 Relative humidity RH measured in closed containers above crushed **samples of mortar REF** after water storage, wet storage (98%), and sealed storage for comparison. (a) Measured curves $RH(t)$. (b) Example of the crushed material.

Only precise RH measuring instruments are to be used for this purpose. The material must be crushed beforehand **according to Figure 4.81b**. It is essential that this fragmentation be carried out in such a way that the water content of the samples is not altered. For this purpose, the samples from preliminary storage in resistant plastic bags were processed by hammer-blow.

After a few hours, the RH value sought in each case is now stably established in the closed bags. As an example, in **Figure 4.81a**, the RH values resulting from a REF mortar (at the boundary conditions given there) are shown. It can be seen that an RH value of 98.5% corresponded to the "wet storage" designation. Also, the lower RH values after "sealed storage" indicate the expected lower water content, **compare Section 4.5.9**.

In an "inverse" sorption experiment using two different mortar samples with different moisture content similarly set, **the mean slope of scanning isotherms can be determined**: For this purpose, Deckelmann+Schmidt-Döhl/TUHH have performed experimental investigations with two separately arranged samples M_0 and M_1 of the same mortar **(screed-concrete),** each corresponding to **Figure 4.81b**, with different water saturations generated by desorption at two different RH values φ_0 and φ_1. Afterwards, these samples were placed together separately in a closed container, and the equilibrium air humidity φ established after a few hours in the container was measured. This equilibrium air humidity was significantly different from the mean values of φ_0 and φ_1. For example, initial values of $\varphi_0 = 54\%$ and $\varphi_1 = 85\%$ resulted in $\varphi = 78\%$.

This experiment is suitable for determining the slope of the scanning isotherms. The following **Eqs.** (4.25) **to** (4.31) **and Figure 4.82** explain the relationship.

The basic fact is that at the end of storage (of the two material samples in the same container), the water content increase Δw_0 of one sample is identical to the water

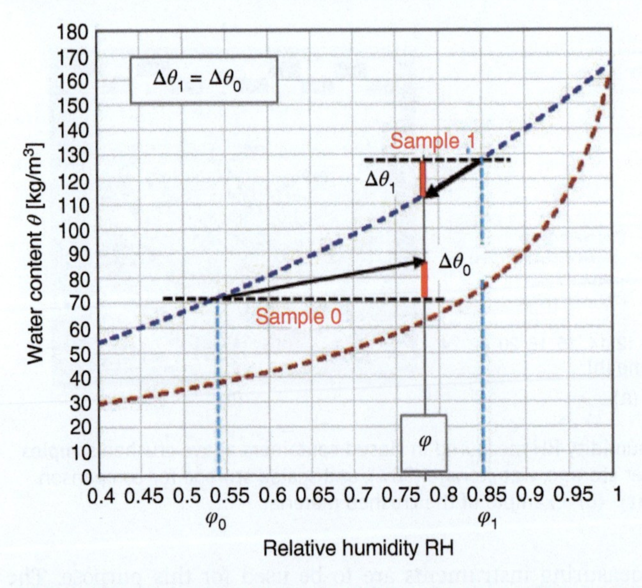

Figure 4.82 Schematic illustration of the moisture exchange up to transport equilibrium between two spatially separated porous material samples with hysteretic moisture storage properties (here REF mortar) with different initial moisture content, stored together in a closed container. The physically true mean slope of a scanning isotherm is obtained in the range of humidity considered.

content decrease Δw_1 of the other material sample, according to Eq. (4.27).

$$M_0 = M_{0(\varphi 0)} = (1 + m_0) \cdot M_{0\ dry} = \text{mass of wet sample 0 with lower water content} \tag{4.25}$$

$$M_1 = M_{1(\varphi 1)} = (1 + m_1) \cdot M_{1\ dry} = \text{mass of wet sample 1 with upper water content} \tag{4.26}$$

Both values are 0 and 1 from adsorption isotherm or both from desorption isotherm.

$$\Delta\theta_0 \cdot V_0 = \Delta\theta_1 \cdot V_1 \left[\frac{\text{kg}}{\text{m}^3} \cdot \text{m}^3\right] \quad \textbf{or} \quad \Delta m_0 \cdot M_{0\ dry} = \Delta m_1 \cdot M_{1\ dry} \left[\frac{\%}{100} \cdot kg = kg\right] \tag{4.27}$$

At the end of the experiment, the resulting relative humidity $\varphi = \varphi_{experiment}$ and the weight or water content changes are measured. Defining the slope of the scanning isotherms according to Eq. (4.28), it follows from $\Delta\theta_0 = \Delta\theta_1$ the Eq. (4.29) and from this with $\varphi_{experiment}$ **the slope a_{Sc} sought at the point φ_0 of the desorption isotherm from Eq.** (4.30).

$$a_{des\ (\varphi)} = \frac{\Delta m_{des\ (\varphi)} \cdot \varrho_{dry}}{\Delta\varphi} = \frac{\Delta\theta_0}{\Delta\varphi} \qquad a_{Sc\ (\varphi)} = \frac{\Delta m_{Sc\ (\varphi)} \cdot \varrho_{dry}}{\Delta\varphi_{Sc}} = \frac{\Delta\theta_1}{\Delta\varphi} \tag{4.28}$$

$$a_{des\ (\varphi 1)} \cdot (\varphi - \varphi_1) \cdot M_{1\ dry} = a_{Sc\ (\varphi 0)} \cdot (\varphi - \varphi_0) \cdot M_{0\ dry} \tag{4.29}$$

$$a_{Sc\ (\varphi 0)} = \frac{a_{des\ (\varphi 1)} \cdot (\varphi_{experiment} - \varphi_1)}{(\varphi_{experiment} - \varphi_0)} \cdot \frac{M_{1\ dry}}{M_{0\ dry}} \tag{4.30}$$

Control experiments were carried out by Deckelmann+Schmidt-Döhl/TUHH for this purpose:

Material corresponding to **Figure 4.81b** was divided in two portions M_0 and M_1 was adjusted to the water content corresponding to $\varphi_0 = 54\%$ and $\varphi_1 = 85\%$. $M_{1\ dry}$ corresponded to $M_{0\ dry}$. The equilibrium moisture content from experiment RH was found to be $\varphi_{experiment} = 78\%$. Thus, according to Eq. (4.30), the slope $a_{Sc\ (\varphi 0)}$ we are looking for can be determined:

For the material REF, the slope $a_{des\ (\varphi 1)}$ can be taken from **Figure 4.46b** to $\Delta\theta_1/\Delta\varphi \approx (140\ \text{kg} - 118\ \text{kg})/10\%\ RH = 220\ \text{kg}/100\%\ RH$.

The ratio of the slope of the scanning isotherm at point M_0 to the isotherm slope at point M_1 is $f = \frac{64}{220} = 0.291$ or about 30%.

This slope of the scanning isotherms corresponds to the straight line connection between the value of the desorption isotherms at $\varphi = 54\%$ and the value of the adsorption isotherms at $\varphi = 91\%$.

This result confirms the statement that the average slope of the scanning isotherms corresponds, at least in the middle region of RH, to **the straight-line connection between the isotherms between $\varphi = 50-55\%$ and** $\varphi = 90\%$.

On the other hand, if the slope a_{Sc} of the scanning isotherm is known, the unknown air humidity RH setting can be determined from it by the following Eq. (4.31) (f inserted positively).

$$f = \frac{a_{Sc\ (\varphi 0)}}{a_{des\ (\varphi 1)}}; \qquad f_{Mat} = \frac{M_{0\ dry}}{M_{1\ dry}}; \qquad \varphi = \frac{\varphi_1 + f \cdot f_{Mat} \cdot \varphi_0}{1 + f \cdot f_{Mat}} \tag{4.31}$$

Section 5.2.3 shows that the same result is obtained using computer simulation.

The studies on moisture equilibrium in closed systems make it particularly clear that correct predictions can only be expected when hysteretic sorption behavior is taken into account.

4.6.8 Influence of Carbonation on Moisture Storage and Transport Behavior

In the pore systems of cementitious products, a carbonation of the incorporated calcite and reactions with the C–S–H phases usually lead to a change in the pore size distributions, narrowing of the pores, and slowing down of the transport of liquids and gases. In practice, however, the possible influence of a realkalization must be checked in each case.

Kropp [47] has investigated the influence of carbonation on the diffusion coefficients of cement mortars. According to him, the diffusion coefficient in the water-saturated state falls by about 50% compared to the non-carbonated state.

Houst and Wittmann [48] measured the effect of gas diffusivity to O_2 and CO_2 of HCP with water-cement ratios of 0.30 to 0.80 after storage in different relative humidities. The samples were first completely carbonated and accelerated in an atmosphere containing 80–90% CO_2 at 76% RH. In subsequent diffusion tests, a reduction in diffusivity was observed between 30% and 50% of initial values, depending on the water-cement ratio.

The work of Houst [49] contains additional results in a different presentation compared to [48], in particular measurement results on desorption isotherms on HCP before and after complete accelerated carbonation.

For the investigated HCP made of Portland cement with a W/C value of 0.40, the water storage capacity at complete saturation was reduced after carbonation to 40% of the initial value.

The desorption isotherm of the carbonated state showed a correspondingly lower course below the isotherm of the initial state, as newly drawn by Auroy et al. [50]. A comparison of these desorption isotherms after calibration of the curves to their respective saturation state $S = 1$ at 100% RH showed a largely congruent course of the calibrated curves. This behavior reveals that the pore size distribution after carbonation largely corresponds to the previous course, at least concerning the HCP samples of Portland cement.

Thiery et al. [51] investigate the change of pore systems of HCPs and concrete also with admixtures of fly ash (FA) on HCP and some concretes after accelerated carbonation. They undertake investigations with NMR, porosity measurements, and mainly by MIP. They find that the pore morphology changes can be fully represented by the MIP analyses. The observations show that although the pores are reduced in size in the range of 10–100 [nm], the calcite crystal formation and the attack on the CSH phases simultaneously cause a pore enlargement in the region of the large capillary pores of about 100 [nm].

On non-carbonated and fully carbonated concrete specimens made of normal Portland cement, desorption isotherms were determined from the water-saturated state. The measured relative desorption curves with respect to their water-saturated state showed an almost congruent course, from which it could be concluded that at least the relative pore size distributions before and after carbonation remained the same for this Portland cement-based material. The water storage capacity of the carbonated specimens was about 40% of the initial non-carbonated value.

The authors note that in the presence of pozzolanic CSH phases when FA is used, their decalcification can even lead to an overall increase in porosity due to carbonation.

The work of Auroy et al. [50] (whose aim is to determine the influence of carbonation on transport properties in the unsaturated state) also contains results of investigations on HCPs from four different cements or binders on the influence of carbonation on porosity and desorption isotherms.

The results shall be presented below for the materials PI (of Ordinary Portland Cement CEM I 52.5 R) and PIII [of CEM III/A (39% CEMI + 61% blast furnace slag)] according to the European standard EN-206. The HPC specimens with $W/C = 0.40$ were all initially hydrated under aqueous solution for four months. Thereafter, the specimens for the carbonation experiment part were stored for one month at 55% RH and 25 °C until moisture equilibrium, and then for a further 11 months under the aforementioned climate with 3.0% CO_2 until complete carbonation.

For determination of the porosity via water saturation, the samples were first dried at 80 °C and then water-saturated under vacuum.

In these experiments, complete water saturation of **PI** decreased to 45% of the initial value and to 55% at **PIII** due to carbonation.

The measured desorption isotherms of PI and PIII are shown in **Figures 4.83a and 4.84a**:

The material samples, which were again water-saturated prior to the start of desorption, were tested in the desiccator method, and for comparison also in the Dynamic Water Sorption (DVS) method. Compare Section 4.6.3, where advantages and problems of the DVS-method are addressed.

Figure 4.83b shows a comparison of the desorption isotherms of the Portland cement material PI calibrated to the respective saturation states. For these desorption

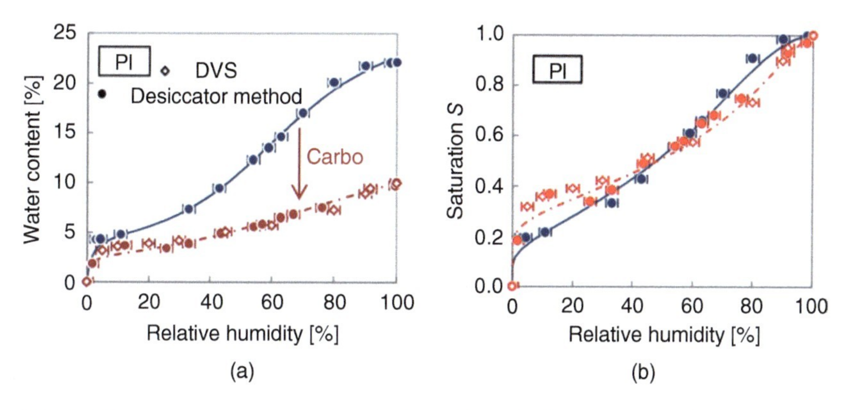

Figure 4.83 Desorption isotherms of HCP PI from Portland cement CEM I 52.5 R, $W/C = 0.40$, before and after complete carbonation from Auroy et al. [50]/with permission of Elsevier. (a) Isotherms determined according to the desiccator method and partially with DVS. (b) Comparison of the curves of the left figure calibrated to $S = 1$.

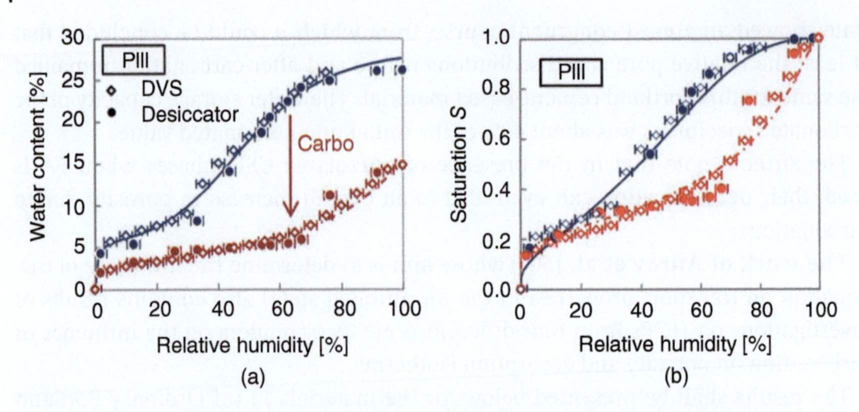

Figure 4.84 Desorption isotherms of HCP PIII (Blast furnace cement CEM III/A), $W/C = 0.40$, before and after complete carbonation from Auroy et al. [50]/with permission of Elsevier. (a) Isotherms determined according to the desiccator method and partially with DVS. (b) Comparison of the curves of the left figure calibrated to $S = 1$.

curves determined by the desiccator method, there results again an extended coincidence of the curve courses before and after carbonation and thus of the relative pore size distributions. This result confirms the results from Houst and Wittmann [48] and Thiéry et al. [51].

Figure 4.84b illustrates the difference of the calibrated desorption curves of HCP PIII (Blast furnace cement). before and after carbonation. The accelerated desorption behavior of the carbonated sample PIII is evidently due to the microcrack formation observed and documented on the disc-shaped specimens in the course of carbonation. This influence was also evident with respect to the permeability, for which an increase was determined after carbonation instead of the expected decrease. In addition, the permeability was inversely calculated from evaporation test results. Critical remarks on this (when using evaporation test results) can be taken from Section 3.1.7.

Ren et al. [52] perform permeability measurements on mortars made of white Portland cement 52.5 with a cement : sand mixing ratio of 1 : 3 with W/C values of 0.45 and 0.55. For an equivalent number of specimens, 10% of the cement content is replaced by FA. The first mentioned mortars have the designation M45 and M55, the mortars with FA-part have the designation MS45 and MS55.

Accelerated carbonation over 56 days is carried out with high CO_2 gas concentration on water-saturated samples at 100% RH. The permeability measurements before and after carbonation are carried out in a triaxial apparatus on thin specimens at hydraulic pressure of 30 bar.

The following permeabilities are measured:

In the initial state: M45: $1.4 \cdot 10^{-19}$ MS45: $0.8 \cdot 10^{-19}$ [m²]
M55: $1.8 \cdot 10^{-19}$ MS55: $1.7 \cdot 10^{-19}$ [m²].

After carbonation, M45 and MS45 showed a decrease of about 30%,
M55 and MS55 (with 10% FA) a permeability decrease of about 50%.

It should be noted that the carbonation of the samples in this work is carried out (to prevent desiccation effects) via dissolved CO_2 in saturation concentration via the pore solution, in contrast to other carbonation methods where the CO_2 enters the pore system largely in gas form.

4.6.9 Cavitation in the Pore System During Desorption or Drying

In Section 1.3.7, the phenomenon of cavitation in free water was discussed via the formation of vapor bubbles, usually favored by impurities, at falling external pressure p_{atm}.

Referring to Section 1.4.6.2 and Figure 1.28a,b, recall that when the pore system is partially saturated with water, the capillary stress p_{cap} in the pores acts as a tensile stress on the contained water. The effective stress state acting on the water molecules in the pore water is, however $\overline{p}_{cap} = p_{atm} - p_{cap}$, which is decisive with respect to possible cavitation.

As the absolute pore pressure $\overline{p}_{cap}$ decreases, the boiling point of the water contained in the pores drops below 100 °C. In the following, we will therefore consider in more detail the risk of cavitation that arises when, as a result of a decrease in pore pressure, the boiling point also decreases.

The **relationship of Clausius–Clapeyron** is needed for the computational estimates. It describes the course of the "separation line" between the liquid phase and the vapor phase in the phase diagram of the substance, here water, on the one hand as a (saturation) vapor pressure curve, on the other hand as a boiling point curve depending on the temperature T and the respective (total) pressure p on the system liquid-vapor.

Since Clausius–Clapeyron is also needed for the explanations in the following Section 4.6.10, the well-known derivation will be shown anyway:

The equilibrium between the liquid phase (L) and its vapor (v) is considered (from 1 mol) by equating the chemical potentials.

$\mu_{(L)} = \mu_{(v)}$ or $-S_{(L)} \cdot dT + V_{(L)} \cdot dp = -S_{(v)} \cdot dT + V_{(v)} \cdot dp$, transformed into
$dp/dT = (S_{(L)} - S_{(v)})/(V_{(L)} - V_{(v)}) = \Delta S/\Delta V$
At constant pressure, for 1 mol of substance
$\Delta S = \Delta H/T$ with $\Delta H =$ **molar vaporization enthalpy** ΔHv.

Since in $\Delta V = (V_{(L)} - V_{(v)})$, the magnitude of $V_{(L)}$ is much smaller than $V_{(v)}$, only $V_{(v)}$ is kept and replaced by the general gas equation $V = R \cdot T/p$. This then gives the initial form of the Clausius–Clapeyron equation to be

$$\frac{dp}{dT} = \frac{\Delta Hv}{T} \cdot \frac{p}{R \cdot T} = \frac{\Delta Hv}{R} \cdot \frac{p}{T^2} \tag{4.32}$$

After separation of variables and integration $\int_{p1}^{p2} \frac{1}{p} dp = \frac{\Delta Hv}{R} \cdot \int_{T1}^{T2} \frac{1}{T^2} dT$ follows.

$$ln\left(\frac{p_2}{p_1}\right) = \frac{\Delta Hv}{R} \cdot \left(\frac{1}{T_1} - \frac{1}{T_2}\right) \tag{4.33}$$

as the most common equation form.

In Figure 4.33 of Section 4.5.4, the distribution of the effective pressure as a function of the water saturation of a pore system open to the environment from $r = 1$ [nm] to $r = 1$ [mm] is schematically shown. It can be seen that below the pore radius of about 2 [μm] and an atmospheric pressure of $p_{atm} = 10^5$ [Pa], the absolute pressure of the water filling is increasingly negative as the saturation decreases, and thus the conditions for cavitation are present or increase with increasing drying in water-filled pores.

Figure 4.86 shall show that at a material temperature of 80 °C at a pressure of $0.48 \cdot 10^5$ [Pa] (according to Clausius–Clapeyron or Magnus from **Figure 4.85**) corresponding to a pore filling up to $r = 2.8$ μm, the water content is set to the boiling state.

For a pore filling $r \leq 1.49$ μm, a potential boiling state already exists at a material temperature of 20 °C with an effective pressure of 2340 [Pa].

The maximum pore radius in conventional cement-bonded materials is about 200 [nm]. Thus, the pore filling is always under effective tensile stress and thus in a potential cavitation state. Associated with this is the formation of water-vapor bubbles in the affected water-filled pores, see Section 1.3.7. This relationship thus describes the "normal state" of a HCP or concrete.

The influence of cavitation, in particular, on the moisture properties is increasingly being investigated by comparative studies with different compositions.

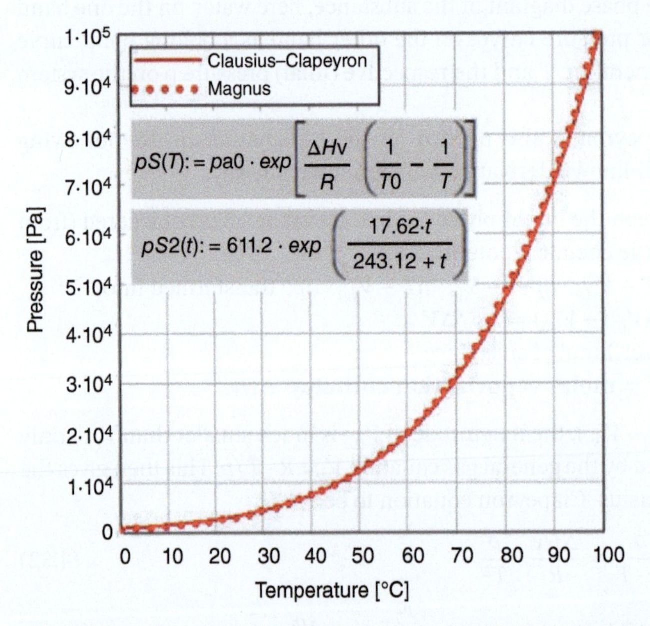

Figure 4.85 Boiling point and vapor pressure curves, respectively, of water below 100 °C. Clausius–Clapeyron curve has been calculated with constant value of $\Delta Hv = 42.50$ [kJ/mol], starting with $T = 373$ K and $pa0 = 1.01325 \cdot 10^5$ [Pa]. The values of ΔHv are dependent from temperature only:ΔHv = 5.05/(0.01 °C), Standard value 43.99/(25 °C), 42.48/(60 °C), 41.59/(80 °C), 40.66/(100 °C), 36.30/(180 °C) [kJ/mol]. The Magnus formula related to t (°C) is empirical.

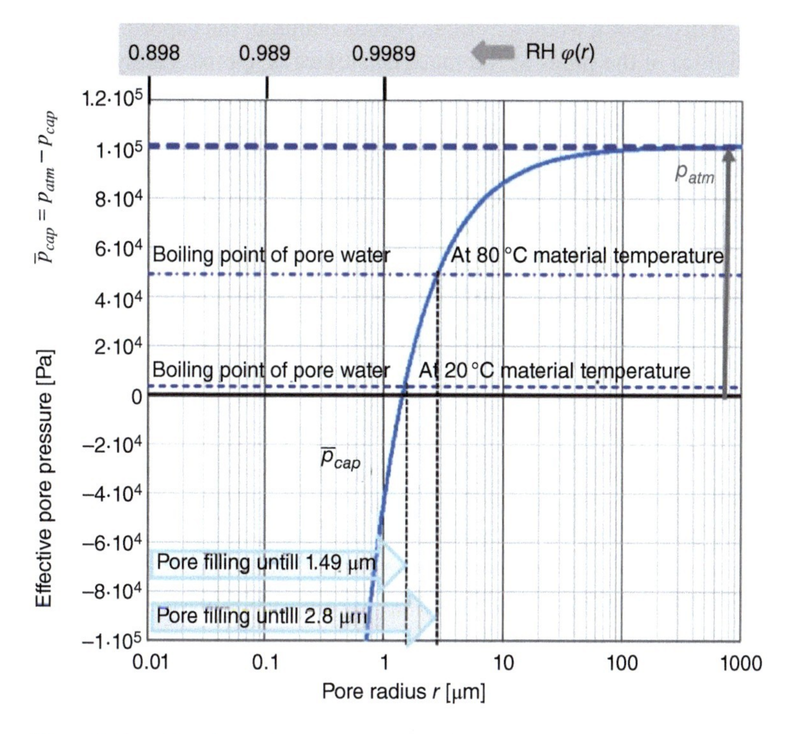

Figure 4.86 Effective water-pressure curve in the pores (of a hardened cement paste) during drying from water saturation, with identification of pore sizes below which dewatering can lead to cavitation in the pore system. The marked pore radii are those at which, for a given atmospheric pressure p_{atm}, the absolute vapor pressure in the pores is reduced to the boiling point pressure at 80 °C and at 20 °C, respectively, by the action of capillary suction.

The relationships can only be clarified indirectly, since the presence of temporary vapor bubbles in the pore system, in particular, defies direct observation.

Only a few studies have so far succeeded in optically monitoring the formation and behavior of vapor bubbles, at least in artificially created structures.

Duan et al./Berkeley [53] succeed in observing the formation and behavior of water vapor bubbles from in artificially created nanochannels of dimensions 20–120 [nm] as the applied vapor pressure is reduced. The structures were created via etching techniques on silicon wafers ("Cavitation experiments were performed on silica nanochannel devices fabricated using sacrificial layer etching and microchannel bonding"). The main result is that generated water-vapor bubbles are not limited by the pore diameter and can expand in the pores, but they do not affect menisci formed at the pore ends. During the formation and enlargement of vapor bubbles, a significantly increased evaporation rate of the system is observed. Unfortunately, a more precise reason for the temporary increase in the evaporation rate could not be given there.

The following reasoning for the increase in evaporation rate is assumed here: (In this description, known relationships are also repeated.)

At the start of drying of a water-saturated porous material, the vapor pressure p_i of the water fillings of the pores at the material surface is approximately equal to the saturation vapor pressure of free water p_S (at given temperature T and atmospheric pressure p_{atm}), while the external vapor pressure is $p_a = \varphi_a \cdot p_S$, so that $p_i > p_a$, which starts the drying process of the material.

For a pore system with decreasing pore size, the drying of the pore system comes to a stop when the pore filling has reached the pore region with smaller pore diameter, where the vapor pressure p_{S_i} (see Kelvin equation) is applied to the associated meniscus curvatures, which corresponds to the external vapor pressure p_a, so that the vapor pressure outside/inside is in equilibrium.

The evaporation rate $\dot{m}$ on the path between drying start and this equilibrium is given by the equation

$$\dot{m} = \frac{D_v}{x} \cdot (p_i - p_a) \tag{4.34}$$

where the vapor diffusion coefficient D_v depends on the current water saturation degree and temperature, x is the distance between the inner water filling of the pores and the outer surface. A vapor transfer coefficient at the surface was not considered. **Independent of these parameters, the internal vapor pressure p_i controls the evaporation rate.**

If the boundary conditions for cavitation are given during the drying of the pore system and steam bubbles form behind the menisci of pores, this leads to a corresponding spontaneous increase in the filling behind the menisci concerned because of the large volume difference between the liquid and the vapor state, i.e. the bubble volume and the volume of the evaporated water quantity. This dynamic process leads to a flattening of these menisci, compare **Figure 1.42** and the explanations on this in Section 1.4.8, which continues as long as the vapor-volume arises from cavitation. The flattening of the menisci in question leads to a corresponding increase in $p_{S(i)}$ and therefore to an increase in $\dot{m}$ or in evaporation rate according to Eq. (4.34).

For conventional cement-bonded materials, the conditions for cavitation are at least theoretically given and one can assume that there is an influence during drying tests.

The work by Vincent et al. [55] also presents results of optical studies on a synthetic pore system.

Rastogi et al. [54] report in 2022 on possible influence of cavitation on cement paste with shrinkage reducing agent (SRA) additives, compare **Figure 4.87**.

In particular, these results seem to confirm those obtained by **Maruyama et al.** 2018 [56] of an influence of cavitation on the desorption isotherms in the pore region between $\varphi = 0.50$ and $\varphi = 0.30$. Results from the work [56], which also reports extensive sorption isotherms as a function of temperature, will be discussed in more detail in Section 4.6.10.

Nevertheless, there is still a need for investigation. Due to the reduction of the surface tension in the pore system when using SRA additive, the capillary suction in the pore system is reduced, so that the tendency to cavitation decreases. With increasing relative humidity, this tendency also decreases. It would therefore be necessary to examine why, compared to the control mixture, desorption of appreciably

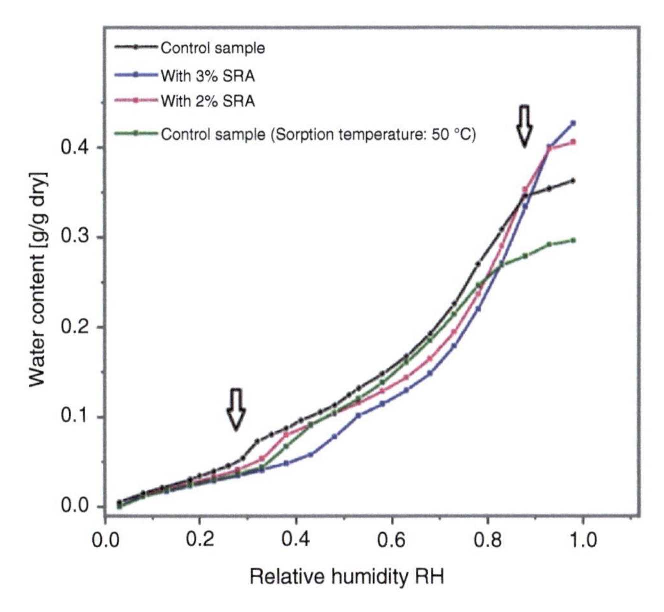

Figure 4.87 Results of desorption studies on the influence of cavitation on the desorption behavior of hardened cement paste. DVS as a desorption method. Result diagram of Rastogi et al. [54]/with permission of Elsevier on HCP of modified Portland cement with 10% silica-fume addition, $W/B = 0.7$ (Control samples) and corresponding blends with addition of 2% and 3% SRA (Shrinkage Reducing Agent), respectively.

higher RH due to cavitation starts in the mixture with 3% SRA. This question arises similarly with the results of Maruyama et al. [56], where the desorption jump in the same vapor-pressure range was determined with increasing temperature.

It should be examined whether possibly the water-molecule layers adsorbed in the water-filled region of the pores during desorption released the transport pathways at the relevant pore necks and in this way led to the observed desorption jumps, compare Section 1.4.6.5.

However, such an influence can also be questioned, since the pressure considered as the driving force is determined via the capillary pressure. This can however only act as long as the menisci required for this are actually present in the micropore system. According to Section 1.3.2, the validity of the Kelvin relationships is limited to actual pore radii between 0.7 and 1.5 [nm]. According to Section 1.4.5, this corresponds to a pore filling corresponding to a relative vapor pressure of about 20% to 50%. In the finer nanopores, the water molecules are present in the adsorbed state in the fiber mesh of the CSH phases. The effect of capillary pressure from the larger pores into this nanopore region should therefore be questioned.

The increase in desorption curves observed in **Figure 4.87** by Rastogi et al. falls within the aforementioned pore size range and may therefore indeed suggest the effects of cavitation.

The influence of cavitation becomes distinct and concrete when additionally the external pressure (p_{atm}) is lowered by applying a vacuum, so that even at high degrees of water saturation in the total water content of the material, the boiling point is exceeded.

This takes place, for example, in standard drying with 40 °C/1 mbar. In this type of drying, which is gentle on the material, drying thus takes place at $10^5/1000 = 100$ [Pa] external pressure. All the water contained in the material is in the boiling state. The boiling point present here is still well below the boiling point for 0 °C, which is present at a pressure reduction of 611 [Pa] or 6.1 mbar. It should be noted that at the selected 40 °C, the surface tension of the water decreased only insignificantly compared to 20 °C, so that essentially **a strongly accelerated evaporation** occurs due to the standard drying.

4.6.10 Dependence of Sorption Isotherms on Temperature

There are also results in the literature of sorption isotherms on HCP or concrete measured as a function of temperature available. It should be noted that clearly different dependencies result from DVS measurements and desiccator measurements.

Temperature-dependent desiccator measurement results are from the publications of Hundt and Kantelberg [57], Ishida et al. [58], Poyet [59], Poyet and Charles [60], Poyet [61], and de Burgh and Foster [42].

Temperature-dependent measurements using the DVS method were published by Maruyama et al. [56].

Climatic chamber measurements by Brue et al. [62] and from vapor-pressure measurements at specified water contents and temperatures Radjy et al. [63] and Schneider and Goss [64] on different soils.

Hundt and Kantelberg [57] determined the desorption behavior of cement paste, cement mortar, and concrete in the temperature range from 20 to 70 °C as early as 1978 in an extensive test program.

Figure 4.88a shows the results for cement blocks of different water–cement ratios. These results already show the strong influence of temperature on the equilibrium desorption moisture contents in the material as a function of relative humidity, as determined by the desiccator method.

In Figure 4.88b are desorption results of Poyet [59] reproduced for a concrete of 400 [kg/m^3] CEMI 52.5 R and $W/C = 0.43$ at temperatures of 30 and 80 °C, determined by the desiccator method after 82 weeks of preliminary water storage. The obviously high influence of the temperature on the desorption equilibrium is also confirmed here.

However, the marked difference in the extent of desorption at 100% RH is remarkable. Later results by Poyet and coworkers 2015 [65] also show only a small difference at 100% RH and HCPI.

Our own control tests at 100% RH also show no significant decrease in water content:

Results of these control tests are shown in Figure 4.89. The figure contains the results of desorption measurements at 100% RH and 85% RH on 40 °C and 60 °C water-stored samples and results of adsorption measurements at the temperatures mentioned after standard drying. It can be seen that no significant decrease in water

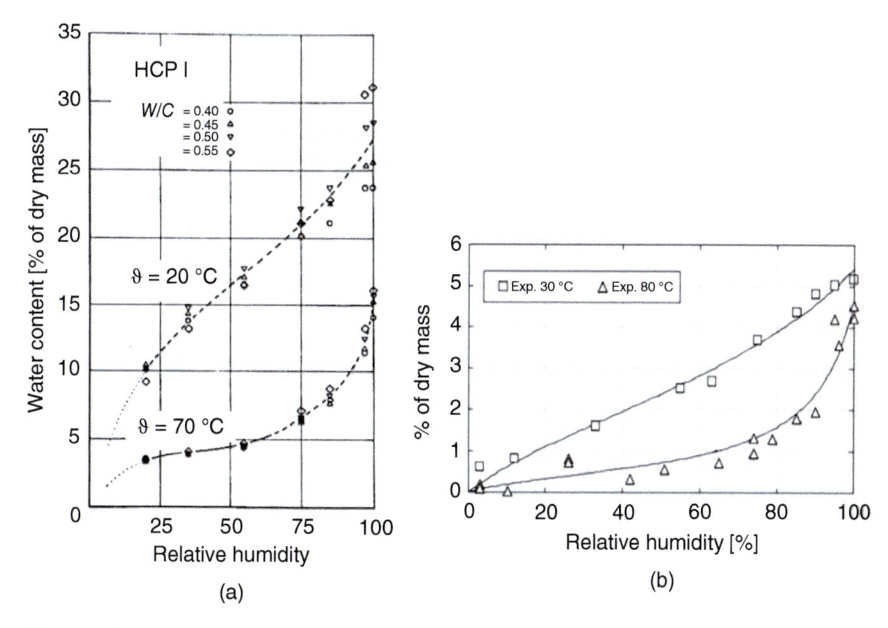

Figure 4.88 Results of **desorption tests** using the desiccator method at elevated temperatures. (a) Results of Hundt and Kantelberg [57] on water-stored HCP of CEM I with different W/C values at 20 and 70 °C. (b) Measurement results of Poyet and Charles from [60]/with permission of Elsevier on High Performance Concrete $W/C = 0.43$ at 30 and 80 °C after year of water storage.

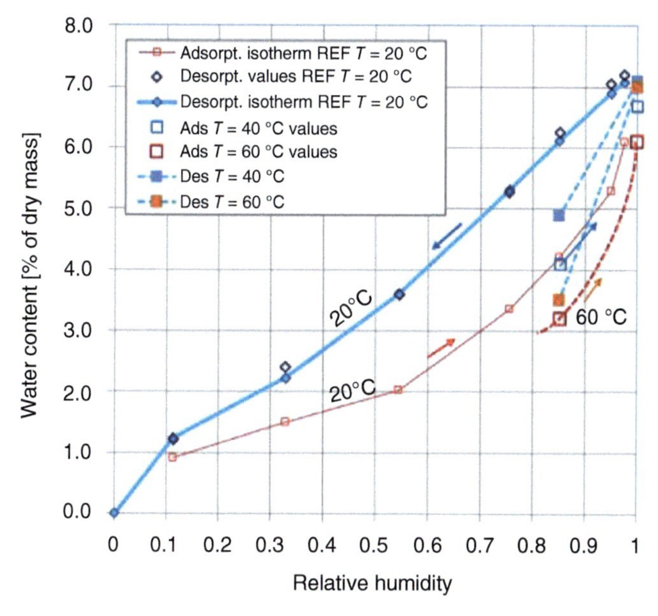

Figure 4.89 Tests on the temperature influence on the desorption and absorption behavior **of the material REF** at high relative humidities, desiccator tests after 15 months of water storage. For comparison, the desorption and adsorption isotherms were measured again at 20 °C. The arrows at the large open squares indicate the expected adsorption progress between 85% and 100% RH **at 40 and 60 °C**.

content was measured at desorptions from 100% RH in contrast to 85% RH. In the adsorption tests, the measured results at 40 °C were only slightly smaller than the comparative values at 20 °C, while at 60 °C, the water contents were at equilibrium reached only approx. 75% of the comparative values at 20 °C.

Further results of temperature-dependent desorption measurements and adsorption measurements have since been published by Drouet et al. [65] and de Burgh and Foster [42], which are discussed below.

Figure 4.90a gives desorption curves based on data from de Burgh and Foster [42]. These results confirm the significant temperature dependence of the desorption curves of cementitious materials, in this case HCP CEMI. The curves start from water storage at the given temperatures. Desorption values at 100% RH were obviously not determined in this project.

To handle the measured values from Figure 4.90a, regression curves were created using the van Genuchten equation, **Figure 4.90b**. The associated equation parameters can be obtained from figure caption.

The equation proposed in 1980 by van Genuchten [66] to describe the relationship between water saturation and capillary pressure can also be used in many cases to describe the measured values of sorption isotherms, since it can be fitted to many of the occurring curve progressions (in addition to prespecified parameters) simply by the two variables $m1$ and m, compare results of Drouet et al. [65].

The equation between water content and capillary pressure $\theta(p_{cap})$ is

$$\theta(p_{cap}) = \theta_{sat} \cdot \left[1 + \left(\frac{p_{cap}}{m1}\right)^{\frac{1}{1-m}}\right]^{-m} \tag{4.35}$$

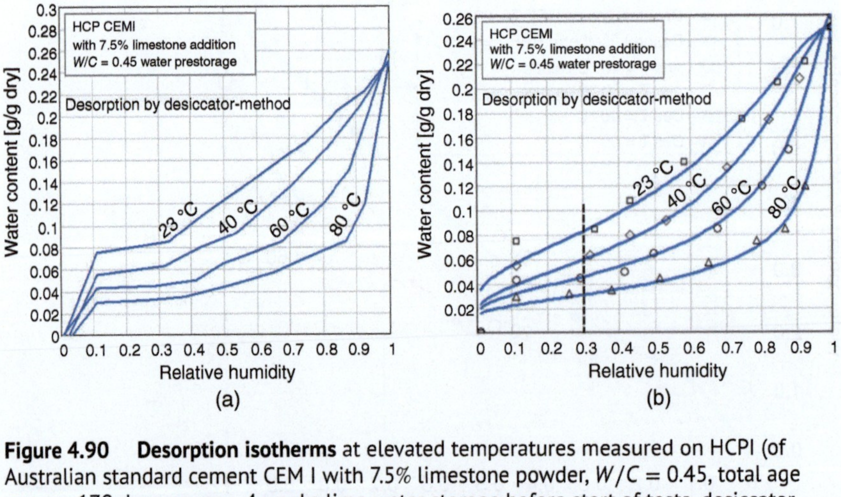

Figure 4.90 **Desorption isotherms** at elevated temperatures measured on HCPI (of Australian standard cement CEM I with 7.5% limestone powder, $W/C = 0.45$, total age approx. 170 days, approx. 4 weeks lime water storage before start of tests, desiccator method). **(a)** Curves generated based on the measurements from de Burgh and Foster [42]/with permission of Elsevier. **(b)** Regression curves to the measured values of the left image, created with the van Genuchten equation (4.36). The dashed line is to delineate the region below $\varphi \approx 0.30$, where the measured values are not reliably captured by the regression. Van Genuchten-parameters m, m1: for 23 °C-curve: 0.3959, 33.64; 40 °C: 0.4159, 23.75; 60 °C: 0.3776, 10.04; 80 °C: 0.3397, 2.985; $\theta_{sat} = 0.255$.

With $p_{cap} = -(R \cdot T)/18 \cdot ln(\varphi)$ it follows an equation to describe the measured values $\theta(\varphi)$ as a function of RH

$$\theta(\varphi) = \theta_{sat} \cdot \left[1 + \left(\frac{-R \cdot T}{18 \cdot m1} \cdot ln(\varphi)\right)^{\frac{1}{1-m}}\right]^{-m} \tag{4.36}$$

A simple derivation and an inverse function can also be given for this equation. For the following tasks the **inverse function** $\varphi(\theta)$ of Eq. (4.36) is needed, here also parameters given in **Figure 4.92b** are valid:

$$\varphi(\theta) = exp\left[-\left[\left(\frac{\theta}{\theta_{sat}}\right)^{-\frac{1}{m}} - 1\right]^{1-m} \cdot \left(\frac{m1 \cdot 18}{R \cdot T}\right)\right] \tag{4.37}$$

From Figure 4.90b it can be seen that the measured values below $\varphi \approx 0.30$ cannot be accurately represented with the van Genuchten regression equation. On the other hand, this range of values is of less importance from a practical point of view.

By de Burgh and Foster [42] **adsorption isotherms were also measured** for the same material as in **Figure 4.90a**. The results are shown for temperatures 23, 40 and 60 °C in **Figure 4.91**. It can be seen that at temperatures above 40 °C there is already a significant slowdown in the rise of the adsorption isotherm. The measured values were also approximated by the van Genuchten equation, and the parameters can be obtained from **Figure 4.91**.

The physical causes of the different temperature dependence of the desorption and adsorption isotherms were described as early as 1972 by Kast and

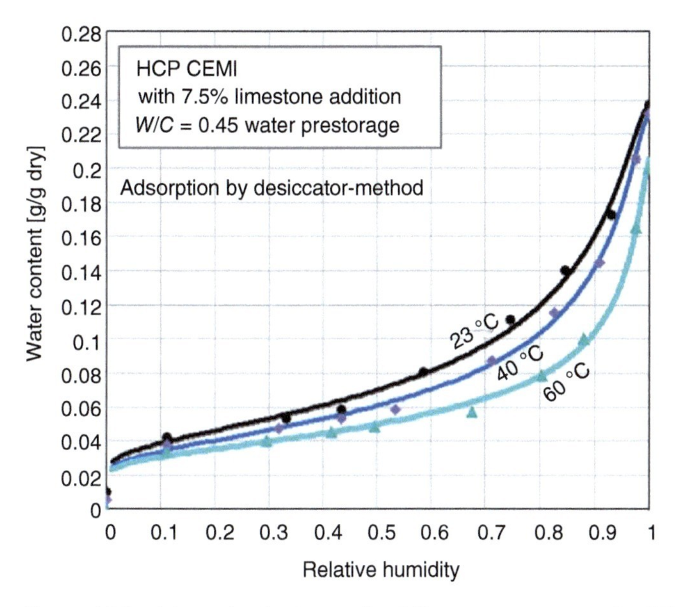

Figure 4.91 Adsorption isotherms for different temperatures for CEM I (material and storage see **Figure 4.90a**): Values measured from de Burgh and Foster [42]/with permission of Elsevier, regression with the Van Genuchten equation, Van Genuchten-parameters $m, m1, \theta_{sat}$: for the 23 °C-curve: 0.335, 9.0, 0.237; 40 °C: 0.330, 6.6, 0.232; 60 °C: 0.290, 3.1, 0.205.

Jokisch [67], in particular the thermodynamic basis of sorption in porous solids. From this paper it is clear that the binding energy is the main approach to describe the sorption processes. The terms sorption enthalpy or heat of sorption may also be used for the binding energy. See also Poyet [59] and Poyet and Charles [60].

Although the data presented were fortunately measured for comparatively close temperature spacings, **question arises as to the prediction of the magnitude of $\theta(\varphi, T)$ in general or for arbitrary intermediate states,** respectively. This is addressed below. The desorption enthalpies play a decisive role here.

The desorption enthalpies depend on the pore structure or internal surface area of the material. They must be determined empirically as a function of water saturation, vapor pressure, and temperature. For this purpose **the Clausius–Clapeyron equation** (4.33) presented in Section 4.6.9 was used and transformed into the following equation for the calculation of desorption enthalpies.

$$\Delta hmol_{T1}(\varphi_1) = R \cdot ln\left(\frac{\varphi_1 \cdot pS(T1)}{\varphi \cdot pS(T0)}\right) \cdot \left(\frac{(T1+273) \cdot (T0+273)}{T1 - T0}\right) \quad [J/mol] \tag{4.38}$$

where $R = 8.31$ [J/(mol· K)], temperatures $T0, T1$ in [°C]; $pS(T\,°C)$ are the saturation vapor pressures; since in the 2nd part of the equation absolute temperatures in Kelvin are to be entered, the partial addition of 273 is necessary.

The temperature $T0$ (for example 20 °C) is the temperature of the reference isotherm, the temperature $T1$ (for example 40 °C) refers to the associated isotherm for which the desorption enthalpies are to be calculated. The result $\Delta hmol_{T1}(\varphi_1)$ is the sought desorption enthalpy of one point on the desorption isotherm for $T1$. To calculate this point with (4.38), an associated point $(\varphi, \theta(\varphi))$ must be preselected on isotherm $T0$. The relative humidity of the point sought on isotherm $T1$ is φ_1 for the same water content $\theta(\varphi)$. Further enthalpy values are determined via the two isotherms analogously.

If one chooses in the computer program for the formulation of the sorption isotherms an analytic function , for which also a closed inverse function exists, for example the function (4.36) with the inverse function (4.37), then $\Delta hmol_{T1}(\varphi_1)$ can be represented continuously for the whole range $\varphi = 0 \ldots 1.0$.

Figure 4.92a shows the determined $\Delta hmol(T, \varphi)$ curves for the temperature-dependent desorption isotherms from **Figure 4.90b**, generated for measured values from de Burgh and Foster [42] with Eq. (4.36).

The curves of enthalpy values depend on the temperature range, i.e. on the temperature level and the temperature spreading $T1 - T0$. This temperature dependence must be taken into account when calculating any values $\theta(\varphi, T\,°C)$.

Therefore the enthalpy curves were calculated starting from the given 23 °C-isotherm for the isotherms 40, 60 and 80 °C given curves. The resulting enthalpy curves are shown in **Figure 4.92b**. The enthalpy values are naturally largest at low φ values and exhibit the evaporation enthalpy of pure water at $\varphi = 1.0$. One can see that there are relevant differences between the curves according to the temperature ranges chosen.

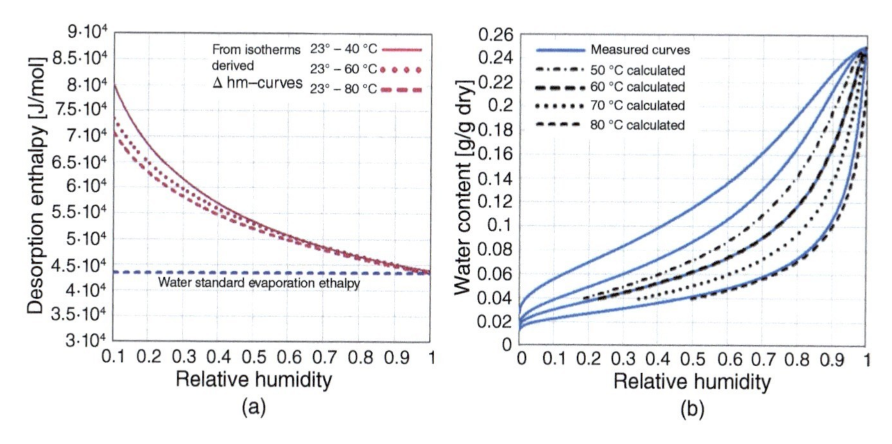

Figure 4.92 The computational modeling results presented are based on the material and the curves from **Figure 4.90a,b**. (a) Progress of the different desorption-enthalpy curves for the temperature ranges indicated in the figure. (b) Examples of calculated temperature-dependent desorption curves, temperature-ranges according to the legend.

If one wants to obtain reliable isotherms for intermediate temperatures in the temperature ranges given, the corresponding enthalpy values can be obtained by a kind of interpolation. The procedure is shown below for the ranges 40°–60° and 60°–80°:

For the 2nd range $T1 = 40\,°C$ to $T1 = 60\,°C$ with $\Delta hmol1(\varphi)$ for 40 °C and $\Delta hmol2(\varphi)$ for 60 °C (determined with Eq. (4.38)) one can obtain the enthalpy intermediate value named $\Delta hm2(\varphi)$ for the intermediate temperature $T[°C]$ as:

$$\Delta hm2(\varphi, T) = \Delta hmol1(\varphi) - \frac{\Delta hmol1(\varphi) - \Delta hmol2(\varphi)}{60 - 40} \cdot (T - 40). \quad (4.39)$$

For the 3rd range $T1 = 60\,°C$ to $T1 = 80\,°C$ with $\Delta hmol2(\varphi)$ for 60 °C and $\Delta hmol3(\varphi)$ for 80 °C (determined with Eq. (4.38)) one can obtain the enthalpy intermediate value named $\Delta hm3(\varphi)$ for the intermediate temperature $T[°C]$ as:

$$\Delta hm3(\varphi, T) = \Delta hmol2(\varphi) - \frac{\Delta hmol2(\varphi) - \Delta hmol3(\varphi)}{80 - 60} \cdot (T - 60) \quad (4.40)$$

On the basis of the desorption enthalpies, determined as described above, the corresponding relative humidities φ_1 can be determined for any intermediate temperatures T in the total temperature range between 23 and 80 °C, leading to a given water saturation resp. the corresponding water content θ in the material due to desorption at φ_1 and T. After choosing a content θ resp. φ belonging to θ from the basic isotherm (here for $T0 = 23\,°C$), the relative humidity φ_1 sought results from the following Eq. (4.41) for a temperature T [°C] given :

$$\varphi 1(\varphi, T) = \varphi \cdot \frac{pS(T0)}{pS(T)} \cdot exp\left[\frac{\Delta hm(\varphi, T)}{8.31} \cdot \left(\frac{T - T0}{(273 + T) \cdot (273 + T0)}\right)\right] \quad (4.41)$$

where $\Delta hm(\varphi, T)$ is the enthalpy value to be taken from the temperature range matching T (for example, for T between 60 and 80 °C from Eq. (4.40)).

Figure 4.92b shows the desorption isotherms calculated in this way for the intermediate temperatures indicated in the figure.

If, on the other hand, it is asked what water saturation θ of the material occurs at a given temperature of, for example, $T1 = 74\,°\text{C}$, and a relative humidity of, for example, 85%. The water content that occurs is obtained **with very good approximation** from the "neighboring" desorption isotherms for $T1 = 60\,°\text{C}$ and $T1 = 80\,°\text{C}$ as follows:

$$\theta_{(74\,°\text{C},85\%)} = \theta_{(60\,°\text{C},85\%)} - \left[\theta_{(60\,°\text{C},85\%)} - \theta_{(80\,°\text{C},85\%)}\right] \cdot \frac{74-60}{80-60} = 0.103\ [\text{g}/\text{g}_{dry}] \tag{4.42}$$

The possibility of an even better fitting of the measured values for example of the curves in **Figure 4.90b** is illustrated in the following. From this figure it is evident that the easy-to-use van Genuchten regression curves cannot adequately reproduce the measured values below $\varphi \approx 0.30$.

A more accurate fit to the given measurement results is obtained by **using the following Pickett equation** (4.43), who in his 1945 publication [68] supplemented the BET theory with respect to a better fit to measured values at larger RH, in which he brings a constant g used by Brunauer–Emmett–Teller into dependence on φ, whereby also a convex curvature of the behavior at high φ values can be modeled. This equation is also propagated by Drouet et al. [65], among others. The resulting equation is

$$\theta(\varphi, n, C, b, \theta_{mono}) = \frac{\theta_{mono} \cdot C\left[\varphi \cdot (1-\varphi^n) + b \cdot n \cdot \varphi^n \cdot (1-\varphi)\right]}{(1-\varphi) \cdot \left[1 - \varphi + C \cdot (\varphi + b \cdot \varphi^n)\right]} \tag{4.43}$$

The four parameters n, C, b, θ_{mono} to be inserted all have a "physical" meaning, so fitting to different measured values is comparatively easy.

C is decisive for the slope of the curve or a possibly existing "plateau" in the value range $\varphi \leq 0.30$ and behaves as indicated in **Figure 1.17** Section 1.4.3.1.

θ_{mono} is the water content of the system at mono-molecular occupancy of the inner surface and corresponds to the measured water content at $\varphi \approx 0.20$.

n is the assumed number of molecular layers and controls the slope of the curve, b controls any concave curvature of the curve when approaching $\varphi = 1.0$.

The formation of a closed inverse function $\varphi(\theta)$ of (4.43) has not been tested, but at least at 1st glance appears to be impossible, which limits its ease of use. If necessary, however, the reversal points can be obtained over a mathematical step wise regression algorithm.

The Pickett equation (4.43) **was used here as a regression equation** for measurement results of Drouet et al. [65]. These are results of desorption tests at 20, 50 and 80 °C on HCPI from CEM I 52.5 R (according to EN-206), $W/C = 0.40$, three months water-stored.

The resulting desorption curves are shown together with the corresponding Pickett parameters in Figure 4.93. These results also demonstrate the strong dependence of the desorption curves on temperature and only a slight decrease in saturation at 100% RH at 80 °C.

The desorption curves for intermediate temperature values were calculated according to the previously described procedure on the basis of the data resp. the

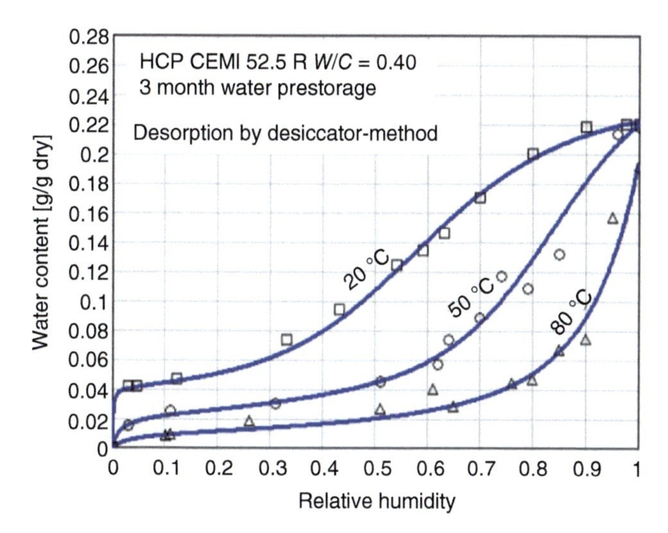

Figure 4.93 Desorption isotherms at elevated temperatures describing measurement results of Drouet et al. [65]/with permission of Elsevier by the Pickett equation (4.43). Pickett equation parameters n, C, b, θ_{mol} : 20 °C-curve: 5.55, 1000, 5.9, 0.040; 50 °C-curve: 10.0, 70, 2.5, 0.022; 80 °C-curve: 19.0, 29, 0.0, 0.0103.

curves from **Figure 4.93** using Pickett regression curves for 20, 50, and 80 °C respectively. For this purpose, the $\Delta hmol50(\varphi)$ and $\Delta hmol80(\varphi)$ progression curves had to be determined first using Eq. (4.38) analogously.

The inverse values necessary were determined step wise via an interpolation algorithm. The Δhm values for T [°C] in the temperature range 50–80 °C are then:

$$\Delta hm(\varphi, T) = \Delta hmol50(\varphi) - \frac{\Delta hmol50(\varphi) - \Delta hmol80(\varphi)}{80 - 50} \cdot (T - 50) \; [J/mol] \tag{4.44}$$

Using the Eq. (4.41), the courses of the temperature-dependent desorption isotherms for arbitrary temperatures can now be given for the material of the measured curves of **Figure 4.93**. Results of these calculations can be seen in **Figure 4.94a,b** for selected values $T1$.

On the question of the course of scanning isotherms at elevated temperatures, reference is made to investigation results of de Burgh and Foster [42] on HCP from CEM I with 7.5% limestone flour, $W/C = 0.45$. The following **Figure 4.95** from de Burgh and Foster [42] gives the measured results for 40 and 60 °C. It can be seen that the observed behavior of the scannings at these temperatures corresponds in principle to the behavior of the material at 20 °C, compare **Figure 4.67** and Section 4.6.4.

The suitability of DVS measurements should be verified, especially at elevated temperatures, for the determination of different sorption isotherms and minimum standards for the time management of the apparatus should be defined. For example, the results published by Maruyama et al. 2018 [56] in a paper on the influence of cavitation on adsorption and desorption isotherms and scanning

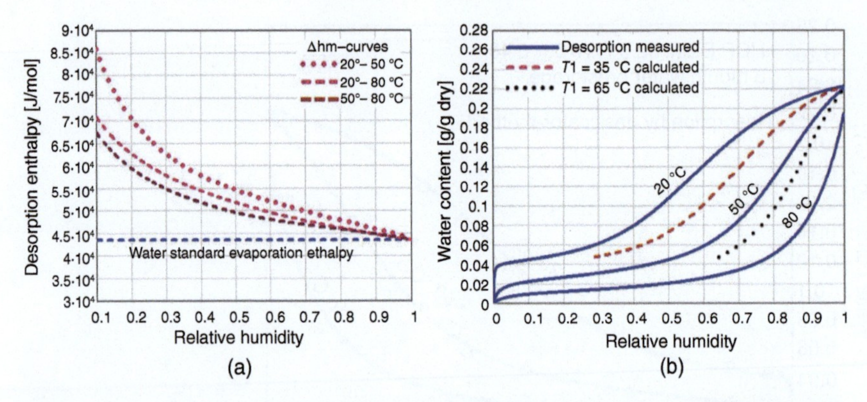

Figure 4.94 Calculated curves based on the material and the curves from **Figure 4.93**. (a) Progress of the different desorption-enthalpy curves for the temperature ranges indicated in the figure. (b) examples of calculated intermediate temperature-dependent desorption curves temperature-ranges according to the legend.

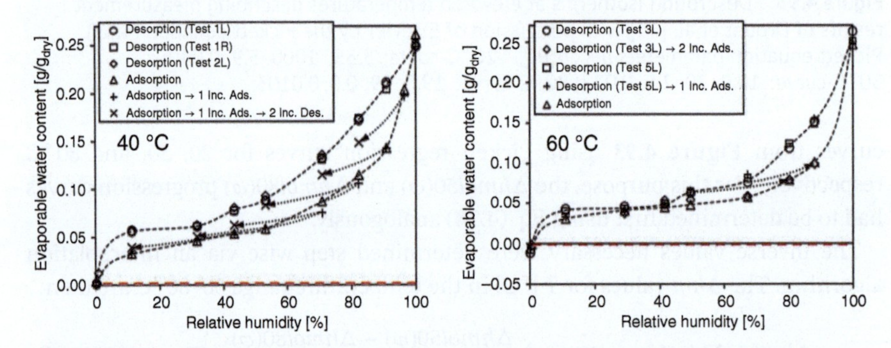

Figure 4.95 Measured values of scanning isotherms of HCPI (CEMI+7.5% limestone powder), $W/C = 0.45$ (Designations in the figure: 1 Inc. Ads or 1 Inc. Des corresponds to 1 incremental step of 10% RH in adsorption or desorption). Source: Diagrams from de Burgh and Foster [42]/with permission of Elsevier.

curves at different HCP, which show only a minor influence of temperature up to 80 °C, should be questioned. Hygric equilibria apparently did not exist completely in the DVS measurements.

4.6.11 MIP Curve as a Boundary Storage Function at Elevated Temperature

Mercury pressure porosimetry generates cumulative volume curves of the given pore system in the range of pore radii from 151 000 [nm] or 0.151 [mm] to 1.8 [nm]. The cumulative MIP curve is generated starting from the upper boundary at initially low pressure, thus capturing the large pores of the system first. With increasing pressure, the pore volume is then increasingly captured up to the pore radius of 1.8 [nm] or the maximum pressure. The initial range of the cumulative MIP curve therefore refers to the pore area filled at humidities close to 100%. The upper endpoint of the sorption isotherms therefore corresponds in principle to the starting point of the

MIP curve. Based on the pore volume or water content of the material (for example, concrete) at the pore size corresponding to the starting pore size of the mercury porosimeter (151 000 [nm]), subtracting the MIP volume measurements from the aforementioned water content of the material yields an associated MIP-based moisture storage function.

A relative humidity of 99.9993% corresponds to the starting radius of the porosimeter according to Kelvin. However, the measured cumulative pore volume also includes the volume of processing-related air pores and the volume of any artificially inserted air pores embedded in the HCP. The fact that air pores created by air pore formers are also included has been demonstrated, for example, by Y. Chen et al. [69] measured.

Since there is usually no superhygroscopic region in the cementitious material, this value is in principle equivalent to the water saturation value θ_{sat}.

An MIP-based moisture storage function can therefore be calculated in principle by subtracting from the water saturation value θ_{sat} of the material the volume of the cumulative MIP curve for given relative humidities <100%. For this purpose, however, the pore radii of the MIP measurements must first be converted into corresponding relative humidities RH.

On this basis, it was now investigated whether the MIP-based "desorption curve" allows a usable statement to describe desorption under temperature influence.

For this purpose, on the one hand the results of Drouet et al. [65] from **Figure 4.93** were used, on the other hand a MIP curve of the cumulative pore volume of the reference mortar REF converted for this purpose, where the pore radii were first expressed in φ values using Eq. (1.38).

The Drouet material used is a HCPI with $W/C = 0.40$. The REF material has a high cement paste content with a W/C value of 0.45. The MIP measured values for the MIP curve (**Figure 4.1a**, curve 40 °C/1 mbar) were therefore first converted to the pure HCP volume of the REF by multiplying the volume measured values by the volume relation $k_{V,HCPI}$ = total volume to HCP volume fraction = 1/0.430, further converted to the W/C value of 0.40 of the reference material with $k_{V,W/C} = 0.93$ from Eq. (4.24).

The MIP "desorption curve" related to the dry bulk density ϱ_{HCPI} of the HCPI is then obtained as:

$$m_{MIP}(\varphi) = \left[\theta_{sat,HCPI} - \left(V_{REF,MIP}(\varphi) \cdot \frac{0.93}{0.430} - \Delta V_{Air} \cdot \varphi\right)\right] \cdot \frac{1}{\varrho_{HCPI}} \tag{4.45}$$

$\theta_{sat,HCPI}$ is therein the initial water saturation of the HCPI curves, $V_{REF,MIP}(\varphi)$ the cumulative volume measurements of the MIP measurement of the REF, ΔV_{Air} the difference of the air content of the REF mortar to the lower content of the HCPI from **Figure 4.93**, considered in Eq. (4.45) with rough approximation.

The result of this comparison with the MIP desorption curve determined according to Eq. (4.45) is shown in **Figure 4.96**. This comparison shows that MIP curves can be suitable to delineate the limit curves of desorption at high temperatures. This can be justified by the fact that the MIP curves arise in the case of completely dried pore systems, a condition which also increasingly arises during desorption at high material temperatures of HCP due to desorption resp. drying.

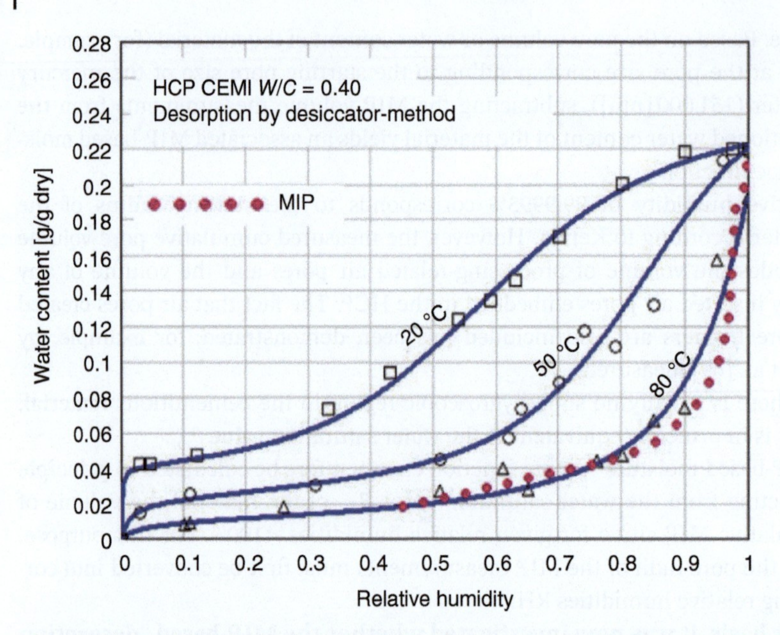

Figure 4.96 Comparison of applicability of MIP-measurements as bordering "desorption curves" for desorption isotherms of HCP or concretes at high temperatures. The figure shows a test comparison, using the desorption isotherms from **Figure 4.93** and a MIP-curve deduced from results of measurements on material REF.

4.6.12 Influence of Salt Contents on Moisture Storage

Increased salt contents in the pore system lead to a lowering of the (saturation) vapor pressure above the pore solution and can therefore permanently enhance the effect of capillary condensation, depending on the salt content, salt type, and temperature boundary conditions.

Therefore, a significant change in the sorption isotherms takes place due to additional salt contents. This is obviously of practical relevance in all cases in which ions are additionally introduced into the pore system by continuous exposure or corrosion processes and are dissolved there or remain soluble with sufficient moisture supply, if necessary, after chemical reactions with parts of the cement paste matrix.

The so-called deliquescence moisture of the salt compounds present or formed is of essential importance. Below the deliquescence moisture (for NaCl corresponding to approx. 75% RH), salt solutions dry out and then obviously have no or only a subordinate influence on the sorption isotherms in a fine-pore system with capillary condensation. Above the deliquescence humidity, a salt solution is generated by water vapor uptake, which dilutes as long as the external air humidity is higher than the deliquescence humidity value. Pore systems are filled with water in the process.

This physico-chemical behavior is also used to keep the relative air humidity constant in desorption experiments with the **exsiccator method**.

The influences on surface energy, which is also relevant for transport, have already been discussed in Section 1.2.2. Villani et al. investigate in [70] the influence of the properties of the pore solution on hygric properties of the HCP such as viscosity,

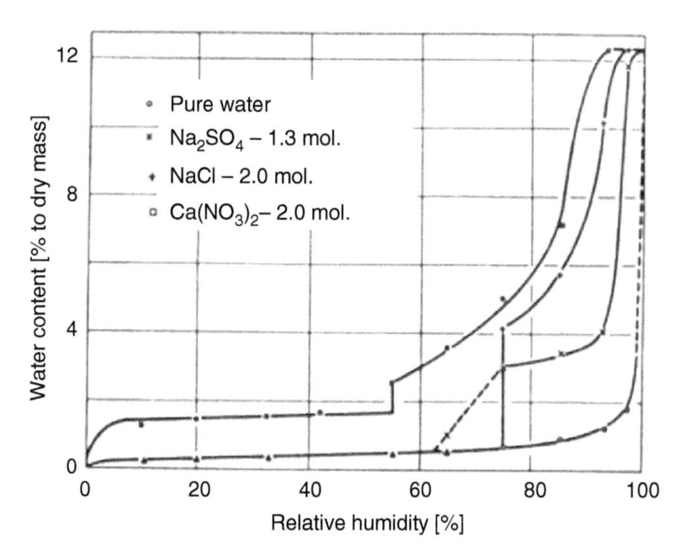

Figure 4.97 Desorption isotherms of sandstone material, previously doped with salt solutions of different concentrations according to the legend, investigation results of Garrecht [72]/Harald.

surface tension, relative humidity depending on the salt concentration, the temperature, and partly on the pore size, especially with regard to diffusion coefficients and the drying behavior of the HCP.

When evaluating the investigation results on the influence of salts, it should be taken into account that the concentration of sodium, potassium, and sulfur in the hydrated HCP does not exceed about 0.2 mol/L, compare Lothenbach et al. [71].

The nature of the influence of elevated salt contents on desorption behavior is indicated in **Figure 4.97**, taken from Garrecht [72]. The curves were determined on sandstone material previously doped with salt solutions of different concentrations.

One can see the large influence on the desorption curves at the given high salt contents, although the result is not directly transferable to cement paste or concrete.

4.7 Results on Capillary Water Absorption Depending on Initial Water Content and Time

4.7.1 Experimental Studies on the Behavior of HCP and Concrete

Extensive capillary water uptake measurements were performed on prisms made of REF mortar. A main objective of this part of the project was to investigate **the dependence of the water uptake curves and the associated water uptake coefficients** on the initial water content of the mortar prisms used. The results have a decisive influence on the formulation of the moisture transport coefficients.

In a subproject, the prisms were first specially adjusted to homogeneous unsaturated water contents until the start of the intended water uptake tests. This was done

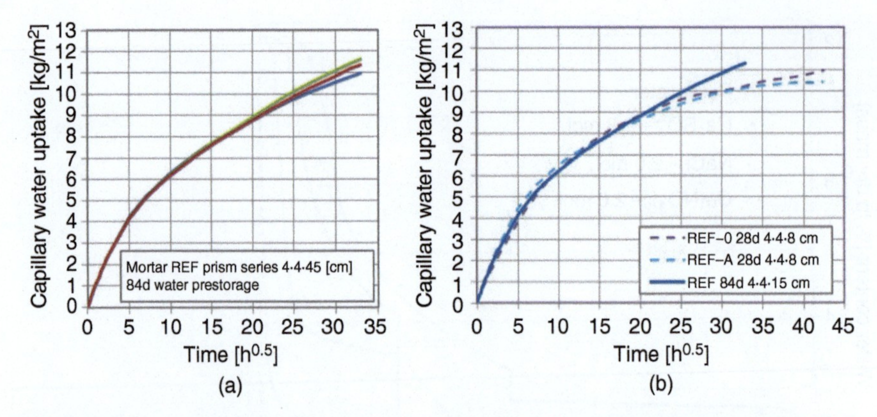

Figure 4.98 Capillary water uptake curves measured in two independent projects at a time interval of 5 years with newly produced mortar REF batches. (a) Contains the individual water uptake curves (measured over 1000 hours) of a series of **three laterally sealed prisms of length 150 [mm]**, cured by 84 days of water storage and standard dried water uptake via sawn face. (b) Shows for comparison the mean value curve from the left figure (a) and the later measurement results of two series of prisms 40·40·80 [mm³] from Goedeke [44]/TUHH. Up to a measuring duration of approx. 400 hours, largely equivalence of the curves is observed, although the specimens were cured in water for only 28 days. After approx. 400 hours, the decelerating effect of the shorter prism lengths becomes obvious.

by desorption from initial water saturation or by adsorption from previous standard dry state or by alternate climate storage until intended unsaturated moisture equilibrium. The desiccator method was always used to establish the moisture equilibria.

The reproducibility of the water absorption tests carried out here was checked, for example, using the results from two independent projects carried out several years apart on new mortar REF batches. **Figure 4.98a and b** show a comparison of test results.

In addition to the good reproducibility found, the curves show the pronounced curvature, which is always observed on cement-bound materials in water absorption tests with low initial water contents.

The water absorption coefficient ww(24h) related to approx. 5 hours in the root scale is approx. 0.90 [kg/(m^2 · h$^{0.5}$]). An ambient climate of 20 °C/65% RH was generally selected during the water uptake experiments.

The slope and shape of the water uptake curves depend on the initial water content, even with a homogeneous initial water distribution and a regular unidirectional execution.

Figure 4.99 shows a series of capillary water uptake curves measured on prisms of dimensions 40·40·147 [mm³] made of mortar REF with different (homogeneously distributed) initial water contents of approx. 0 to 127 [kg/m³], corresponding to a water saturation degree of approx. S_0 = 0–0.85 related to a maximum capillary water content of 0.150 [kg/m³].

Figure 4.99 makes it clear **that with increasing initial water saturation, a decrease of the slope** of the curves or of the slope parameter ww(24h) takes place and obviously also a linearization of the curves in the sense of a clear decrease of

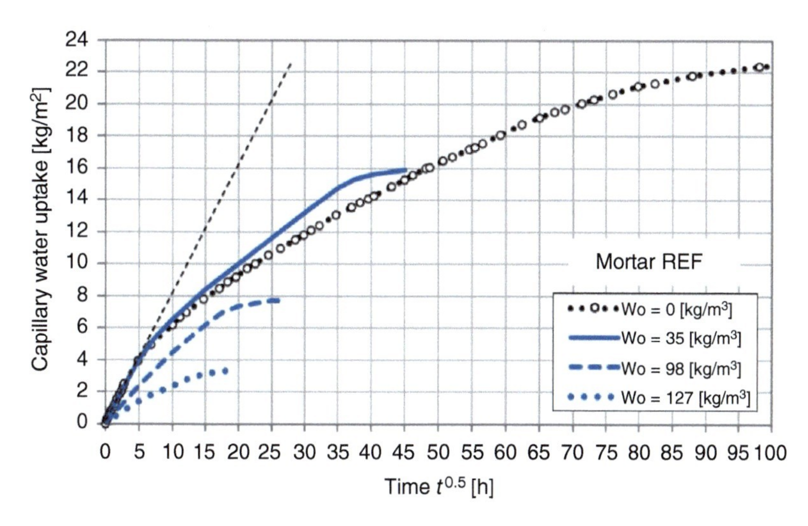

Figure 4.99 Plot of the shape of experimental water uptake curves as a function of $t^{0.5}$ of mortar REF for different degrees of (homogeneous) water saturation S at the beginning of capillary water uptake. (Water content of ≈149 [kg/m³] corresponds to capillary saturation $S = 1.0$).

the curvature. The transition of the curves to the horizontal when water saturation is reached, naturally depends on the free pore space at the beginning of the experiment. It should also be noted that there is no decrease in the slope ww(24h) at initial saturation degrees from 0 to about 0.25.

In view of a modeling of the capillary water uptake over the material surface and the water transport at different water saturations in the material interior, it is necessary to know this dependence in more detail. Therefore, the numerous experiments carried out at 20–23 °C were evaluated in more detail in this regard.

Figure 4.100 summarizes the investigation results. The curves of the capillary water uptake experiments continued over long periods of time are plotted in the figure, starting from the different initial moisture contents of the prism series. The drawn curves result from the curves of the unidirectional water absorption tests by division by the prism length of approx. 150 [mm]. It can also be seen that, despite of striving for high precision and reliability, visible differences between measurement curves can occur, especially with long test times, which can have various causes.

A systematic influence by the way of setting the initial moisture equilibrium by adsorption or by desorption could not be found.

Possible error influences are preparation of the suction surface, carbonation effects in the area of the suction surface at the foot of the test prisms during long-term tests, and the lateral sealing of the test specimens, which on the one hand must be reliably vapor-tight, while on the other hand, a "creeping up" of water components must be prevented at the transition, for example, of bonded aluminum coatings into the water plane at the foot point of the prisms. For this reason, test series were also repeated for verification purposes.

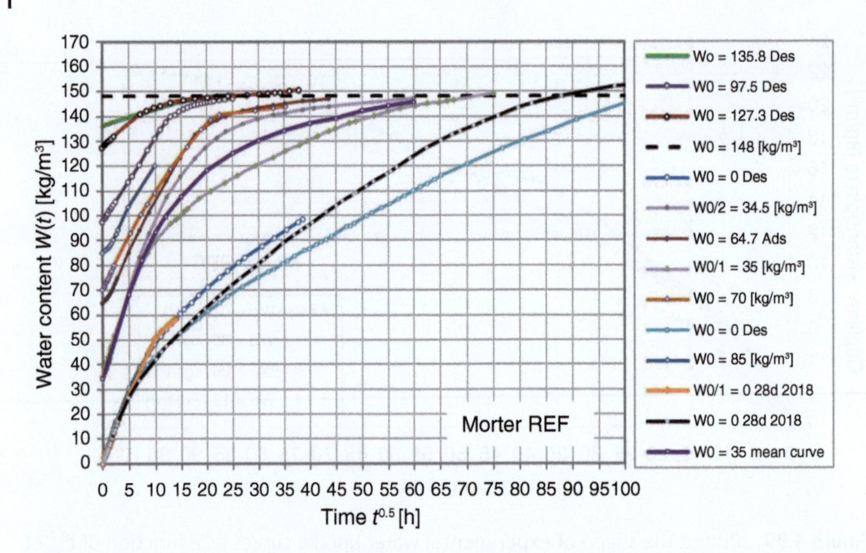

Figure 4.100 Water content curves $W(t)$ in [kg/m^3] of mortar REF derived from measured capillary water uptake curves [kg/m^2] at 20 °C as a function of $t^{0.5}$, all related to the constant volume of the 150 [mm] long test specimens, starting from the different degrees of water saturation at the beginning of capillary water uptake. The homogeneous initial water content of the prisms is indicated on the right. [Franke/TUHH].

Overall, however, the test results in **Figure 4.100** confirm the findings made in **Figure 4.99.**

A look at Figure 4.100 shows that the initial slope of the curves at initial water content of 35 [kg/m^3] was systematically a little larger than at initial water content of about 0 [kg/m^3].

With respect to the programming modeling of the storage and transport processes, the dependence of the curve slopes or coefficients $ww(t0)$ in [kg/(m^2 · h$^{0.5}$] is needed.

Figure 4.101 shows **the measured values ww(5h) and ww(24h) in [kg/(m^2 · h$^{0.5}$]** for the mortar REF as a function of the water saturation level, where $\theta_{REF,sat}$=149 [kg/m^3] was used to produce the diagram. It can be seen that water uptake tests at an initial water content above about 50% of the maximum saturation result in largely linear curves (at the root scale) and only show the curved shape when the maximum saturation or "top" of the prisms is reached. Water absorption tests in the capillary saturation range **above θ_{gel} thus always have a linear course as a function of** $t^{0.5}$ (compare Section 5.2.1 for the definition of θ_{gel}).

On the other hand, it is already clear from **Figure 4.100** that, at least **at lower initial filling levels, there is a considerable slowing down of the moisture transport** with increasing time in the present REF mortars and corresponding cement-bonded materials compared with the root-t function or the initial phase. This also has serious consequences for the modeling of moisture transport.

Further conclusions with respect to the simulation of the capillary transport or the associated transport coefficients are made in Chapter 5.

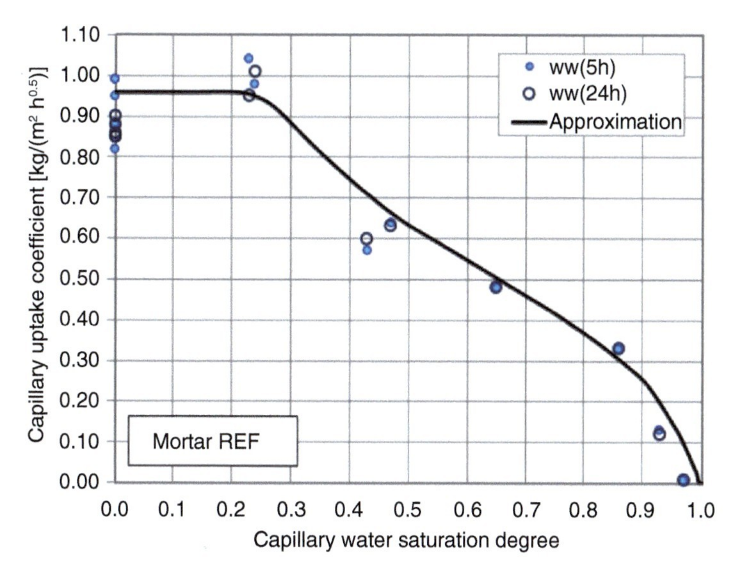

Figure 4.101 Capillary water uptake coefficients, respectively, slope of the measured water imbibition curves in [kg/(m^2 · h$^{0.5}$] of material REF at 20 °C (determined from the curves of **Figure 4.100**) in dependence of the initial saturation degree. The maximum capillary water saturation degree $S = 1.0$ was set to a water content of 149 [kg/m^3].

It should be noted that the capillary water absorption coefficients of concrete and mortars are also affected by temperature.

For comparison, the unidirectional **capillary water absorption of 2 UHPC products is given in Figure 4.102**. One can also see the strong curvature of the curves in $t^{0.5}$ scale. The capillary water absorption coefficients related to 1 hour are 1/20 to 1/10 of the corresponding value of the mortar REF.

4.7.2 Self-Sealing: Conclusions from NMR Analysis and Computational Results

Gräf and Grube [75] **and Jacobs [76]** have performed water permeability tests on hydrated concrete specimens. The specimens were first dried to equilibrium and then water-saturated. Immediately thereafter, a flow was generated over a series of days with time interruptions via an external water pressure, from which a time-dependent water permeability evolution of the specimens was determined. There was observed a marked decrease in permeability as a function of time.

Hearn and Morley [77] **conducted similar tests** in 1997 on 26-year-old fully hydrated concretes. Within about 100 hours of testing after the specimens were dried and re-saturated with water, permeability reduced to about 20% of the initial value immediately after drying and re-saturation. **This behavior was termed self-sealing by Nataliya Hearn**. A number of causes for this behavior were considered. Experimental flow reversal was ruled out as a cause. Ultimately, it was suggested that solution processes and precipitation of material in the pore system could be the cause of the behavior.

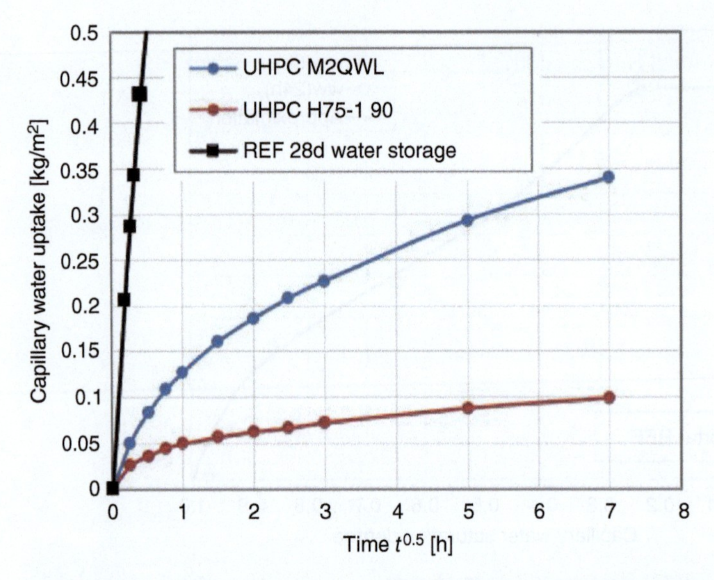

Figure 4.102 Capillary water uptake after standard drying of 2 UHPC products in comparison with mortar REF. 28-day compressive strength: UHPC M2QWL: 181 [N/mm^2]; UHPC H75 90: 178 [N/mm^2]. Compositions: M2QWL: $W/B = 0.17$, 830 [kg/ m^3] CEMI 42.5 R + micro-silica; H75 90: $W/B = 0.24$, 220 [kg/ m^3 CEMI 42.5 R + 600 [kg/ m^3] ground blast furnace slag + micro-silica, plus quartz flour and quartz sand in both products, grain size ≤ 0.5 [mm]. Source: Adapted from Franke et al. [73] and Schmidt [74].

This slowing down of the water uptake compared to the $t^{0.5}$-function discussed in Section 4.7 **does not occur** (on the same material) when the suction test is carried out with solvents such as hexanes instead of water. Numerous tests by different authors have shown this, e.g. compare the results of Krus et al. [78] on HCP with different W/C values from 0.20 to 0.45. The same effect is obtained in comparative tests with different concretes. From this, it must be deduced that the polarity of the water molecules (compared to the nonpolar solvents) plays a significant role in the measured behavior.

With regard to the modeling of water uptake and transport, especially in cement-bound products, this "nonlinear" behavior is of great importance, so its causes will be discussed in more detail below.

There are now a large number of publications that use NMR measurement techniques specifically for the analysis of cement-bound products. Therefore, the NMR procedure is also described here, at least in outline.

Zimmerman et al. [79] make remarks in fundamental papers in 1956 and 1957 on the behavior of "hydrogen nuclei of H_2O adsorbed on silica" and NMR, respectively. They make clear the presence of adsorbed water-molecule layers on silica surfaces, for example, and refer to their $T2$ relaxation times in the millisecond range, while in free water the relaxation times are in the seconds range.)

Petra Rucker-Gramm [80] has already dealt in detail with the question of self-sealing in her dissertation 2008. Here, she specifically observed water uptake and time-dependent water storage in cement mortar prisms and, in particular, followed them metrologically with 1H NMR.

Thiel and Gehlen [81] **and Rucker-Gramm** [80] **describe the NMR** measurement technique used in a very comprehensible and concentrated way, so that their description is adopted in parts in the following:

The method is based on the angular momentum of the positively charged protons (H1) and the resulting magnetic moment. **In a constant external magnetic field,** hydrogen nuclei behave like gyrating magnetic dipoles. The orientation of the magnetic moments is manipulated by additional electromagnetic radio pulses at certain frequencies. The electromagnetic energy is absorbed by the protons 1H, who are put into phase-coherent oscillations and generate a signal. After termination of the radio-frequency pulse, the excited nuclei return with a time response described by the relaxation times *T*1 and *T*2. These relaxation times are associated with an energy release into the environment.

*T*1 refers to the time it takes for the spins to recover after an excitation pulse along the main magnetic field (*z*-axis). The transverse relaxation time *T*2 (or *T*2, eff for measurements in inhomogeneous magnetic fields) corresponds to the loss of magnetization in the transverse plane). Both relaxation times *T*1 and *T*2 are characteristic of the molecular mobility of the nuclei. *T*1 is mainly dependent on the physical bonding of the hydrogen nucleus; *T*2 measurements are commonly used to identify water in different pores.

In the measurement, the excitation pulse is irradiated in a broadband manner, and then the total signal is recorded, **Figure 4.103a**. The amplitude of the NMR signal S_0 is an indicator for the number of hydrogen nuclei present in the screened volume and is thus proportional to the total water content within a sample. The sum signal is then broken down by a Laplace inversion calculation module into the relaxation times contained in the signal in the form of peaks, **Figure 4.103b**. These relaxation times then allow a conclusion to be drawn about the respective proportion of water molecules bound to different extents.

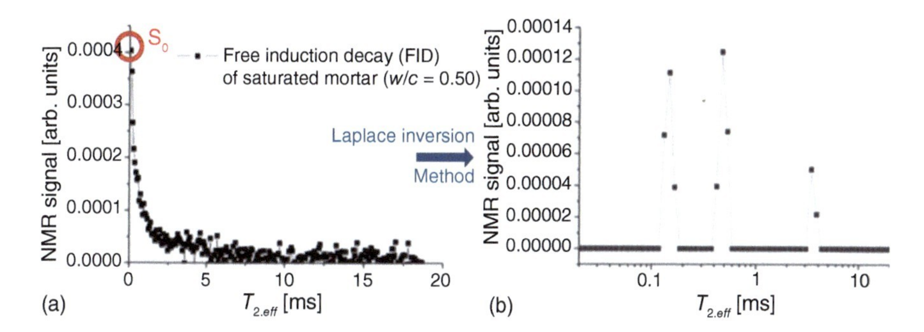

Figure 4.103 Base curve (echo train intensity decay) of an NMR measurement on HCP from Portland cement ($W/C = 0.50$) (plot a) and breakdown by Laplace inversion calculus module into the corresponding *T*2 relaxation peaks (plot b). Source: Figure from Thiel and Gehlen [81]/Springer Nature.

In Rucker-Gramm's apparatus [80], within the permanent magnetic field, a carriage controlled by a stepper motor pushes the specimen under investigation into the measuring coil in steps (apparatus see **Figure 3.9**).

The coil used serves both to radiate the excitation pulse sequence and to receive the useful signal. By switching from transmitting to receiving, the apparatus has a certain dead time. For strongly surface-bound molecules the relaxation times are very short. In the measuring apparatus of [80], these particularly short relaxation times could not be detected or **only partially detected due to the existing dead time** and thus also not the fraction of water adsorbed on the inner surface.

In the papers by Maruyama et al. [82], **Valori et al.** [83], **Muller et al.** [84], **or McDonald et al.** [85], **the method is established for deriving pore-size distributions in cement-bound products by NMR.**

The method was derived from measurements on HCP test specimens, which were subjected to selective moisture desorption and drying, respectively, whereby smaller pore sizes were water-filled in each case according to the drying condition. Since the volume/surface area ratio of the respective pores decreases with decreasing pore size, these pores contain an increasing number of strongly surface-adsorbed water molecules with a correspondingly high proportion of the total water content of the pore group or of the sample saturation state. This results in a correspondingly high fraction of very short $T2$ relaxation times in NMR analysis of the samples. With the knowledge of the parallel verified pore-size distribution of the material, a procedure has now been worked out for the size-ordered determination with NMR of the pore-size distribution present in each case.

Valori et al. give in [83] a comprehensive overview of the current possibilities and limitations of NMR analysis on HCP in particular. According to the authors, three pore-size reservoirs can be detected in cement paste with NMR in the nano-size range: interlayer spaces of size circa 1–1.5 [nm], gel pores of size circa 2–5 [nm], and inter-hydrate pores of circa 10 [nm].

The authors also point out the difficulty or uncertainty of reliably detecting the nature of the water in the interlayer range with MNR.

Muller et al. [84] describe, in terms of interlayer porosity, that in such small spaces, the concept of mobile pore water above an adsorbed layer that underpins the model starts to break down.

Also in the scientific works characterizing the pore systems on the basis of NMR analyses, the porosity is divided into three pore groups with the previously mentioned mean pore radii.

This is also largely consistent with the previous view of a subdivision as given in **Figure 5.2a**, also with a view to modeling moisture transport.

On the basis of the described tool, statements are also made with respect to the rearrangement of water molecules, specifically between gel pores and interlayer pores.

Thus, **Wyrzykowski et al.** [86] **conducted NMR studies** on HCP from white cement with different W/C values with the aim of measuring any water redistribution between pores with temperature changes.

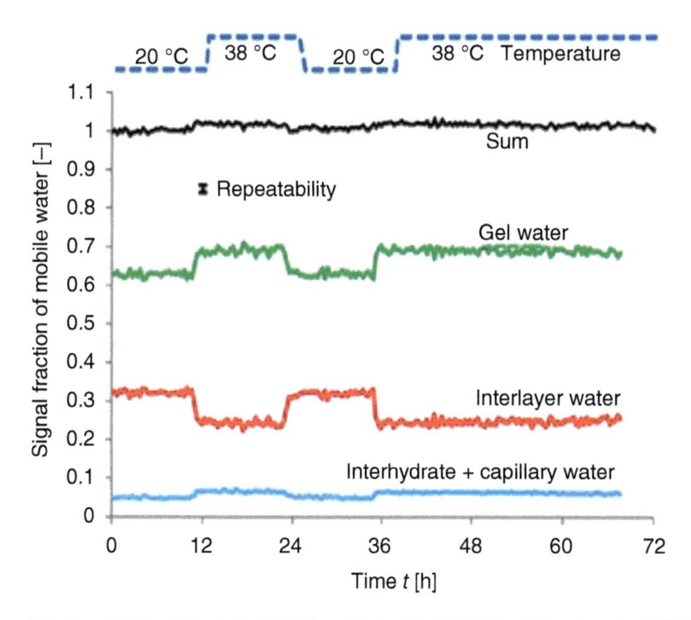

Figure 4.104 Results of NMR analyses on HCP of white cement ($W/C = 0.40$), 28d of water prestorage. The water-saturated test specimens were exposed alternately to temperatures 20 °C and ≈ 40 °C. The determined relaxation times $T2$ were assigned to the pore types "gel water" and "interlayer water." Source: Wyrzykowski et al. [86]/American Chemical Society.

The results are shown in Figure 4.104. From the signal-sum curve, it can be seen that, as expected, during the measurements, the water content remained unchanged to saturation. The measurement signals reacted to the temperature changes with only a slight delay. It is evident that when the temperature is increased from 20 to 38 °C, the $T2$ signal change means an increase in the gel-water content, while a corresponding decrease results for the "interlayer water."

The authors speculate that this result can be interpreted as water transport from the fine "interlayer pores" to the gel pores. However, this is unlikely because there cannot be any appreciable free pore space due to water saturation. Furthermore, the repeated measurements showed complete reversibility. Furthermore, a reversal of the transport from smaller to larger pores would in principle contradict the Kelvin notion.

From this point of view, the explanation for the behavior is that the water fraction adsorbed on the inner pore surface, which was classified as "interlayer water," is partially released by the temperature increase, whereby the "interlayer water" fraction decreases and the fraction classified as gel water increases accordingly. The reverse is true for a temperature decrease. Water transport is not necessary for this purpose.

From these results, it is also possible to derive **conclusions, for example, for the behavior of desorption isotherms at elevated temperature**.

The question of the causes of self-sealing, which is in the foreground here, was specifically investigated by McDonald et al. [87] by considering a larger number of possibly effective causes. A satisfactory explanation could not be found, according to McDonald et al. Conclusions of McDonald et al.: Macroscopic swelling is considered a possible partial cause, microscopic rearrangement of porosity with unchanged overall porosity may play a more significant role.

Fischer et al. [88] **perform NMR analyses on test specimens of concrete** and mortar. These test specimens were previously stored under water for at least three months and alternatively sealed. The composition of the mortar used corresponds to the standard REF mortar used here. The integral echo curves determined in an outer surface layer of 10 [mm] are evaluated. The initial signal (after water storage), the signal after drying of the bodies at 60 °C for one month, and the signal as a function of time after renewed water saturation were measured first.

The evaluations show that after drying of the test specimens and subsequent renewed capillary water uptake, the NMR sum signal indicates a higher capillary water content than in the saturated initial state for up to about two days, but that this saturated initial state or the original ratio of capillary water to gel water is regained after only four days.

Since no inverse Laplace transformation of the integral echo curves could be performed and the short $T2$ relaxation times of adsorbed water were neglected, it can rather be assumed that strongly adsorbed water could not be detected in the pore system during the measurements in the initial state. This volume of adsorbed water, which is temporarily released during drying, is temporarily replaced by detectable water when water is absorbed again. However, it then resumes the sorbed molecular state and thus the original initial state over a period of time. The measurements also reveal that restructuring of the water in the pore system after drying and resaturation is a time-dependent process that can last at least several days.

The authors attribute the process to a temporary decrease in gel porosity and increase in capillary porosity, and the swelling of CSH phases upon reabsorption of water. The probable relationship between swelling in the micro-gel pore region after rewetting and the time dependence of water restructuring will be further discussed below

Rucker-Gramm in [80] first evaluates on the basis of literature results possible causes for the phenomenon of self-sealing (Bingham's flow, reduction of water mobility by formation of ion-solvate shells, osmotic pressure, air entrapment, deposition of soluble substances, hydration, clogging by loose particles, **swelling plus surface forces**). Rucker-Gramm concludes that only the latter effect is likely to be the cause.

In [80], NMR analyses are performed on prisms of a large number of mortars and concretes of different compositions, stepper motor-controlled over the prism lengths to measure the time-dependent water distribution curves within the prisms.

The NMR apparatus used has been described previously.

Figure 4.105a shows the results of control measurements where the NMR signal was characterized as a function of the water content of the samples. It is evident that the proportionality between the signal and the water content is lost below a moisture content corresponding to about 65% RH and below about 50% RH no water content

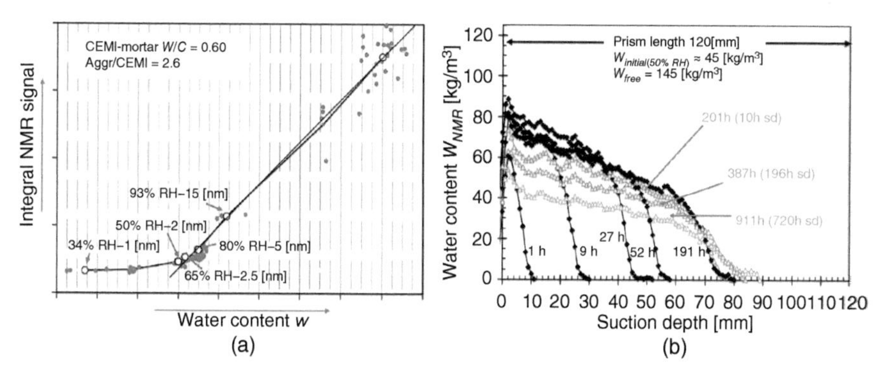

Figure 4.105 Evidence of change in water structure with decreasing pore sizes by NMR analysis. (a): Integral NMR signal as a function of water content *w* due to prestorage in the equilibrium moisture contents 34, 50, 65, 75, 80, 93% RH, respectively, and after water storage (equivalent filled pore radii in nanometer are also indicated); (b): The black curves show the water content distribution during 191h water uptake, the gray curves the following "self distribution"(sd) curves after sealing the water uptake (CEMI-mortar, $W/C = 0.50$, $CEM = 578$ [kg/m^3]). Source: Adapted from Rucker-Gramm [69].

is registered anymore. This result highlights the restructuring of the water within the pore system below the mentioned water contents, resulting from the strong adsorption of the water molecules on the inner surface of the pores. Although this phenomenon occurs on the entire inner surface, due to the surface-to-volume ratio of the micro-capillary pores with radius $\lesssim 2.5$ [nm], these pores contain a high proportion of adsorbed molecules or are completely filled with them.

In Figure 4.105b, the consequences of this relationship are shown. The figure documents the measured distribution curves when water is taken up into a prism on one side over 190h, as well as the water distributions that occur over the course of 700 hours after sealing the absorbent surface. Again, the question arises as to the time delay of the phenomenon.

Figure 4.106 reproduced from Rucker-Gramm [80] shows schematically the derived behavior in the course of a water absorption, respectively, the influence on the "water structure" within the material.

In summary, it can be stated that the previously described results and presentations of Rucker-Gramm [80] are confirmed by the Sections 1.4.6.4 and 1.4.6.5 presented derivations and computational results of [Franke] for water-molecule sorption in the pore region corresponding to $\varphi \leq\approx 65\%$ RH at complete saturation. The diagram obtained in this section is shown again as **Figure 4.107**. Parallels with **Figure 4.106** are striking.

The results of molecular dynamic studies described in Section 1.4.6.5 as well as the results of experimental studies using Atomic-Force Microscopy AFM, described in Section 1.4.7, prove the solid adsorption of water molecules in the nanopores in question and a time influence of the processes, which is also found in [82].

Maruyama et al. [82] find a time-dependent formation of adsorbed water molecules in the micro-gel pore or interlayer pore region in the course of stepwise

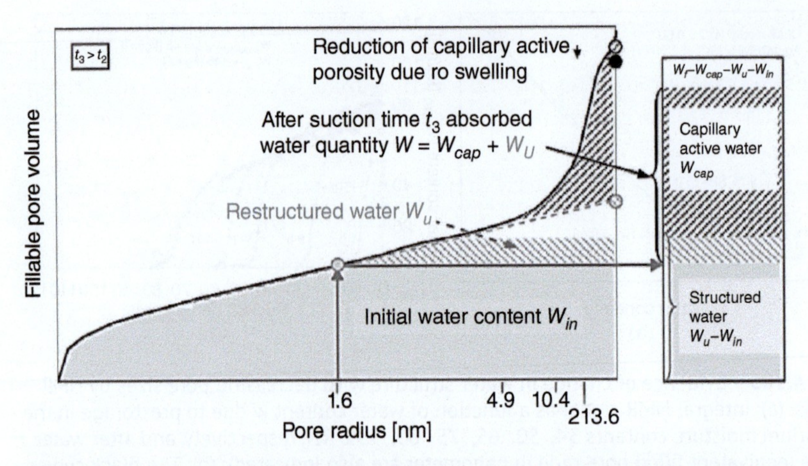

Figure 4.106 Schematic description of water storage in HCP material at an initial water content corresponding to 50% RH **with special reference to the water fraction structured by adsorption**. Source: From Rucker-Gramm [69]/Technische Universität München.

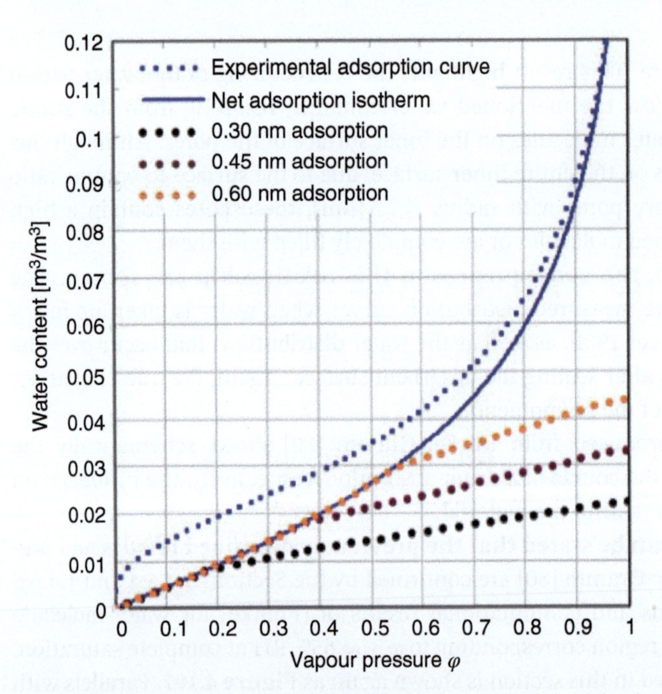

Figure 4.107 Adsorption isotherm of the material REF and net storage function and **volume of adsorbed water molecules at the inner surface in the water-filled pore region** with film thicknesses of 0.30, 0.45, and 0.60 [nm]. The material is considered fully saturated. The RH dependence is used to characterize the pore areas in question. **The net adsorption isotherm** shows the water content of the pore system up to the considered value φ, without the moisture content adsorbed on the inner surface in the non-filled pore area above. **Results from Section 1.4.6.5.**

moisture desorption on the basis of NMR analyses on cement paste, associated with a structural densification.

Still to be answered is the question of the longer time delay of this process: In this regard, reference is made to the results of [89] and, in particular [90], addressed below.

Setzer et al. [89] show by means of new fine measurements that HCP material, in addition to mass-related sorption isotherms, also exhibits strain-related swelling and shrinkage curves between ≈ 0% RH and ≈ 99% RH, respectively, with also a hysteretic course. They also explain the importance of "disjoining pressure" and its physical effect components on the pore structure during drying and rewetting.

Schulte-Holthausen and Raupach [90] perform NMR measurements on the internal swelling of cement-bound material. **They note the following important results:**

"Upon drying, a considerable amount of water bound in C-S-H interlayer spaces is evaporated. Looking closer at the porosities immediately after rewetting, a reduction of fine pore space is seen up to a drying temperature of 40 °C causing the opening of larger cavities. After rewetting of mortars dried at 40 °C, three processes happen simultaneously, though subsequently seen in the measurements. Water reinvades empty pores immediately, a fast reinvasion of beforehand shrunk gel pores causes swelling and reduces the overall coarser porosity and **a much slower reinvasion of consolidated C-S-H interlayer spaces proceeds over several weeks, further reducing gel and capillary pore space.**"

It is thus clear from this section **that self-sealing** in the sense for example of **Figure 4.98a is caused by the delayed incorporation of water molecules in the micro-gel pore region resp. in the interlayer pore region and the restructuring or surface sorption of this water during the internal swelling process.** The pore region up to a filling degree corresponding to 50–65% RH is affected.

The macroscopically expressed time dependence of the water transport behavior is modeled for relevant materials in Section 5.2.2 on the basis of the results of the performed water uptake or water transport experiments in Section 4.7.

References

1 Thommes, M., Kaneko, K., Neimark, A.V. et al. (2015). Physisorption of gases, with special reference to the evaluation of surface area and pore size distribution (IUPAC Technical Report). *Pure and Applied Chemistry* 87: 160.

2 Feldman, R.F. and Sereda, P.J. (1968). A model for hydrated Portland cement paste as deduced from sorption-length change and mechanical properties. *Materiaux et Constructions* 1: 509–520.

3 Ahlgren, L. (1972). Moisture Fixation in Porous Building Materials. *Report 36*. The Lund Inst. of Technology.

4 Mualem, Y. (1974). A conceptual model of hysteresis. *Water Resources Research* 10: 514–520.

5 Funk, M. (2012). Hysteresis der Feuchtespeicherung in porösen Materialien. Dissertation. Technische Universitt Dresden.

6 Espinosa, R.-M. (2005). Sorptionsisothermen von Zementstein und Mörtel. Dissertation. Technische Universität Hamburg.

7 Gätje, B. (2004). Nachträgliche Ermittlung betontechnologischer Parameter in Zementstein, Mörtel und Betonen unbekannter Zusammensetzung. Dissertation. Technische Universität Hamburg.

8 Molendowska, A., Wawrze, J., and Kowalczyk, H. (2020). Development of the measuring techniques for estimating the air void system parameters in concrete using 2D analysis method. *Materials* 13: 482.

9 Liu, Z., Hansen, W., Wang, F., and Zhang, W. (2018). Simulation of air-void system in hardened concrete using a geometrical model. *Magazine of Concrete Research* 71 (13): 680–689.

10 Bollmann, K. (2000). Ettringitbildung in nicht wärmebehandelten Betonen. Dissertation. University Weimar (Germany).

11 Liu, Z. and Hansen, W. (2016). Effect of hydrophobic surface treatment on freeze-thaw durability of concrete. *Cement and Concrete Composites* 69: 49–60.

12 Snoeck, D., Velasco, L.F., Mignon, A. et al. (2014). The influence of different drying techniques on the water sorption properties of cement-based materials. *Cement and Concrete Research* 64: 54–62.

13 Gluth, G.J.G. (2011). Die Porenstruktur von Zementstein und seine Eignung zur Gastrennung. Dissertation. Technische Universität Berlin.

14 Jennings, H.M., Kumar, A., and Gaurav, S. (2015). Quantitative discrimination of the nano-pore-structure of cement paste during drying: new insights from water sorption isotherms. *Cement and Concrete Research* 76: 27–36.

15 Espinosa, R.M. and Franke, L. (2006). Influence of the age and drying process on pore structure and sorption isotherms of hardened cement paste. *Cement and Concrete Research* 36: 1969–1984.

16 Cohan, H.C. (1938). Sorption hysteresis and the vapor pressure of concave surfaces. *Journal of the American Chemical Society* 60: 433–435.

17 Wu, M., Johannesson, B., and Geiker, M. (2014). A study of the water vapor sorption isotherms of hardened cement pastes: possible pore structure changes at low relative humidity and the impact of temperature on isotherms. *Cement and Concrete Research* 56: 97–105.

18 Zandavi, S.H. and Ward, C.A. (2015). Characterization of the pore structure and surface properties of shale using the zeta adsorption isotherm approach. *Energy Fuels* 29: 3004–3010.

19 Zandavi, S.H. (2015). Vapours adsorption on non-porous and porous solids: Zeta adsorption isotherm approach. Dissertation. University of Toronto.

20 Schulte Holthausen, R., Raupach, M., Merkel, M., and Breit, W. (2020). Auslaugungswiderstand von Betonoberflächen in Trinkwasserbehältern. *Bautechnik* 97 (6): 368–376.

21 Schwotzer, M., Scherer, T., and Scherer, A. (2010). Protective or damage promoting effect of calcium carbonate layers on the surface of cement based materials in aqueous environments. *Cement and Concrete Research* 40: 1410–1418.

22 Schwotzer, M. (2008). Zur Wechselwirkung zementgebundener Werkstoffe mit Wässern unterschiedlicher Zusammensetzung am Beispiel von Trinkwasserbehälterbeschichtungen. Dissertation. Karlsruhe: Universität Fridericiana zu Karlsruhe.

23 Eriksson, D., Gasch, T., and Ansell, A. (2018). A hygro-thermo-mechanical multiphase model for long-term water absorption into air-entrained concrete. *Transport in Porous Media* 20: 573.

24 Mercury, L., Azaroual, M., Zeyen, H., and Tardy, Y. (2003). Thermodynamic properties of solutions in metastable systems under negative or positive pressures. *Geochimica et Cosmochimica Acta* 67 (10): 1769–1785.

25 Miller, R.D. (1994). Comment on Paradoxes and realities in unsaturated flow theory. *Water Resources Research* 30 (5): 1623–1624.

26 Schulze, G. and Schluender, E.U. (1985). Absorption von einzelnen Gasblasen in vorbeladenem Wasser. *Chemie Ingenieur Technik* 57: 233–235.

27 Scheffler, G.A. (2008). Validierung hygrothermischer Materialmodellierung unter Berücksichtigung der Hysterese der Feuchtespeicherung. Dissertation. Technische Universität Dresden.

28 Wu, Q., Rougelot, T., Burlion, N., and Bourbon, X. (2015). Representative volume element estimation for desorption isotherm of concrete with sliced samples. *Cement and Concrete Research* 76: 1–9.

29 Brue, F., Liang, Y., Burlion, N. et al. (2010). Influence of temperature and composition upon drying of concretes. In: *Fracture Mechanics of Concrete and Concrete Structures*, 745–750. Seoul: Korea Concrete Institute. ISBN: 978-89-5708-181-5.

30 Chen, W., Liu, J., Brue, F. et al. (2012). Water retention and gas relative permeability of two industrial concretes. *Cement and Concrete Research* 42: 1001–1013.

31 Plagge, R., Scheffler, G., and Nicolai, A. (2007). Experimental methods to derive hygrothermal material functions for numerical simulation tools. *Buildings X Conference*, Clearwater Beach, FL.

32 Plagge, R., Funk, M., Scheffler, G. et al. (2006). Experimentelle Bestimmung der hygrischen Sorptionsisotherme und des Feuchtetransportes unter instationären Bedingungen. *Bauphysik* 28: 81–87.

33 Plagge, R. and Teutsch, A. (2003). *Water Retention Transfer Functions of Ceramic Bricks of the Dresden Building Stock*. Institute of Building Climatology, University of Technology Dresden. Institutsmitteilung. Technische Universität Dresden.

34 Jaynes, D.B. (1984). Comparison of soil-water hysteresis models. *Journal of Hydrology* 75: 287–299.

35 Pham, H.Q., Fredlund, D.G., and Barbour, S.L. (2005). A study of hysteresis models for soil-water characteristic curves. *Canadian Geotechnical Journal* 42: 1548–1568.

36 Espinosa, R.M. and Franke, L. (2006). Inkbottle Pore-Method: prediction of hygroscopic water content in hardened cement paste at variable climatic conditions. *Cement and Concrete Research* 36: 1954–1968.

37 Deckelmann, G. and Schmidt-Döhl, F. (2018). Das Feuchte-Aufnahme- und Abgabeverhalten zementgebundener Estriche und Konsequenzen für die Bestimmung der KRF. TKB-Bericht 4. Köln.

38 Baroghel-Bouny, V. (2007). Water vapour sorption experiments on hardened cementitious materials: Part I: Essential tool for analysis of hygral behaviour and its relation to pore structure. *Cement and Concrete Research* 37: 414–437.

39 Saeidpour, M. and Wadsö, L. (2015). Moisture equilibrium of cement based materials containing slag or silica fume and exposed to repeated sorption cycles. *Cement and Concrete Research* 69: 88–95.

40 Zhang, Z., Thiéry, M., and Baroghel-Bouny, V. (2014). A review and statistical study of existing hysteresis models for cementitious materials. *Cement and Concrete Research* 57: 44–60.

41 de Burgh, J.M., Foster, S.J., and Valipour, H.R. (2016). Prediction of water vapour sorption isotherms and microstructure of hardened Portland cement pastes. *Cement and Concrete Research* 81: 134–150.

42 de Burgh, J.M. and Foster, S.J. (2017). Influence of temperature on water vapour sorption isotherms and kinetics of hardened cement paste and concrete. *Cement and Concrete Research* 92: 37–55.

43 Jiang, Z., Xi, Y., Gu, X. et al. (2019). Modelling of water vapour sorption hysteresis of cement-based materials based on pore size distribution. *Cement and Concrete Research* 115: 8–19.

44 Goedeke, H.K. (2018). Lösender Angriff auf zementgebundene Baustoffe - Veränderungen der Porenstruktur und Folgen für Transport-und Korrosionsprozesse. Dissertation. Hamburg University of Technology.

45 Åhs, M.S. (2008). Sorption scanning curves for hardened cementitious materials. *Construction and Building Materials* 22 (11): 2228–2234.

46 Deckelmann, G. and Schmidt-Döhl, F. (2015). *Überprüfung von marktüblichen Feuchte- und Temperaturmessgeräten für die Beurteilung der Belegreife von Estrichen*. Inst. für Baustoffe, Bauphysik und Bauchemie. Forschungsbericht. Hamburg University of Technology TUHH.

47 Kropp, J. (1983). Karbonatisierung und Tansportvorgänge in Zementstein. Dissertation. Universität Stuttgart.

48 Houst, Y.F. and Wittmann, F.H. (1994). Influence of porosity and water content on the diffusivity of CO_2 and O_2 through hydrated cement paste. *Cement and Concrete Research* 24 (6): 1165–1176.

49 Houst, Y.F. (1992). Diffusion de gas, carbonatation et retrait de la pâte de ciment durcie. Ph.D. thesis. Ecole Polytechnique Fédérale de Lausanne.

50 Auroy, M., Poyet, S., Le Bescop, P. et al. (2015). Impact of carbonation on unsaturated water transport properties of cement-based materials. *Cement and Concrete Research* 74: 44–58.

51 Thiéry, M., Faure, P., Morandeau, A. et al. (2011). Effect of carbonation on the microstructure and moisture properties of cement-based materials. *International Conference XII DBMC*, Porto, Portugal.

52 Ren, F., Chen, X., Zeng, Q., and Zhou, C. (2022). Effects of pure carbonation on pore structure and water permeability of white cement mortars. *Cement* 9: 100040.

53 Duan, C., Karnik, R., Lu, M.-C., and Majumdar, A. (2012). Evaporation-induced cavitation in nanofluidic channels. *Proceedings of the National Academy of Sciences of the United States of America* 109 (10): 3688–3693.

54 Rastogi, M., Müller, A., Haha, M.B., and Scrivener, K.L. (2022). Role of cavitation in drying cementitious materials. *Cement and Concrete Research* 154: 106710.

55 Vincent, O., Sessoms, D.A., Huber, E.J. et al. (2014). Drying by cavitation and poroelastic relaxations in porous media with macroscopic pores connected by nanoscale throats pores connected by nanoscale throats. *Physical Review Letters* 113: 134501

56 Maruyama, I., Rymes, J., Vandamme, M., and Coasne, B. (2018). Cavitation of water in hardened cement paste under short- term desorption measurements. *Materials and Structures* 51: 159.

57 Hundt, J. and Kantelberg, H. (1978). Sorptionsuntersuchungen an Zementstein. *Schriftenreihe DAfStB H* 297: 25–39.

58 Ishida, T., Maekawa, K., and Kishi, T. (2007). Enhanced modeling of moisture equilibrium and transport in cementitious materials under arbitrary temperature and relative humidity history. *Cement and Concrete Research* 37: 565–578.

59 Poyet, S. (2009). Experimental investigation of the effect of temperature on the first desorption isotherm of concrete. *Cement and Concrete Research* 39: 1052–1059.

60 Poyet, S. and Charles, S. (2009). Temperature dependence of the sorption isotherms of cement-based materials: heat of sorption and Clausius-Clapeyron formula. *Cement and Concrete Research* 39: 1060–1067.

61 Poyet, S. (2016). Describing the influence of temperature on water retention using van Genuchten equation. *Cement and Concrete Research* 84: 41–47.

62 Brue, F., Davy, C.A., Skoczylas, F. et al. (2012). Effect of temperature on the water retention properties of two high performance concretes. *Cement and Concrete Research* 42: 384–396.

63 Radjy, F., Sellevold, E.J., and Hansen, K.K. (2003). Isosteric vapour pressure - temperature data for water sorption in hardened cement paste. Report Departement of Civil Engineering MTNU, Trondheim.

64 Schneider, M. and Goss, K.-U. (2011). Temperature dependence of the water retention curve for dry soils. *Water Resources Research* 47: https://doi.org/10.1029/2010WR009687.

65 Drouet, E., Poyet, S., and Torrenti, J.-M. (2015). Temperature influence on water transport in hardened cement pastes. *Cement and Concrete Research* 76: 37–50.

66 van Genuchten, M.Th. (1980). A closed-form equation for predicting the hydraulic conductivity of unsaturated soils. *Soil Science Society of America Journal* 44: 892–898.

67 Kast, W. and Jokisch, F. (1972). Überlegungen zum Verlauf von Sorptionsisothermen und zur Sorptionskinetik an porösen Feststoffen. *Chemie Ingenieur Technik* 44: 556–563.

68 Pickett, G. (1945). Modification of the Brunauer-Emmett-Teller theory of multimolecular adsorption. *Contribution from the Portland Cement Association*, Chicago, Illinois, Vol. 67.

69 Chen, Y., Al-Neshawy, F., and Punkki, J. (2021). Investigation on the effect of entrained air on pore structure in hardened concrete using MIP. *Construction and Building Material* 292: 123441

70 Villani, C., Spragg, R., Pour-Ghaz, M., and Weiss, W.J. (2014). The influence of pore solutions properties on drying in cementitious materials. *Journal of the American Ceramic Society* 97: 386–393.

71 Lothenbach, B., Winnefeld, F., Alder, C. et al. (2007). Effect of temperature on the pore solution, microstructure and hydration products of Portland cement pastes. *Cement and Concrete Research* 37: 483–491.

72 Garrecht, H. (1992). Porenstrukturmodelle für den Feuchtehaushalt von Baustoffen mit und ohne Salzbefrachtung und rechnerische Anwendung auf Mauerwerk. Dissertation. Universität Karlsruhe.

73 Franke, L., Deckelmann, G., and Schmidt, H. (2009). Behavior of ultra high performance concrete with respect to chemical attack. *ibausil Conference*, Weimar.

74 Schmidt, H. (2010). Korrosionsverhalten von Normalmörtel sowie UHPC - Experimente und numerische Simulation. Dissertation. Technische Universität Hamburg.

75 Graef, H. and Grube, H. (1986). Einfluss der Zusammensetzung und der Nachbehandlung de Betons auf seine Gasdurchlässigkeit. *Beton* 36 (11+12): 426-9–473-6.

76 Jacobs, F.P. (1994). Permeabilität und Porengefüge zementgebundener Werkstoffe. Dissertation. ETH Zürich.

77 Hearn, N. and Morley, C.T. (1997). Self-sealing property of concrete - experimental evidence. *Materials and Structures/Matériaux et Constructions* 30: 404–411.

78 Krus, M., Hansen, K.K., and Künzel, H.M. (1997). Porosity and liquid absorption of cement paste. *Materials and Structures/Materiaux et Construction* 30: 394–398.

79 Zimmerman, J.R. and Brittin, W.E. (1957). NMR-Studies in multiple phase systems: lifetime of a water molecule in an adsorbing phase on silica gel. *Journal of Physical Chemistry* 61: 1328–1333. Dallas, TX: Magnolia Petroleum Company, Field Research Laboratory.

80 Rucker-Gramm, P. (2008). Modellierung des Feuchte- und Salztransports unter Berücksichtigung der Selbstabdichtung in zementgebundenen Baustoffen. Dissertation. Technische Universität München.

81 Thiel, C. and Gehlen, C. (2018). Nuclear Magnetic Resonance and Magnetic Resonance Imaging. State-of-the-Art Report of the RILEM Technical Committee 248-MMB.

82 Maruyama, I., Ohkubo, T., Haji, T., and Kurihara, R. (2019). Dynamic microstructural evolution of hardened cement paste during first drying monitored by ^{1}H NMR relaxometry. *Cement and Concrete Research* 122: 107–117.

83 Valori, A., McDonald, P.J., and Scrivener, K.L. (2013). The morphology of C–S–H: lessons from 1H nuclear magnetic resonance relaxometry. *Cement and Concrete Research* 49: 65–81.

84 Muller, A.C.A., Scrivener, K.L., Gajewicz, A.M., and McDonald, P.J. (2013). Densification of C-S-H measured by ^{1}H NMR relaxometry. *Journal of Physical Chemistry C* 117: 403–412.

85 McDonald, P.J., Rodin, V., and Valori, A. (2010). Characterisation of intra- and inter-C–S–H gel pore water in white cement based on an analysis of NMR signal amplitudes as a function of water content an analysis of NMR signal amplitudes as a function of water content. *Cement and Concrete Research* 40: 1656–1663.

86 Wyrzykowski, M., McDonald, P.J., Scrivener, K.L., and Lura, P. (2017). Water redistribution within the microstructure of cementitious materials due to temperature changes studied with ^{1}H NMR. *Journal of Physical Chemistry C* 121: 27950–27962.

87 McDonald, P.J., Istok, O., Janota, M. et al. (2020). Sorption, anomalous water transport and dynamic porosity in cement paste: a spatially localised ^{1}H NMR relaxation study and a proposed mechanism A spatially localised ^{1}H NMR relaxation study and a proposed mechanism. *Cement and Concrete Research* 133: 106045.

88 Fischer, N., Haerdtl, R., and McDonald, P.J. (2015). Observation of the redistribution of nanoscale water filled porosity in cement based materials during wetting. *Cement and Concrete Research* 68: 148–155.

89 Setzer, M.J., Duckheim, C., Liebrecht, A., and Kruschwitz, J. (2006). The solid-liquid gel-system of hardened cement paste. *Conference Paper*. www.researchgate.net/publication/301746898 (accessed 29 December 2023).

90 Schulte-Holthausen, R. and Raupach, M. (2018). Monitoring the internal swelling in cementitious mortars with single-sided ^{1}H nuclear magnetic resonance. *Cement and Concrete Research* 111: 138–146.

5

Modeling of Moisture Transport Taking into Account Sorption Hysteresis and Time-Dependent Material Changes

5.1 Preliminaries

The following chapter deals with the modeling of moisture transport, mainly in cement-bound materials, taking into account and applying the results presented in Chapters 3 and 4 and the theoretical relationships discussed. Also in the following chapter, additional measurement results are included for water absorption and especially for drying behavior under different boundary conditions, which can be better represented on the basis of the simulation program. Among other things, the chapter deals with

- the formulation of the moisture transport coefficients for liquid water as well as for vapor and their dependence on the pore size distribution and on the degree of water saturation in the considered material elements as well as on time (self-sealing)
- the influence of the hysteretic storage behavior on the transport as well as the modeling of the interrelationships
- the drying behavior depending on different boundary conditions, water content, temperature, air velocity, and influence of superficial films from carbonation.

5.2 Modeling of Capillary Transport

5.2.1 Water Content Dependence of the Water Transport Coefficients

With respect to modeling moisture transport in concrete materials, knowledge of **measured, typical, or expected water content distributions** within the material is important.

Figure 5.1a shows such water distributions obtained from NMR analyses on unidirectional capillary water uptake experiments as a function of suction time. The measurements in **Figure 5.1a** were made on mortar prisms with a uniform initial water content corresponding to 50% RH, the measurement results in **Figure 5.1b** up to 72 hours refer to mortar prisms dried at 50 °C. Of importance

Moisture Storage and Transport in Concrete: Experimental Investigations and Computational Modeling, First Edition. Lutz H. Franke.

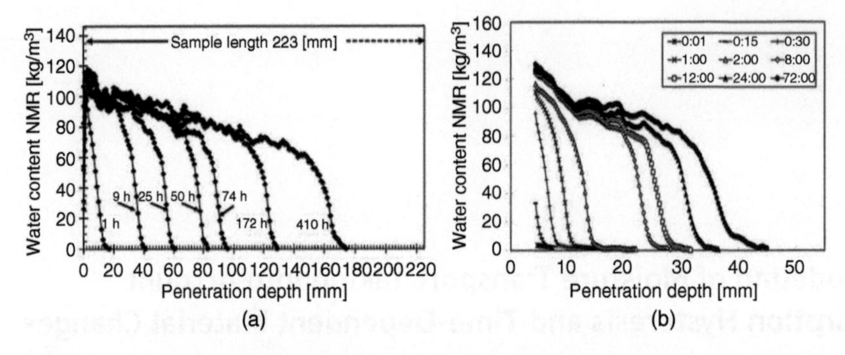

Figure 5.1 Water content distribution and penetration depth curves after one-side water adsorption, measured with NMR on Portland cement mortar samples. **(a)** Mortar of CEMI 42.5, $W/C = 0.60$, $CEM = 516$ [kg/m^3], age > 12 month. Water content at measuring start at approx. 45 [kg/m^3] corresponding to content after constant storage at 50% RH. Free pore volume (including air pores) at adsorption start at approx. 146 [kg/m^3]. Source: Results from Rucker [1]/Technische Universität München. **(b)** Similar curves measured until 72h adsorption on mortar of CEMI 42.5, $W/C = 0.60$, $CEM = 500$ [kg/m^3] after 14d water curing and 4d drying at 50 °C. Source: Adapted from Wittmann et al. [2] and Zhang et al. [3].

for the modeling of the water transport properties of cement-bound products (but also of ceramic bricks and natural stone) is the division of the pore sizes into three groups, which differ significantly in their mean pore size and, in the case of cement-bound materials, in their characteristic properties. The usefulness of this distinction is underlined by numerous recent papers that perform porosity studies based on NMR analyses, compare Section 4.7.2.

A crucial influence in the formulation of transport is played by the water-saturation transitions or boundaries θ_{min}, θ_{gel}, and θ_{sat}, referred to in the explanations of **Figure 5.2a,b**.

The boundaries between pore groups **Figure 5.2a,b**, which are justified from Chapters 2, 3 and Sections 4.6 and 4.7, were defined at 50% and about 92% with respect to the adsorption isotherm. These limits of relative air humidity correspond to pore radii of ca. 1.6 [nm] and ca. 12.5 [nm].

Figure 5.2b contains relevant specifications with regard to the modeling of moisture transport. There, depending on the water saturation, three water content zones are defined, which differ significantly with respect to their water and vapor transport properties. These three zones are limited in terms of modeling by the values θ_{min}, θ_{gel} and θ_{sat}. In addition to the θ_{sat} value, the θ_{sat0} value is given, which occurs as the maximum water saturation in water storage. However, in capillary water uptake experiments, only saturation up to θ_{sat} is observed. The difference largely corresponds to the air-pore content, compare, for example, the Sections 4.3.3, 4.4.1, or 4.5.9.

The basis of the formulation of the capillary diffusion coefficients $D(\theta_{(i)})$ is the transport coefficient *ksat* from Eq. (3.9) in Chapter 3. *ksat* **describes the transport of water at complete saturation** θsat **of the material**. *ksat* can be approximated from pressure plate experiments. For reliable transport calculations, however, *ksat* should be checked via the transport calculation program used, so that

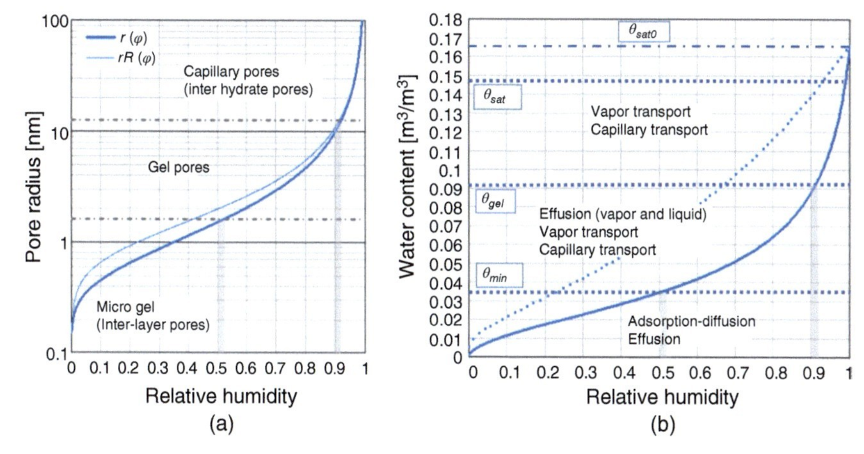

Figure 5.2 Definition of water transport and vapor transport properties as a function of pore size distribution and water saturation degree. **(a)** Subdivision into the effective pore groups at 50% RH and approx. 92% RH related to the adsorption isotherm. Pore radii of ca.1.6 [nm] and ca.12.5 [nm] correspond to these limits of relative air humidity.
(b) Depending on the water saturation, three water content zones are defined, which differ significantly with respect to their water and vapor transport properties. These three zones are bound in terms of modeling by the values θ_{min}, θ_{gel}, and θ_{sat}. In addition to the θ_{sat} value, the θ_{sat0} value is given, which occurs as the maximum water saturation in water storage. However, in capillary water uptake experiments, only saturation up to θ_{sat} is observed. The difference largely corresponds to the air-porosity content.

the coefficient of the root-*t* function of a capillary suction test of the material under consideration must be correctly reproduced with the entered value *ksat*.

An additional coefficient for the consideration of the so-called **tortuosity** is not to be inserted here, since this effect is considered integrally on the one hand by an adjustment to the initial slope of the measured root-t function and on the other hand – starting from this value – by the functional dependence on the respective degree of water saturation.

For the mortar REF investigated here in detail experimentally, the value of *ksat* at 20 °C (after a calculation of acceleration due to gravity) is according to Eq. (3.9).

$$ksat = 3.0 \cdot 10^{-13} [s]$$

For the calculation of moisture transport within the material, **the transport coefficients for partial water saturation of the material must be known.** In the following formulation of the transport coefficients for liquid-water, reference is made to **Figure 5.3**.

A crucial influence in the formulation of the transport is played by the water-saturation transitions or limits θ_{min}, θ_{gel}, and θ_{sat}, which were justified in the explanation of **Figure 5.2a,b**.

For the material REF, the following saturation values were used: $\theta_{min} = 0.035$, $\theta_{gel} = 0.090$, $\theta_{sat} = 0.152$ taken from the adsorption isotherm shown.

The following Eqs. (5.1) to (5.3) refer to the material under the condition that there is no time variation of the transport coefficient *ksat*.

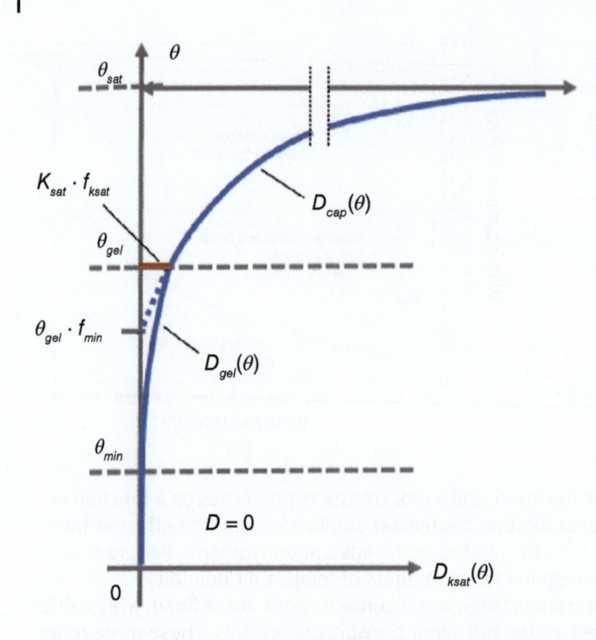

Figure 5.3 Schematic picture of the dependence of the liquid water transport properties on water saturation degree according to Eqs. (5.1) to (5.3).

For a relative water content $\theta_{(i)}$ between θ_{gel} and θ_{sat}:

$$D_{cap}(\theta_{(i)}) = ksat \cdot \left[\frac{\theta_{(i)} - \theta_{gel} \cdot f_{min}}{\theta_{sat} - \theta_{gel} \cdot f_{min}}\right]^{expo1} \quad [s] \tag{5.1}$$

The subscript (i) refers to the location within the material body or the number of the volume element.

For a relative water content $\theta_{(i)}$ between θ_{min} and θ_{gel}:

$$D_{gel}(\theta_{(i)}) = ksat \cdot fksat \cdot \left[\frac{\theta_{(i)} - \theta_{min}}{\theta_{gel} - \theta_{min}}\right]^{expo2} \quad [s] \tag{5.2}$$

Compared to the capillary diffusion coefficient $D_{cap}(\theta)$, the diffusion coefficient $D_{gel}(\theta)$ is much smaller, which should be taken into account via the **factor** *fksat* **as multiplier** of *ksat*. The meaning of *fksat* can be taken from **Figure 5.3** at $\theta_{(i)} = \theta_{gel}$.

For the mortar REF, $fksat = 0.50 \cdot 10^{-3}$ **was used**. This value corresponds to the section of the line drawn in **Figure 5.3** at the level of θ_{gel} where the diffusion coefficients $D_{cap}(\theta)$ and $D_{gel}(\theta)$ or the corresponding *k*-functions meet. For this meeting point to be exact, the function for $D_{cap}(\theta)$ must reach the θ axis at $\theta_{gel} \cdot f_{min}$ and be accounted for accordingly in Eq. (5.1).

For a given value *fksat*, the corresponding value f_{min} is obtained from Eq. (5.3).

$$f_{min} = \frac{1 - fksat^{(1/expo1)} \cdot (\theta_{sat}/\theta_{gel})}{1 - fksat^{(1/expo1)}} \tag{5.3}$$

From **Figure 5.2a,b** emerges, **that at water saturation degrees of less than θ_{min}, capillary water transport can no longer occur**, so that according to **Figure 5.3**

$$D(\theta_{(i)} \leq \theta_{min}) = 0$$

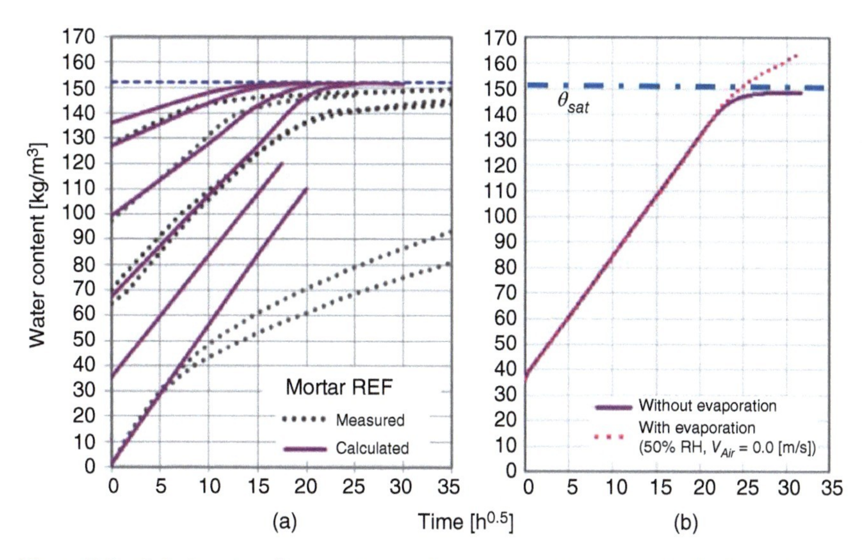

Figure 5.4 Calculated capillary **water uptake curves** (red) into longitudinally sealed prisms of length 150 [mm] from mortar REF compared to measured curves at different initial water contents $\theta oo(i)$. Calculations using the Eqs. (5.1) and (5.2) with time-independent transport coefficients, but with account of internal contribution of vapor transport. Curves were volume normalized to [kg/m^3] by division with the prism length 0.15 [m]. Parameter used: see text.

[In the calculation program, depending on the elementization, the transport parameter $kx_{i-1,i}$ is then used according to Eq. (3.34) or (3.35) in Chapter 3.]

Figure 5.4 now shows a comparison of measurement results for capillary water uptake of mortar REF at different initial water contents and the calculated results based on Eqs. (5.1) to (5.3) for **time-constant transport parameters**.

The measurement results on prisms of length 150 [mm] are taken from Section 4.7. Volume element thicknesses of $\Delta x_i = 3.0$ [mm] were used in the transport calculation. For θ_{min}, θ_{gel}, and θ_{sat}, the previously mentioned values were used based on the adsorption isotherm.

For all calculations on the material REF, **the exponents** $expo1 = 3.0$ **and** $expo2 = 1.3$ (determined empirically by previous comparison calculations) and $ksat = 3.0 \cdot 10^{-13}$ [s] and $fksat = 0.50 \cdot 10^{-3}$ [–] were used. Thus, the capillary transport coefficient for saturation degree corresponding to θ_{gel}, for example, is then $D(\theta_{(i)} = \theta_{gel}) = 1.5 \cdot 10^{-16}$ [s].

Figure 5.4a shows the curves for the capillary water uptake as a function of h$^{0.5}$ in [kg/m^3] according to Section 4.7, whereby the initial water content can be included in the presentation and its influence becomes more visible. It becomes clear that with the mentioned set of parameters, at least the initial slopes up to a suction time of 24 hours are accurately reproduced for all different initial water contents.

From an initial water content of 68 [kg/m^3] (corresponding to about 45% of θsat or about 80% RH of the adsorption isotherm), the calculated curves were completely plotted in **Figure 5.4a**. It can be seen that **with the time-independent**

transport parameters above this water content, the water uptakes are calculated well, especially above θ_{gel}.

Below these initial water contents, the measured curves are much flatter from suction times of about 24 hours, especially for dry initial conditions or initial water contents below θ_{min} or 50% RH.

At least in these cases, it is necessary to perform the calculations according to the modeling in the following Section 5.2.2.

Figure 5.4b shows the water uptake curve for the initial water content of 36 [kg/m^3] complete with special attention to the parallel vapor transport.

All curves reproduced in Figure 5.4a,b also include the transport fraction from vapor. However, for the curves in **Figure 5.4a**, the calculation was done without moisture exchange from vapor at the upper surface, since the influence appears negligible. For verification, the dotted curve, calculated with evaporation at 50% RH and air velocity 0.0 [m/s], is also shown in **Figure 5.4b**. It can be seen that only shortly before reaching the capillary water uptake equilibrium an influence from evaporation becomes visible, caused by the fact that the evaporating fraction at the top must be further supplied at the lower prism side.

5.2.2 Water Content- and Time-Dependence of the Transport Coefficients

A number of publications have addressed the possible causes of the "abnormal" behavior of capillary water uptake, according to **Figure 4.98a**, for example.

Zhang and Angst [4] investigate methods for modeling the retardation of water uptake specifically based on a "dual-permeability model." They find that this model fails to model the effect of drying and rewetting on permeability under different conditions. McDonald et al. [5] also investigate possible causes for the abnormal transport behavior based on NMR analyses. They also discuss three different diffusion coefficients related to varying saturation limits. However, they refer to diffusivity rather than permeability. They find that the diffusion coefficients are time dependent to a certain extent due to "relaxation processes" and slow pore size changes. **Figure 5.5 of Zhang and Angst** [4] is representative of a number of other publications in the model conception for the behavior of the CSH phases in the course of drying and rewetting.

It is obvious that this behavior of the structure results in a time-dependent influence on the permeability, which can be described with the help of the following modeling. The reasons for this behavior were analyzed in more detail, discussed, and elucidated in Section 4.7.2.

Modeling of the behavior assumes the same porosity-dependent water content levels θ_{min}, θ_{gel}, and θ_{sat}, as well as the water content dependence of the transport coefficients with exponents *expo*1 and *expo*2 corresponding to the Eqs. (5.1) to (5.3) in Section 5.2.1.

Figure 5.6 schematically describes the permeability behavior as a function of the saturation degree θ from saturation 0 to θsat. For selected initial saturations

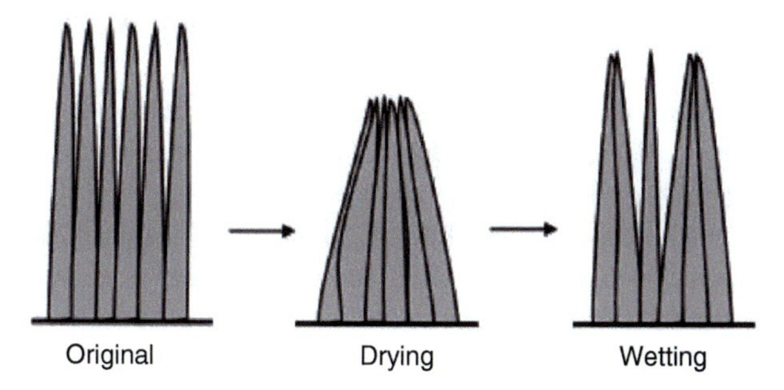

Figure 5.5 Schematic figure about the impact of drying and wetting on the morphology of hydration products. Source: From Zhang and Angst [4]/Springer Nature/CC BY 4.0.

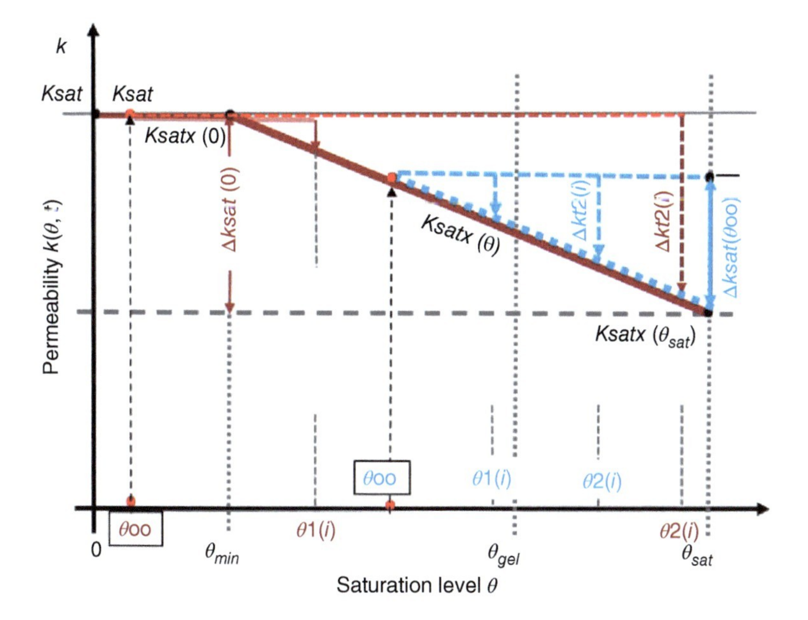

Figure 5.6 Diagram explaining the modeling of the possible time-dependent decreases of the permeability of the mortar REF at the moisture level $\theta(i)$ by Δkt up to a maximum of $\Delta ksat$, as a function of the initial saturation $\theta oo(i)$.

of, for example, θoo (red) or θoo (blue) of a volume element (i), the capillary transport **in the case of non-time-dependent permeability** is symbolized by the upper horizontal red dashed line. It indicates that for the above saturations, the transport coefficients are to be calculated based on the Eqs. (5.1) to (5.3) over *ksat*.

However, **for the case of time-dependent varying permeability**, these transport coefficients of the volume element (i) decrease time-dependently (at unchanged initial saturation θ) down to the solid red line denoted $ksatx(\theta)$ in the diagram, compared to the time-independent case considered before. Inspired by **Figure 5.5**, this state-after-time influence will be **referred to as (complete) conversion** in the following:

The initial saturation level $\theta oo(i)$ can be pictorially imagined as the dark blue water filling in **Figure 3.11** in Section 3.2.

$\Delta ksat(0)$ **is the corresponding maximum possible reduction of the permeability** of a volume element (i) that has been spontaneously fully water-saturated (referred to as a moisture jump) from the dry state $\theta oo(i) \leq \theta_{min}$ and that subsequently reduces its permeability time-dependently maximal by to an amount $\Delta ksat(0)$ to (complete) conversion. [$\Delta ksat(0)$ is visible in **Figure 5.6** at $\theta = \theta_{min}$].

The following time-dependent decrease of the permeability at the moisture level $\theta(i)$ results according to **Figure 5.6**:

$$\Delta kt(\theta(i)) = \Delta ksat(0) \cdot \left[\frac{\theta(i) - \theta_{min}}{\theta_{sat} - \theta_{min}}\right] \cdot fkt(i) \tag{5.4}$$

Determined for the material REF, this maximal possible reduction of permeability $\Delta ksat(0)$ to conversion from the saturation state is approx. 95% of $ksat$ resp.

$$\Delta ksat(0) = ksat \cdot 0.95 \tag{5.5}$$

The permeability (of the volume element (i)) thus decreases time-dependently to about 1/20 of the permeability after spontaneous pore filling from the dry state. This time-dependent process follows the still-to-be defined important time function $fkt(i)$ with the value output 0 to 1.

For initial water saturations $\theta oo(i)$ close to θ_{sat}, there is no more temporal permeability change. According to the results from Section 5.2.1, this is already the case at saturations above $\theta(i) \approx 85\%$ RH.

In Figure 5.6, as an example, the time-dependent permeability decrease in the volume element $VE(i)$ after the jump from the initial water content $\theta oo(i)$(red) to a moisture level $\theta 2(i)$(red) is plotted as $\Delta kt2$(red), occurring at the end of the time influence reaching the conversion (at $fkt(i) = 1$), if the water saturation is maintained at a constant level. The value of $\Delta kt2$(red) is obtained from Eq. (5.4).

A moisture content jump from a given $\theta oo(i)$ is characteristic for near-surface VEs, which spontaneously transform from the dry state to a state close to water saturation in a short time under capillary water uptake.

When modeling an advancing transport process for a non-water-saturated concrete material, it is necessary, in addition to the previously presented background, **to distinguish between the initial saturation $\theta oo(i)$ and the target saturation** $\theta(i)$ in the volume element in question. In order to explain the relationships, we will first assume a spontaneous change (increase) in saturation and refer to this as a moisture jump. This is always significant when $\theta(i) > \theta_{min}$.

With increasing water saturation of the initial state θoo of the volume element toward θ_{sat}, the corresponding permeability of the volume element decreases. The course of the decrease corresponds to the red curve $ksatx(\theta oo)$ approximately modeled in **Figure 5.6.**

When the water saturation is further increased above θoo, the relative decrease in permeability that can still be expected becomes smaller, since part of the maximum possible decrease has already occurred at the initial saturation $\theta oo \geq \theta_{min}$, shown, for example, at $\theta oo(i)$(blue) in **Figure 5.6** resp. the corresponding dashed blue lines:

In Figure 5.6 the time-dependent permeability decrease of the volume element $VE(i)$ after a jump from the initial water content $\theta oo(i)$(blue) to the moisture level $\theta 2(i)$(blue) is labeled as $\Delta kt2$ (blue), the maximal possible reduction is then the maximum $\Delta ksat(\theta oo)$ (blue) when jumping to θ_{sat} according to **Figure 5.6** or the following equations:

$$\Delta ksat(\theta oo(i)) = \Delta ksat(0) \cdot \left[\frac{\theta_{sat} - \theta oo(i)}{\theta_{sat} - \theta_{min}}\right] \tag{5.6}$$

After a moisture jump from $\theta oo(i) \leq \theta_{min}$ to $\theta(i) < \theta sat$, the following time-dependent reduction $\Delta kt(\theta(i))$ of the permeability using the time function $fkt(i)$ at the moisture level $\theta(i)$ results:

$$\Delta kt(\theta(i)) = \Delta ksat(\theta oo(i)) \cdot \left[\frac{\theta(i) - \theta oo(i)}{\theta_{sat} - \theta oo(i)}\right] \cdot fkt(i) \tag{5.7}$$

Using these expositions, **the time-dependent diffusion coefficients $D_{cap}(\theta_{(i)}, t)$ based on the diffusion coefficients and equations from Section 5.2.1** can be formulated using Eq. (5.7) (and Eq. (5.6)) as follows:

For a relative water content $\theta_{(i)}$ between θ_{gel} and θ_{sat} at initial water contents $\theta oo(i) \leq \theta_{(i)}$ given by

$$D_{cap}(\theta_{(i)}, t) = ksat - \Delta kt(\theta(i)) \cdot \left[\frac{\theta_{(i)} - \theta_{gel} \cdot f_{min}}{\theta_{sat} - \theta_{gel} \cdot f_{min}}\right]^{expo1} \quad [s] \tag{5.8}$$

For a relative water content $\theta_{(i)}$ between θ_{min} and θ_{gel} at initial water contents $\theta oo(i) \leq \theta_{(i)}$ by

$$D_{gel}(\theta_{(i)}, t) = ksat - \Delta kt(\theta(i)) \cdot fksat \cdot \left[\frac{\theta_{(i)} - \theta_{min}}{\theta_{gel} - \theta_{min}}\right]^{expo2} \quad [s] \tag{5.9}$$

For $\theta(i)$ values below θ_{min}, capillary water transport no longer occurs, so that

$$D(\theta_{(i)} \leq \theta_{min}) = 0$$

Modeling the time function *fkt*(*i*):

The goal is to define the time function $fkt(i)$ for a total impact period te with a value range between 0 and 1. An exponential function or a special power function would be suitable for this purpose.

As exponential function with $ft1 + ft2 = 1$:

$$f(tm) = \left(1 - ft1 \cdot e^{-\beta 1 \cdot tm} - ft2 \cdot e^{-\beta 2 \cdot tm}\right) \tag{5.10}$$

***tm* is therein the time passed within the time-space 0 to *te*.**

The disadvantage of the exponential function is that it cannot be set exactly to 1 at the end of a time te.

Therefore, in the calculation program, the following easier-to-handle power function was used for $fkt(i)$ in Eq. (5.7):

$$fkt(i, tm) = 1 - \left[1 - \left(\frac{tm}{te}\right)^{c1}\right]^{ct} \tag{5.11}$$

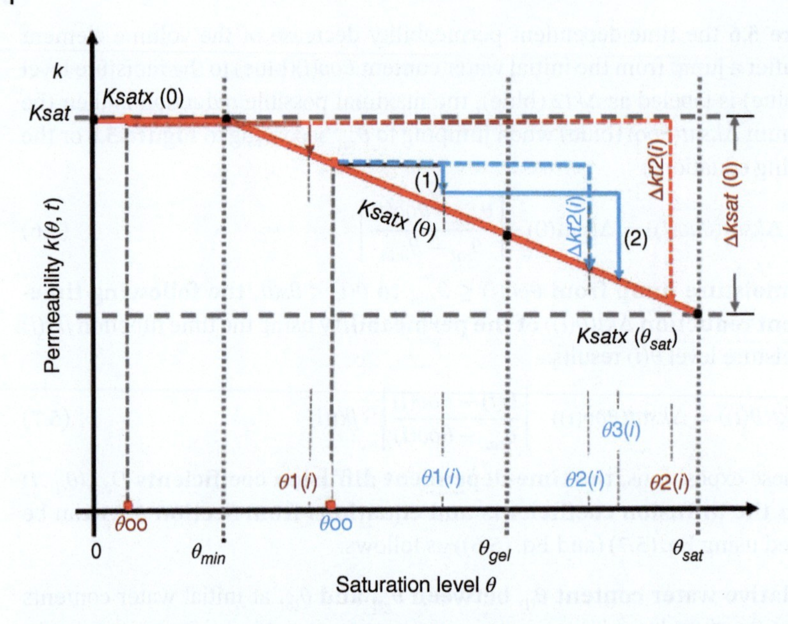

Figure 5.7 Diagram to illustrate the modeling of the possible time-dependent decreases Δkt of the permeability due to steps of saturation up to a maximum of $\Delta ksat$.

With the appropriate constant value $c1 = 0.50$ and with a preselected half-life period th, with *tm* set = *th* and with the half-life value $fkt(tm) = 0.5$ results the exponent $ct = log(0.5) \ / \ log\left(1 - (th \ / \ te)^{c1}\right)$.

The time function *fkt*(*i*, *tm*) (Eq. (5.11)) must be calculated for each *VE*(*i*). The increase of the running time *tm*(*i*) corresponds approximately to the time step Δt specified in the calculation program, but only approximately. Corresponding to Eq. (5.12), **an equivalent action time** ***tm*(*i*) must nevertheless be calculated and inserted into Eq.** (5.11).

The effective time ***tm*(*i*) during water transport in a volume element** *VE*(*i*) must be calculated as **an equivalent action time**. This results (see **Figure 5.7**) when a moisture content jump from θoo (blue in **Figure 5.7**) to $\theta 1(i)$ (blue) occurs, and this saturation level $\theta 1(i)$ is kept constant only for a limited time (1), not reaching the conversion state. If then another moisture jump follows in the sense of a moisture increase (for example, to $\theta 3(i)$ (blue) in **Figure 5.7**) before the time function *fkt*(*i*, *tm*) at $\theta 1(i)$ reaches the value of 1, a new situation to be clarified follows.

In such a case, the effect of the time part (1) (at the saturation level $\theta 1(i)$) on the time still required at $\theta 3(i)$ (part (2)) until conversion is less than its time fraction. In particular, the water content increase does not occur by a single moisture jump but by successive small moisture jumps due to small program time steps of the modeling program.

For the calculation of the change in permeability resp. conversion on a path of changing water saturation, which can only be predicted to a limited extent for the volume elements concerned, an (approximate) method was chosen, which can be described by Eq. (5.12). This is a procedure in which the effect of elapsed time (from start to initial saturation) is weighted over the associated water saturation in relation to the upcoming water saturation after the next time step. In addition, the effect of saturation decreasing due to drying on reaching a conversion state is included.

The equivalent action time $tm(i)$ depends on the time steps $\Delta t(k)$ and the saturation levels is then:

$$tm(i)_{(k)} = tm(i)_{(k-1)} \cdot \frac{\theta(i)_{(k-1)}}{\theta(i)_{(k)}} + \Delta t_{(k)} \tag{5.12}$$

where the index k denotes the current time step and $(k-1)$ the previous time step.

Then the following calculation measures are necessary:

When $tm(i) = te$ is reached, $\theta(i)(te) = \theta oo(i)$new is set.

When water saturation $\theta(i)(k)$ decreases due to drying or capillary removal ($\theta(i)_{(k)} < \theta(i)_{(k-1)}$), $tm(i)$ grows faster toward 1. When falling below $\theta oo(i)$ by this water removal, the corresponding $\theta(i)$ value is set equal to $\theta oo(i)$new, thus starting a new counting of $tm(i)$ from 0.

Figure 5.8 shows the results of capillary water uptake in prisms of length 150 [mm] from mortar REF calculated according to the present modeling in comparison with measured curves at different initial water contents $\theta oo(i)$ as a

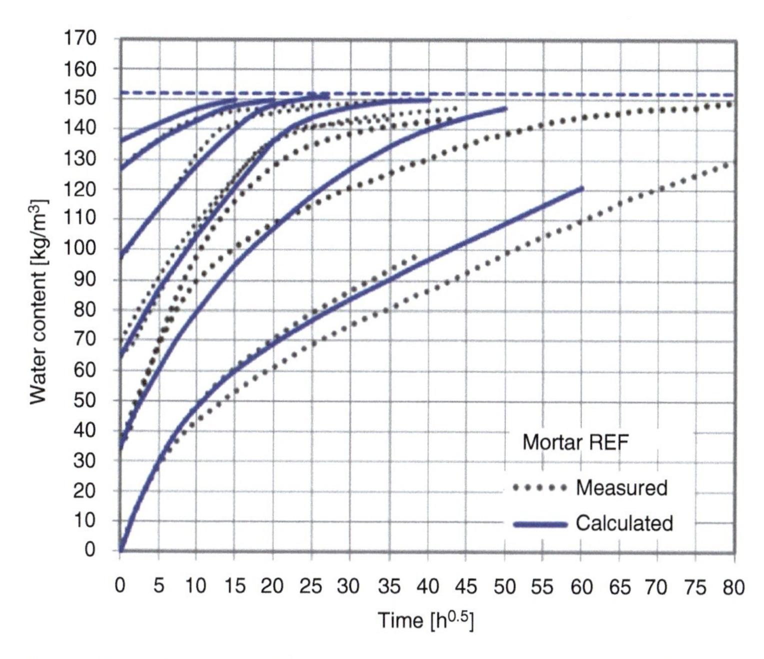

Figure 5.8 Calculated capillary water uptake curves (blue) into longitudinally sealed prisms of length 150 [mm] from mortar REF compared to measured curves at different initial water contents $\theta oo(i)$ as a function of root of time t. For the inclusion of the different initial water contents $\theta oo(i)$ (dimension [kg/m^3]) and the curves of the measured capillary water uptakes (dimension [kg/m^2]) in one diagram, these curves were volume normalized to [kg/m^3] by division with the prism length of 0.15 [m].

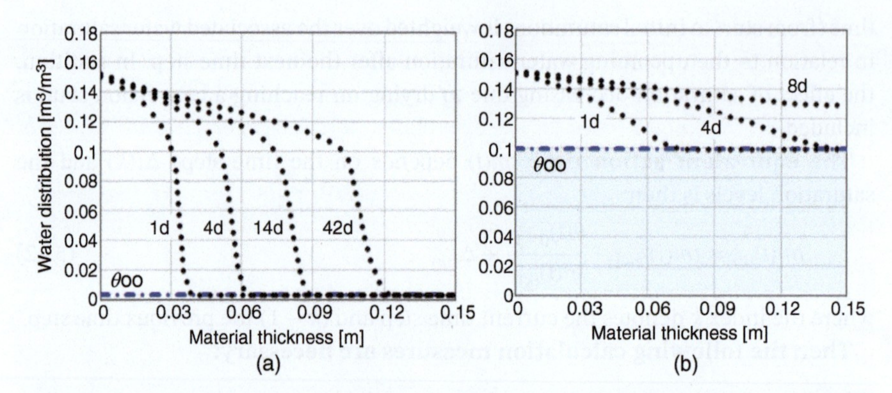

Figure 5.9 Distribution curves of capillary-absorbed water into prisms of length 15 [cm] after different suction times, calculated on the basis of this Section 5.2.2 and corresponding to results in **Figure 5.8**, (with consideration of the fraction of vapor transport). **(a)** At initial water content $\theta oo = 0.003$ [m^3/m^3]. **(b)** At initial water content $\theta oo = 0.099$ [m^3/m^3], [d=days].

function of root t. In addition to the capillary transport corresponding to Eqs. (5.8) and (5.9), the vapor transport fraction was also effective in the specimen (corresponding to the procedure in Section 5.2.1) was considered here. Except for the obvious deviations in the curves with the initial saturation $\theta oo(i) = 0.35$, the comparison yields matching curves from experiment and calculation.

The following parameters were used in the calculations:

The measured initial slope parameter of capillary water uptake in [kg/(m$^2 \cdot$ h$^{0.5}$)] at $\theta oo(i) \approx 0$ was measured at $ww(t_0) = 1.05$ for material REF, compare Section 4.7. Using the basic transport parameters for the Eqs. (5.1) and (5.2) already mentioned in Section 5.2.1, the simulation program provides the measured initial slopes, compare **Figure 5.4**. These transport parameters must then also be used for the simulation of the curves of **Figure 5.8**.

The additional necessary parameters for the time function have been mentioned earlier in this section.

Figure 5.9 shows the distribution curves of capillary absorbed water into prisms of length 15 [cm] after different suction time, calculated on the basis of this Section 5.2.2 and corresponding to results in **Figure 5.8**.

5.2.3 Computational Modeling of Scanning Isotherms and Hysteretic Moisture Transport

The moisture transport and moisture state calculations are basically performed according to the formalisms given in Chapter 3.

To account for hysteretic storage behavior, the balance Eqs. (3.29) and (3.30) are also used. The water-storage functions effective in the individual VEs in the considered time step Δt or their mathematical derivatives result from the "positions" in the θ–φ-diagram or the conversions made from it to the corresponding θ-p_{cap}-dependence. (In Eqs. (3.29) and (3.30), p_{cap} is denoted by p_k).

The water storage functions effective in each VE at the time under consideration may be the desorption isotherm, the adsorption isotherm, or connecting scannings, the exact location in the θ–φ diagram is denoted as "position" for each VE. In addition to the θ–φ value, the instantaneous isotherm or scanning isotherm must also be noted when the "position" is stored.

IAt the following time step Δt a new moisture distribution has been calculated in the test specimen compared to the previous time step, whereby new "positions" are valid for the VEs.

At the end of each time step, check in each VE whether a reversal of the direction of water filling has taken place. For positions that are on the desorption isotherm or adsorption isotherm, in case of a water filling reversal, a position recalculation must take place in a repetition of the time step in the whole material body, where for the respective VEs, the scanning isotherm effective in this position must be calculated and stepped on with a resulting new position.

For each VE with a position on a scanning isotherm, it must be checked after each time step whether a transfer to the desorption resp. adsorption isotherm must be calculated and taken into account in the time step repetition.

The behavior of the scanning isotherms was studied and described in more detail in Sections 4.6.4 and 4.6.5. Approaches for their calculation were given, starting from the desorption and adsorption isotherms of the respective material.

Within the simulation program, a simplified modeling of the scanning isotherms was performed to reduce the computational effort.

A good approximation is obtained when a "master" scanning isotherm is assumed, usually the linear connection between the points $\theta_{Des}(\varphi = 50\%)$ and $\theta_{Ads}(\varphi = 90\%)$, **compare Figure 5.10**. Starting from this scanning isotherm, an intersection point on the horizontal is defined at $\theta = 0$, which is used as the zero point of fan-like outgoing straight lines from which the further scannings are

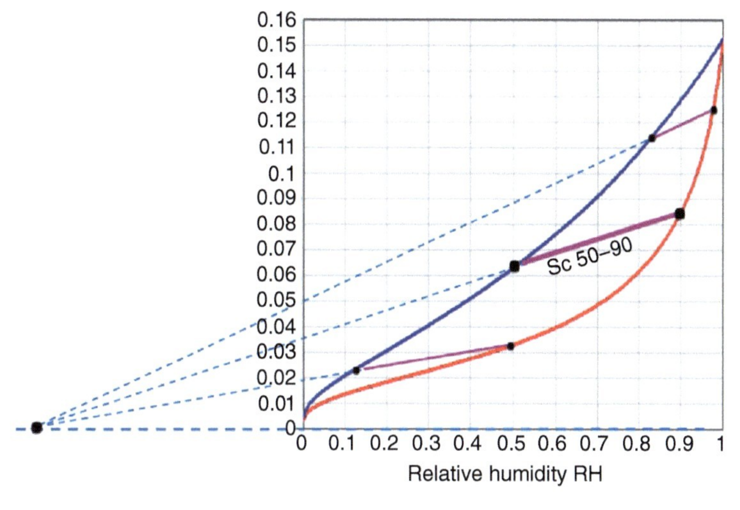

Figure 5.10 Modeling the slope of the scanning isotherms for the mortar REF as slope fan mastered by the scanning Sc 50–90.

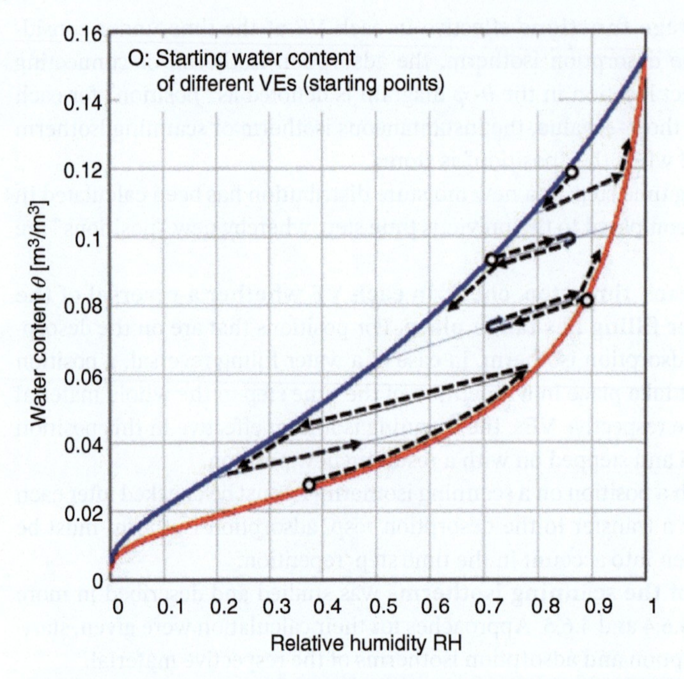

Figure 5.11 Illustration of the possible scanning behavior of differently water-filled volume elements VEs, performed by the present simulation program. Depending on the boundary conditions and the resulting internal moisture states during calculation, the outlined paths can also run simultaneously, provided that the spatial moisture balance requirements are met.

derived. This results in systematically increasing slopes of the scanning isotherms for increasing air humidity intercepts. This modeling is applicable for materials with linear or rather concave desorption isotherms.

Figure 5.11 now shows (as an example for volume elements with different initial water filling) which adsorption and desorption pathways in the θ–φ diagram can result for volume elements due to different moisture content changes or vapor pressure changes. **These different pathways can also occur simultaneously at different locations in a test specimen,** depending on the boundary conditions and moisture content state provided that the spatial equilibrium conditions are satisfied.

The test presented in Section 4.6.7 for the alternative determination of the scanning slope was recalculated by the simulation program for further control, in particular of the modeling of the scanning isotherms. A prismatic mortar body was used as a basis.

In this experiment, two mortar portions (with the same dry mass) of the same mortar (screed-concrete) with different initial water contents were spatially separated but placed together in a closed container. In the container, a redistribution of the water content took place via vapor transport by desorption from the wetter mortar portion and corresponding adsorption in the drier mortar portion.

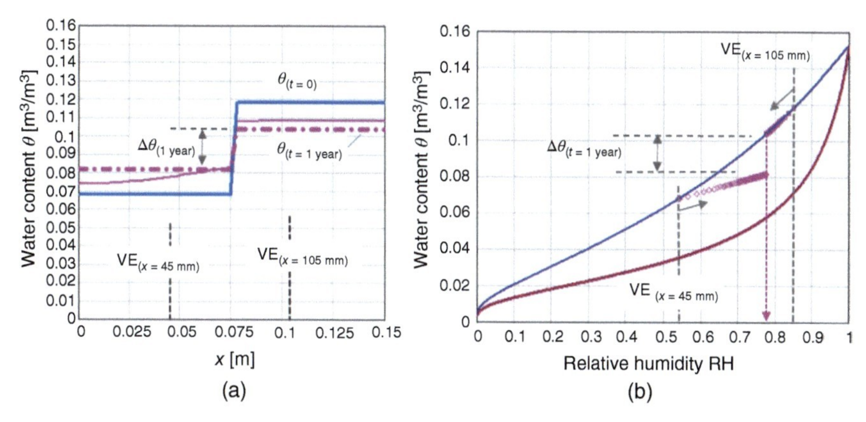

Figure 5.12 Control calculation for reverse sorption experiments on the course of scanning isotherms (compare Section 4.6.7) by means of a mortar prism. (a) Division of the initial water content $\theta_{t=0}$ at calculation start in a mortar prism $L = 0.15$ [m] (with sealed outer sides) into two zones with different water contents corresponding to 54% RH and 85% RH. Subsequent calculation of the water redistribution in the prism. Calculated distribution curves after 10 days and after one year are shown (pink curves), no further subsequent change. (b) Output of the time-dependent moisture content development in the VE ($x = 45$ [mm]) and in the VE ($x = 105$ [mm]) from $t = 0$ to $t = 1$ year by lines of pink-colored diamonds, which show no more progress already before one year. Further, indication of the remaining moisture difference $\Delta\theta_{(t=1year)}$, which occurs after long (redistribution) time.

In the mortar prism used as a basis for the calculated comparison, the initial water content was set to different levels in the longitudinal direction of the prism in the halves.

Figure 5.12a shows the situation. In the two zones in the mortar prism, initial water contents of 54% RH and 85% RH were taken as a basis, **corresponding to the initial water contents for the experiment described in Section 4.6.7.**

The moisture redistribution occurring in the prism was now calculated up to $t = 8800$ hours. From 6600 hours only a negligible change occurred. In accordance with the experimental configuration, all sides of the prism were assumed to be sealed in the calculation.

Figure 5.12a makes it clear that only a part of the different initial water levels is redistributed, and in the computational equilibrium, a difference between the levels of $\Delta\theta$ remains even after 8800 hours.

Figure 5.12b shows the behavior in the $\theta - \varphi$ diagram. Plotted therein are the scanning traces output from the initial water contents of 54% RH and 85% RH on the desorption isotherm by the computational program for two inner volume elements $VE_{(x=45\ [\mathrm{mm}])}$ and $VE_{(x=105\ [\mathrm{mm}])}$, respectively, and the corresponding water contents of these elements, which show no change from about 77.5% RH and are computationally in equilibrium. The remaining moisture difference $\Delta\theta$, which can be read off in the figure, corresponds to the measured value.

It is noteworthy that the comparative test and the simulation calculation lead to the same result, although different cement-bound mortar products with different (but similar) sorption isotherms were considered.

These results show that the scanning modeling method described above obviously leads to accurate predictions. This also proves that **the used method of modeling capillary and vapor transport, described in more detail in Chapter 3, gives reliable results**, even in cases where surface vapor exchange is the main concern.

5.2.4 Examples of the Hysteresis Influence on Water Content Distributions

Examples of water content distributions are discussed below, which were determined based on the computational approaches of Chapter 3 and the computational model described in Section 5.2.3 for scanning behavior using the simulation program.

The transport parameters and moisture storage functions for the REF material were used, taking into account the moisture influence on the parameters and the time influence on the permeability according to Sections 5.2.1 and 5.2.2.

The shown examples of moisture content distributions were calculated for laterally sealed prismatic material bodies of length 150 [mm], divided longitudinally into 45 volume elements VE. The material bodies were each computationally subjected to different successive boundary conditions, using a system temperature of 20 °C.

The examples are particularly intended to illustrate the influence of hysteretic water storage behavior. For the examples with a constant initial water distribution of $\theta = 0.100$ and $\theta = 0.148$ [m^3/m^3], the influence of time on the permeability values has little or no significance anymore.

Figure 5.13a shows the resulting water distributions for the mortar body with an initial capillary water saturation of $\theta = 0.149$ [m^3/m^3]. Let the body then be dried over both faces. Here, the desorption isotherm is decisive. After one year of drying at 50% RH at an air velocity of $v = 0.50$ [m/s], the calculated result is a slightly curved water content distribution equal to about 50% of the initial water content. A unilateral capillary water uptake at $x = 0$ for 24 hours was then performed. The end faces were then also sealed, and the moisture redistribution was calculated. **Figure 5.13a** shows a largely stationary distribution curve after 2000 hours, indicating that even in the long term, there is no uniform water content within the test specimen due to redistribution.

The hysteretic storage behavior responsible for this is illustrated in Figure 5.13b using the moisture content in the selected volume element VE at $x = 45$ [mm], or its "moisture trace" in the θ–φ diagram.

Figure 5.14a gives (using the same boundary conditions as before for **Figure 5.13a**) the moisture distributions when **the hysteretic storage behavior is not considered** and only the normally present adsorption isotherm is used in the calculations.

The resulting significantly different moisture distribution after one year of drying is remarkable. After a further one-sided capillary water supply over 24 hours, the moisture redistribution was then calculated. Above $x = 0.06$ [m], the redistribution runs only slowly in the direction of constant water distribution over the body length.

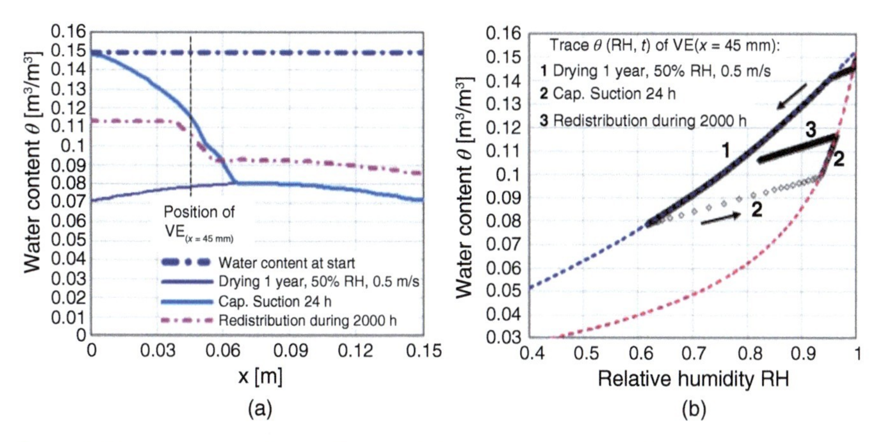

Figure 5.13 Illustration of the influence of the **hysteretic storage behavior** of the material on the water content distributions within a prismatic test specimen. **(a)** The figure shows the distributions after expiration of exemplarily chosen moisture boundary conditions (initial condition water saturation, 12 months of drying over both faces at 50% RH/ v_{Air} = 0.5 [m/s], one-sided capillary water absorption over 24 hours, and then moisture redistribution over 2000 hours with sealed surfaces). In particular, the moisture equilibrium after 2000hours of internal redistribution (pink-colored curve) is affected (calculation based on the measured adsorption and desorption isotherms and the corresponding scanning isotherms, compare **Figure 5.13b**). **(b)** Course of the corresponding water content or scanning behavior in the volume element VE selected as an example at a material depth of 45 [mm] (from the left test specimen face). The numbers refer to the used boundary conditions.

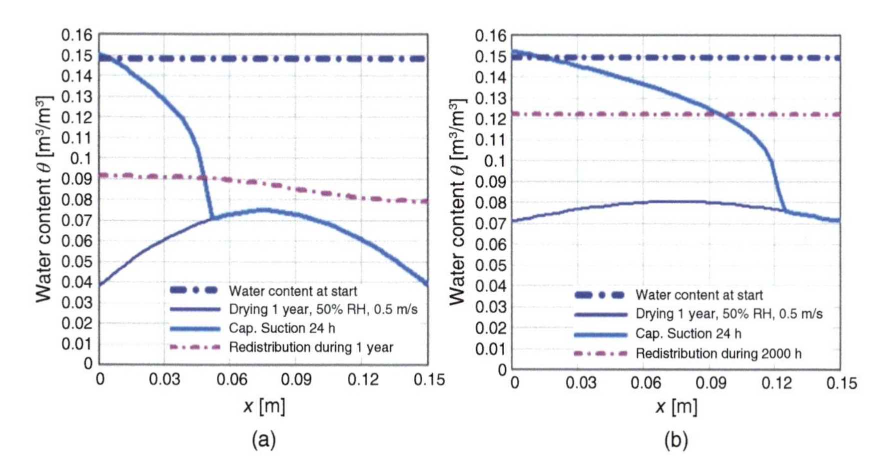

Figure 5.14 Resulting water distributions in the test body of Figure 5.13a, if the hysteretic behavior is neglected. (a) Only the adsorption isotherm of the material is used for the transport calculations. **(b)** Only the desorption isotherm is used.

Even after one year of calculation, computationally, a very slow adjustment still occurs as a result of internal vapor transport.

Figure 5.14b demonstrates how moisture distributions are obtained (under the same boundary conditions as in **Figure 5.13a**) if correctly the desorption isotherm (instead of the adsorption isotherm) is used in the calculation of drying, and also how the following redistribution is determined on the basis of the desorption isotherm without a scanning mechanism. The one-sided capillary water uptake over 24 hours leads computationally to significantly higher water uptake, and the redistribution would already be fully completed after 2000 hours.

Figure 5.15a,b show moisture distributions for material bodies of length $x = 0.15\,[\text{m}]$ starting from an initial moisture of $\theta oo = 0.100\,[\text{m}^3/\text{m}^3]$. It is **assumed that this initial moisture was due to water uptake after a lower initial moisture content**. The water content "positions" of the VEs are then on the adsorption isotherm. After one-sided capillary water uptake for 30 hours, there is then a redistribution within the material bodies according to after sealing the end faces. In the example of **Figure 5.15a** no hysteretic behavior was calculated neglecting the scanning isotherms, resulting in a complete redistribution after 300 hours at the latest. In contrast, **Figure 5.15b** demonstrates that when the scanning behavior is taken into account, even after 600 hours, there is only a slight redistribution in the direction of constant saturation over the material-body length.

In Figure 5.16a,b, it is illustrated which water distributions are obtained in comparison to the results in **Figure 5.15a,b** at unchanged boundary conditions when the moisture transport is calculated on the basis of the desorption isotherm.

The initial water saturation of the material body, amounting to about $\theta oo = 0.100$ $[\text{m}^3/\text{m}^3]$, was determined **for this comparison directly by drying the body** from an initial capillary saturation, (rather than being given as a straight line). For this initial drying process, the physically correct desorption isotherm must be used (if one is known for the material).

In **Figure 2.16a**, the calculation was then continued (without scanning) on the basis of the desorption isotherm, while in **Figure 2.16b**, in the respective VEs in which a water content change in the sense of a directional reversal of saturation takes place as a result of the moisture redistribution, the calculation is switched (via the scannings) to the adsorption isotherm if necessary.

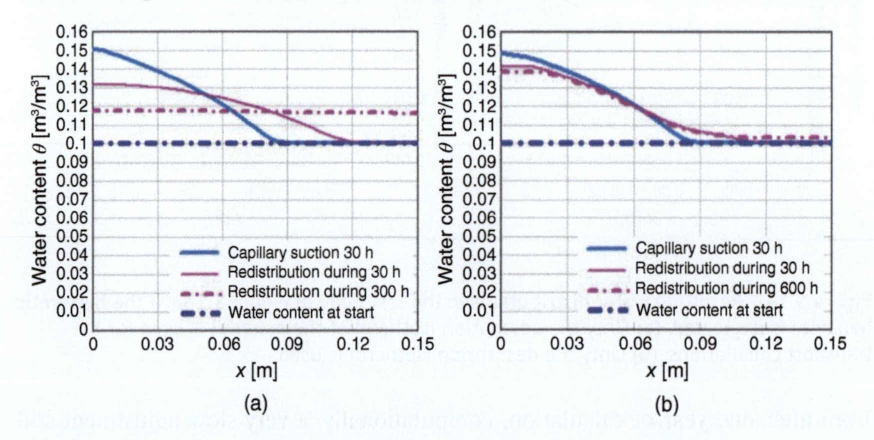

Figure 5.15 Influence of the **hysteretic storage behavior** within the test body of Figure 5.13a at an initial water content of $\theta oo = 0.100$ $[\text{m}^3/\text{m}^3]$, capillary water uptake over 30 hours at $x = 0$ [m], and subsequent redistribution (pink-colored curves). **Adsorption-Isotherm as the starting isotherm. (a)** Only based on adsorption isotherm **without scanning** calculations. **(b)** Consideration of the **hysteretic storage** behavior.

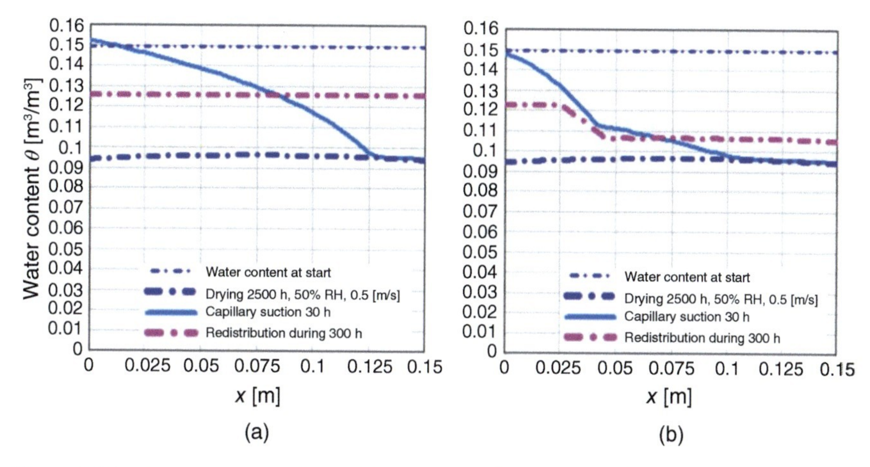

Figure 5.16 Influence of the **hysteretic storage behavior** within the test body of **Figure 5.13a** at an initial water content of $\theta oo = 0.100$ [m^3/m^3], capillary water uptake over 30 hours at $x = 0$ [m], and subsequent redistribution (pink-colored curves). **Desorption-isotherm as the starting isotherm. (a) Without scanning calculations based on desorption isotherm only. (b)** Consideration of the **hysteretic storage behavior with scanning calculations** (starting correctly from the desorption isotherm).

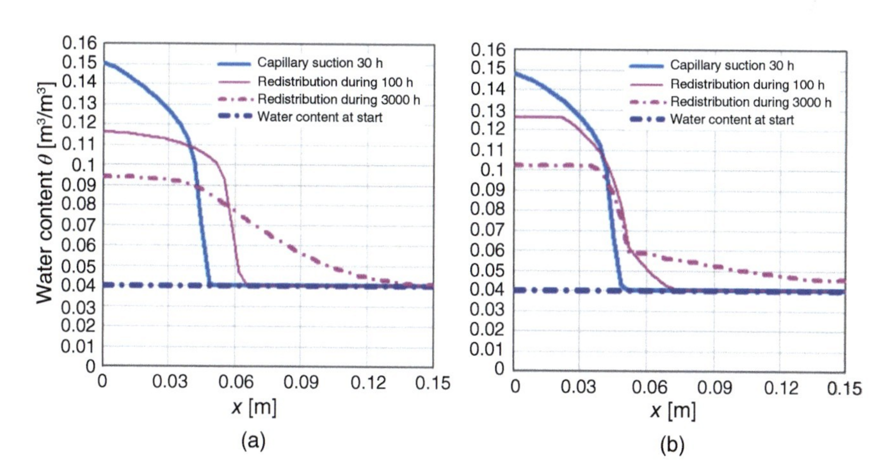

Figure 5.17 Influence of hysteretic storage behavior within the test body corresponding to **Figure 5.13a** at an initial water content of $\theta oo = 0.04$ [m^3/m^3], capillary water uptake over 30 hours at $x = 0$ [m], and subsequent redistribution over 100 hours and 3000 hours (pink-colored curves). **(a) Based on adsorption isotherm only without scanning calculations. (b) With consideration of the hysteretic storage behavior (starting correctly from the desorption isotherm).**

Figure 5.17a,b additionally shows the redistribution behavior at a low initial water content of $\theta oo = 0.040$ [m^3/m^3].

In **Figure 5.17a**, the redistribution was calculated without scanning, assuming only the adsorption isotherm for the material. Even after 3000 hours, no complete redistribution has occurred in this example. If the desorption isotherm were taken as a basis, a linear water distribution in the prismatic body would already have been completed after 700 hours.

Figure 5.17b, on the other hand, shows the behavior of the same body under the same boundary conditions if correctly started from the desorption isotherm and calculated with scanning mechanism.

The calculation results and remarks in the present Section 5.2.4 should make it clear that predictions of water content distributions within material bodies for materials with pronounced hysteretic moisture storage behavior can only be reliable if this behavior is taken into account in modeling.

If only **one** sorption isotherm (usually the adsorption isotherm) is used, serious misestimations of the distributions can result.

Furthermore, it becomes clear that in the computational consideration of the hysteretic storage behavior, the isotherm to be specified at the start of the calculation should correspond as far as possible to **the origin** of the initial moisture distribution, i.e. in the case of an initial condition created by drying, the desorption isotherm.

5.3 Modeling of Vapor Transport and Drying by Evaporation of Concrete

5.3.1 Definition and Measurement of Vapor Diffusion Resistance, Influence of Experimental Boundary Conditions on Nominal Measurand

In the following, first the known but subsequently needed relations are given for the definition of the transport parameters for water vapor through porous material layers. Then, the own measurement results for mortar and concrete as well as some measurement results from the literature are presented.

In the steady-state diffusion condition, the following applies to the water vapor flux density through a homogeneous material layer:

$$\dot{m} = \frac{\delta_{Mat}}{d} \cdot \Delta p \ \ [\text{kg}/(\text{m}^2 \cdot \text{h})] \tag{5.13}$$

where δ_{Mat} is the diffusion coefficient of the material for water vapor in [kg/(h m Pa)], d the material layer thickness in [m], and $\Delta p = p_i - p_a$ **the difference in vapor pressures in Pa.**

In practice, instead of the diffusion coefficient δ_{Mat} for the particular material, it has become common to use the **diffusion coefficient for vapor in air** δ_{Air} and instead of the real thickness d, **the equivalent air layer thickness** d_{Air} of the material layer in [m], i.e:

$$\frac{\delta_{Mat}}{d} = \frac{\delta_{Air}}{d_{Air}} \Rightarrow \ d_{Air} = d \cdot \frac{\delta_{Air}}{\delta_{Mat}} = d \cdot \mu \ \ [\text{m}] \tag{5.14}$$

where μ **is the water vapor diffusion resistance number of the material.** μ indicates how many times the diffusion resistance of a layer is greater than that of a layer of air of the same thickness d.

$$\dot{m} = \frac{\delta_{Air}}{d \cdot \mu} \cdot \Delta p \ [\text{kg}/(\text{m}^2 \cdot \text{h})] \tag{5.15}$$

The vapor diffusion coefficient in air δ_{Air} depends slightly on the air pressure p_{Air} and the absolute temperature T:

$$\delta_{air} = 0.083 \cdot \frac{p_{air0}}{p_{air}} \cdot \left(\frac{T}{273}\right)^{1.81} \cdot \frac{1}{R_D \cdot T} \approx 7.0 \cdot 10^{-7} \ [\text{kg}/(\text{h m Pa})] \text{ at } 20\,°\text{C} \tag{5.16}$$

With $p_{Air} \approx p_{Air0} = 1013$ [hPa] and $R_D = 462$ [Nm/kg K] it follows, for example.

$$\delta_{Air,5\,°\text{C}} = \frac{1}{1.5 \cdot 10^6}, \ \delta_{Air,20\,°\text{C}} = \frac{1}{1.43 \cdot 10^6} \ [\text{kg}/(\text{h m Pa})]$$

The steady-state vapor diffusion flow **measured in the experiment** on samples of thickness d under the set boundary conditions of rel. humidity φ and temperature T is

$$\dot{m}_{ges} = \frac{\Delta m}{\Delta t} \ [\text{kg}/(\text{m}^2 \text{ h})] \tag{5.17}$$

With $\Delta p = \Delta\varphi \cdot p_{sat(T)}$ follows by default from the experiment:

$$\mu = \frac{\delta_{air} \cdot \Delta p}{\dot{m}_{ges} \cdot d} \ [-] \tag{5.18}$$

The diffusion coefficient of the material for vapor as well as the resulting vapor flow over a material length Δx is then (D_V corresponds to δ_{Mat} in Eq. (5.13)):

$$D_V = \frac{\delta_{Air}}{\mu} \ [\text{kg}/(\text{m h Pa})], \qquad \dot{m} = D_V \cdot \frac{\Delta p}{\Delta x} \ [\text{kg}/(\text{m}^2 \cdot \text{h})] \tag{5.19}$$

In the program modeling of the evaporation or drying of the material, the water vapor transfer coefficient β must be taken into account for the surface elements, compare Section 5.3. To capture the influence of air velocity on evaporation, the heat transfer coefficient α_k [W/(m²· K)] can be used.

$$\beta \approx 7 \cdot 10^{-9} \cdot \alpha_k \ [\text{kg}/(\text{m}^2 \cdot \text{s} \cdot \text{Pa})] \tag{5.20}$$

The influence of the vapor transfer coefficient is discussed in Sections 5.3.4 and 5.3.6, which deal with additional influences and modeling.

Vapor diffusion measurements are carried out in the laboratory according to DIN EN ISO 12572. One side of the material discs is exposed to a lower relative humidity, and the other side to the upper relative humidity at a specified ambient temperature. Since in the case of porous materials the transported amount of moisture in the diffusion test increases with the mean value of the applied relative humidity or the diffusion resistance number μ decreases, it has become common practice to carry out so-called **Dry-Cup experiments**, where the upper humidity is approx. 50% and the lower humidity is, for example, 3% RH. In addition, so-called **Wet-Cup tests** can be performed, where the lower relative humidity on one test specimen side is 50% and the upper relative humidity on the second test specimen side is 93%.

In the Wet-Cup diffusion experiment, lower μ values are always obtained compared to the Dry-Cup experiment. In all diffusion measurements on cement-bound products, the diffusion resistance increases with decreasing porosity. Furthermore, there is a strong dependence on the upper value of the selected humidity boundary conditions. This is due to the well-known fact that, with increasing relative humidity, the amount of water transported through the material layer increasingly results from capillary transport in addition to vapor transport.

The DIN EN ISO 12572 standard prescribes preconditioning of the disc-shaped specimens, either by prestorage at 23 °C/50% or alternatively, in the case of specimens wet in the initial state, by pre-drying the specimens at 105 °C or 40 °C and subsequent storage at 23 °C/50% RH. It is shown below that **these different prestorage conditions systematically lead to different μ values when cement-bound material is tested.**

For this purpose, extensive series of REF mortar were investigated in diffusion tests according to DIN EN ISO 12572 with dry and wet boundary conditions, respectively, in order to check an **influence of preconditioning**. DIN EN ISO 12572 stipulates as preconditioning either storage at 23 °C/50% or also, alternatively, for specimens wet in the initial state, pre-drying of the test specimens at 105°C or 40°C and subsequent storage at 50% RH/23 °C. Prestorage only at 50% RH /23°C will be referred to below as **prestorage A**, and prestorage with pre-drying as **prestorage B**. The diffusion tests with the REF mortar were carried out on 7 [mm] discs, which were water-stored for 56 days until the start of prestorage. Composition of the REF material, see Section 4.2.

Prestorage A consisted of 21 days at 23 °C/50%, and prestorage B of 7 days of drying at 40°C/10% RH followed by 14 days storage at 23 °C/50% RH, each in a climate-controlled chamber.

In the Dry-Cup experiments, the relative air humidity on one side of the sample was always 11%, on the other side (upper air humidity) of one series 44%, on another series 54%.

In the Wet-Cup investigations, the relative air humidity on one sample side was always 54%, while on the other sample side of the 4 series of experiments, the upper air humidity was set to 83%, 93%, 97%, and 99% for comparison, respectively.

The results of these tests are shown in Figure 5.18a,b.

In the Dry-Cup experiment (lower RH = 11%), the highest μ value results at the upper test RH of 44%, as expected. The decrease in μ value at the standard humidity setting of 50% (54% RH in the experiment) is approximately 15% relative to the μ value at 44%.

It is noteworthy that **the same μ-value is obtained for the material REF in the Dry-Cup experiment regardless of the storage type A or B.**

However, this is completely different **in the Wet-Cup experiment. There is a systematic difference in the μ-values as a result of the different preconditioning of the samples by storage A or storage B.** According to this, the μ value measured when the samples are preconditioned by storage B (pre-storage with pre-drying) is about 50% higher than the measured value for storage A (pre-storage

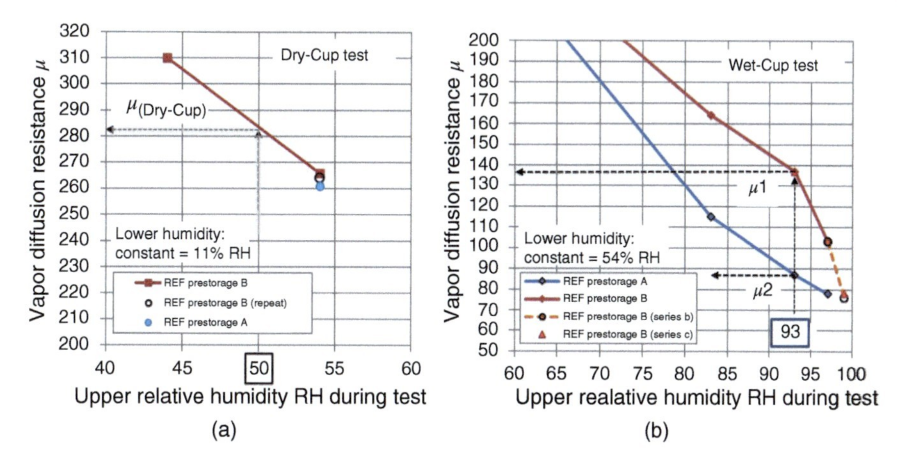

Figure 5.18 Results of the vapor diffusion measurements according to DIN EN ISO 12572 on the mortar REF for the dry range (Dry-Cup) and the wet range (Wet-Cup) after different prestorage methods, A (storage at 50% RH/23 °C only) and B (storage at 50% RH/23 °C after initial drying at 40 °C).

only at 50% RH/23 °C) with the air humidity spread set according to the standard (50% to 93% RH) in the Wet-Cup test.

It can be concluded that **the specifications of the above-mentioned test standard for pre-storage of the test specimens** prior to determination of the water vapor diffusion resistance for cement-bonded products **are not suitable in the present form for the Wet-Cup test**. The reason for the difference in the measurement results depending on the preconditioning is the hysteretic storage behavior of the products. This was discussed in more detail in Section 5.2.3 with the aid of simulation calculations.

Additional measurements on two cement mortars (MA and MAB) made of different Portland cements CEM I 42.5 R were carried out to verify the influence of the W/C value on the diffusion resistance in the Dry-Cup and Wet-Cup experiments. As expected, the results for the Dry-Cup and Wet-Cup conditions show a strong influence of the W/C value and the capillary porosity, respectively.

On two normal concretes produced in a concrete plant, the water vapor diffusion resistances were also determined at W/C values of 0.45 and 0.60 in the Dry-Cup and the Wet-Cup experiments. Table 5.1a shows the composition of the concretes, Table 5.1b contains the results of the Dry-Cup and Wet-Cup tests using preconditioning B (pre-storage with pre-drying).

Measurement results for the temperature dependence of the μ-value for different concretes are presented by Jooss and Reinhardt [6] and Jooss [7]. According to these, the influence in the temperature range between 20 and 70 °C is small in Dry-Cup experiments, while in the Wet-Cup range, there is a decrease in transport resistance of about 25%.

In this Section 5.3.1, the determination of the water vapor diffusion coefficients, especially for cement-bonded materials, was examined in more detail on the basis of measurement results, and it was shown that when the Wet-Cup method is used,

Table 5.1 Results of vapor diffusion measurements using **pre-drying condition B** on normal concretes.

(a) Concrete composition

Mixture	W/C	Cement	Cem. Content	max.aggreg.size
PC C35/45	0.45	CEM I 42.5 R	350 [kg/m³]	8 [mm][a]
PC C25/30	0.60	CEM I 42.5 R	350 [kg/m³]	8 [mm]

(b) Measured μ values

W/C	Dry-Cup	Wet-Cup
0.45	173	136
0.60	146	100

a) Plus plasticizer.

the resulting coefficients depend on the pre-drying of the test specimens. For this purpose, the known associated basic equations of water vapor diffusion have been compiled again.

In the following section, the modeling of the drying of the material by evaporation is investigated and described, using the results of numerous experimental results.

A precondition for a correct description or prediction of the time-dependent evaporation and the corresponding reduction of the water content in the material is also the correct modeling of the vapor transport within thicker material bodies as part of the overall moisture transport.

Therefore, in the following, the vapor transport within the material is formulated first.

5.3.2 Modeling of Vapor Transport Within the Material

In Section 3, **it was described that the vapor transport fraction** is included in Eq. (3.27) or (3.30) by means of the "source term" $\overset{0-1}{\Delta\theta}_{(i)}$.

This vapor transport fraction in the "source term" for a *VE*(*i*) at time step Δt is:

$$\frac{\overset{0-1}{\Delta\theta_i}}{\Delta t} = \left(\overset{0}{J}_{V_{i-1,i}} - \overset{0}{J}_{V_{i,i+1}}\right) = \frac{D_{Vi-1,i}}{\Delta x_{i-1,i}} * \left(p_{V_i} - p_{V_{i-1}}\right) - \frac{D_{Vi,i+1}}{\Delta x_{i,i+1}} * \left(p_{V_{i+1}} - p_{V_i}\right) \quad (5.21)$$

The one-dimensional formulation describes the difference in vapor flows between the *VE*(*i*) under consideration and the neighboring elements ($i-1$) and ($i+1$) over the course of a time step Δt, expressed by the associated vapor pressures $p_{V(i)}$, each multiplied by the associated vapor transport coefficients $k_{V(i)}$. In simplified terms, this results in:

$$\overset{0-1}{\Delta\theta}_{V(i)} = \left[+k_{V(i-1)} \cdot (p_{V(i)} - p_{V(i-1)}) - k_{V(i)} \cdot (p_{V(i+1)} - p_{V(i)})\right] \cdot \Delta t \quad (5.22)$$

The required transport coefficients k_V and vapor pressures for the upcoming time step Δt are calculated from the results of the previous time step.

The vapor pressures are determined with the following Eq. (1.39):

$$p_{V(i)} = pS(T_{(i)}) \cdot exp\left[\frac{-p_{cap_{(i)}}}{Rd \cdot T_{(i)} \cdot \varrho_w}\right] \tag{5.23}$$

where $pS(T_{(i)})$ = saturation vapor pressure in $VE(i)$ at the locally existing temperature $T_{(i)}$. The capillary pressure p_{cap}, (i) is present again in each $VE(i)$ after a time step for all "positions" in the θ–φ-diagram rep. θ-$p_{cap}(i)$-diagram, **especially for the case of hysteretic transport calculations** (Section 5.2.3).

The vapor transport coefficients $k_{V(i)}$ are obtained from Eq. (5.24) as the weighted average of neighboring VEs, where $(i, i+1)$ is truncated to $(i+1)$ (Compare **Figure 5.21**):

$$k_{V(i)} = 2 \cdot \frac{\left(D_{V(i+1)} \cdot \Delta x_{i+1} + D_{V(i)} \cdot \Delta x_i\right)}{\left(\Delta x_{i+1} + \Delta x_i\right)^2} \tag{5.24}$$

With constant Δx values, this gives the mean Eq. (3.34).

Vapor transport takes place only in the pores, which are not filled with water. **Hence, the moisture transport fraction from water vapor depends on the degree of water saturation.**

From Section 5.3.1, it can be seen that the vapor diffusion resistance numbers μ decrease when exposed to higher relative air humidities RH or higher water saturation degree in the test. The apparently higher vapor transport is due to the increasing proportion of liquid water transported along with the vapor. In contrast, the proportion of water transported in form of water vapor decreases instead.

From Section 5.2.1 and **Figure 5.2a,b**, it is clear that below a water-filling degree corresponding to $\varphi \approx 50\%$, no vapor transport takes place, since there is only absorbed water in micro-gel pores or inter-layer pores. This moisture limit is defined here as θ_{min}. In the Dry-Cup, the RH boundary conditions are $\leq 50\%$. In this experiment, therefore, the largest overall open pore space for vapor transport in the pore system is present, and thus **the maximum water-vapor diffusion coefficient D_V of the material, denoted here as $D_{V(Dry-Cup)}$**. As the degree of water saturation increases, the water-vapor diffusion coefficient decreases non-linearly and is 0 at water saturation θ_{sat}. A correct modeling of the moisture transport within the material is a precondition for a correct modeling of the drying or evaporation behavior of the material under consideration. From the extensive experimental program carried out for this purpose, **the dependence of the diffusion coefficient on the water saturation degree $\theta_{(i)}$ of the volume elements VE** can be derived to

$$D_{V(\theta)} = D_{V(DryCup)} \cdot \left(\frac{\theta_{sat} - \theta}{\theta_{sat} - \theta_{min}}\right)^{exp_V} \tag{5.25}$$

$D_{V(Dry-Cup)} = \delta_{Air}/\mu_{(Dry-Cup)}$ [kg/(m h Pa)], compare Eq. (5.19).

For $\theta_{(i)} \leq \theta_{min}$, $D_{V(\theta)} = D_{V(Dry-Cup)}$.

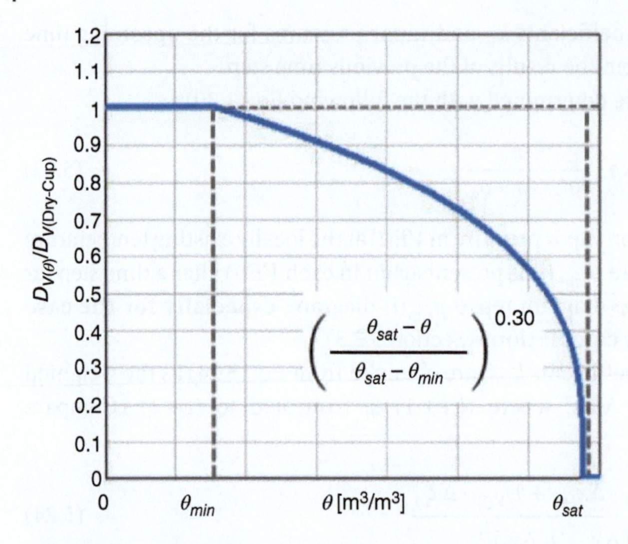

Figure 5.19 Dependence of the water vapor diffusion coefficient D_V on the degree of water saturation of the volume elements, derived from desorption tests on mortar REF and concrete.

Figure 5.19 shows the dependence of the water vapor diffusion coefficient D_V on the degree of water saturation of the volume elements concerned (derived from desorption tests on mortar REF and concrete).

For mortar REF and concrete, $exp_V = 0.30$ **is derived for Eq.** (5.25) from simulation/experiment comparison.

For ceramic brick products, calcium silicate products, porous concrete, and similar products, due to their special pore size distribution, the exponent is usually $exp_V = 1.0$ to be used, compare Section 5.4.

5.3.3 Importance of Vapor Transfer Coefficient for Vapor Transport Through Wall Elements

When modeling the evaporation or drying of the material, the transfer of water in the form of vapor to the surrounding atmosphere occurs at the surface. Here, the first question is whether, similar to heat transfer, a water vapor transfer coefficient β must be taken into account in calculations of the vapor transfer at the surface elements. For this purpose, we shall briefly consider **a material layer (wall) of thickness d with vapor diffusion resistance** μ at 20 °C. The partial pressure difference of the adjacent air volumes is $(p_{Va} - p_{Ve})$. Then, considering the vapor transfer coefficients β_a and β_e, the water mass transported in a steady state between the two air spaces (*a*) and (*e*) through the wall in vapor form is (with the transport resistances *R* of the layers):

$$\dot{m} = \frac{1}{R_1 + R_2 + R_3} \cdot \Delta p_V = \frac{1}{\frac{1}{\beta_a} + \frac{d \cdot \mu}{\delta_{Air}} + \frac{1}{\beta_e}} \cdot (p_{Va} - p_{Ve}) \tag{5.26}$$

Let the values of the vapor transfer coefficients be calculated via the following relation, often mentioned for material surfaces, compare for example Worch [8] and Matiasovsky and Mihalka [9]:

$$\beta = 7 \cdot 10^{-9} \cdot 3600 \cdot \alpha \ \ [\mathrm{kg/(m^2 \cdot h \cdot Pa)}] \tag{5.27}$$

In this, α is the heat transfer coefficient in [W/(m²·K)], with value 8 for unmoved air at indoor wall surfaces, with value 25 for outdoor atmosphere.

For $\alpha = 8$ follows from Eq. (5.27) $\beta = 2.0 \cdot 10^{-4}$ in [kg/(m²· h· Pa)], for $\alpha = 1.0$ would follow $\beta = 2.5 \cdot 10^{-5}$.

Multiplying the resistances in Eq. (5.26) by the coefficient δ_{Air} (Eq. (5.16)), we obtain (after inserting the values for δ_{Air} and β from Eq. (5.27)) the following relation:

$$\dot{m} = \frac{\delta_{Air}}{\frac{\delta_{Air}}{\beta_a} + d \cdot \mu + \frac{\delta_{Air}}{\beta_e}} \cdot \Delta p_V = \frac{\delta_{Air}}{\frac{0.028}{\alpha(a)} + d \cdot \mu + \frac{0.028}{\alpha(e)}} \cdot \Delta p_V \ \ [\mathrm{kg/(m^2 \cdot h)}] \tag{5.28}$$

For a $d = 10$ [cm] thick wall of brick material with a μ value of 10, the equivalent air layer thickness for the wall material is 1.0 [m]. The equivalent air layer thickness $0.028/\alpha(a)$ of the vapor transfer (e.g. also S_{ext} in **Figure 5.21**) when choosing $\alpha = 8$ is 0.0035 [m] or 3.5 [mm]. Therefore, the contribution of vapor transfer to the total vapor transmission resistance of the considered wall is less than 1%. From this example, it can be seen **that in vapor transmission calculations, the influence of vapor transfer is usually neglected.**

The precondition is that vapor pressures on the two surfaces of the material layer under consideration are different and that there is a low water saturation of the layer under consideration in the vapor transport region, which is usually assumed for ceramic brick material and materials of similar porosity.

In the case of increased moisture saturation of the material in the initial state or equal external vapor pressures $p_{V(e)}$ on both sides or, for example, in the case of vapor-tight closure of one side, **the mechanism of drying by evaporation** becomes decisive.

To describe this mechanism, vapor transfer coefficients must be considered because liquid-water transport is involved then. This concerns both brick and concrete materials.

As an example, consider a 100 [mm] thick wall made of concrete material REF, which is capillary water-saturated at the beginning of the observation: At 20 °C the wall is exposed to a relative humidity of 90% at the surface (a) and 30% RH at the surface (e). Let the drying by evaporation be calculated, using the modeling of the evaporation behavior described below for the material concrete.

Figure 5.20a shows **the vapor water absorption** $\dot{m}(a)$ at the surface (a) and **the vapor water release** $\dot{m}(e)$ as a function of time. It can be seen that there is (instead of a release) no vapor water absorption at the side (a) until about 2000 hours, while there is a strong release at the side (e). Only from about 4000 hours on, the one-sided vapor uptake at (a) corresponds approximately to the delivery at (e), in order to run in similar to running out from about 8000 hours on.

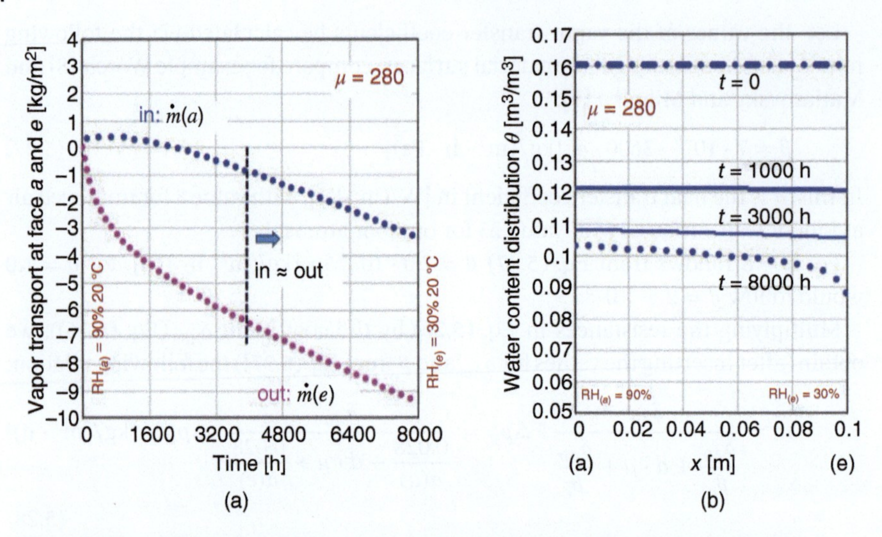

Figure 5.20 **(a)** Vapor transport through a 10 [cm]-thick (initially water-saturated) concrete wall of mortar REF ($w/c = 0.45$) between air humidities RH(a) = 90% and RH(e) = 30% (20 °C): a description of the behavior in the first approx. 4000 hours is only possible reliably by considering vapor transfer coefficients $\beta_{a/e}$. **(b)** Water saturation of the wall resulting from the transport process of figure (a).

A calculation of the behavior in the first 4000 hours is not possible reliably without consideration of a vapor transfer coefficient. Only at a water saturation below approx. θ_{gel} can the course of the curve be determined only by the vapor diffusion coefficient of the material, compare Section 5.3.6.

The corresponding distribution of water content in the wall material can be obtained from **Figure 5.20b**. It can be seen in this that at a degree of saturation above $\theta_{gel} = \theta(\varphi \approx 91.5\%)$, there is a largely constant moisture distribution across the cross-section despite largely one-sided evaporation only.

5.3.4 Influence of Vapor Transfer on Realistic Modeling of Drying Based on Experimental Results

In contrast to vapor transfer through a wall element with specified and, in particular, constant temperature and vapor pressure boundary conditions, the process of drying a porous material by evaporation initially involves only the near-surface zones, with the initial near-surface water content and associated vapor pressure in the pore system dropping sharply as evaporation progresses.

Realistic modeling of nonsteady-state drying by evaporation requires consideration of a vapor transfer coefficient at the surfaces involved in this process. This also results from the comparison of the calculation results with test results, as will be shown in the following.

The source term $\overset{0-1}{\Delta\theta}_{V(nn)}$ in the boundary element (nn) at the surface (e) (compare **Figure 5.21**) is then instead of Eq. (5.22):

$$\overset{0-1}{\Delta\theta}_{V(nn)} = \left[k_{V(nn-1)} \cdot (p_{V(nn)} - p_{V(nn-1)}) - k_{V(e)} \cdot (p_{V(e)} - p_{V(nn)})\right] \cdot \Delta t \tag{5.29}$$

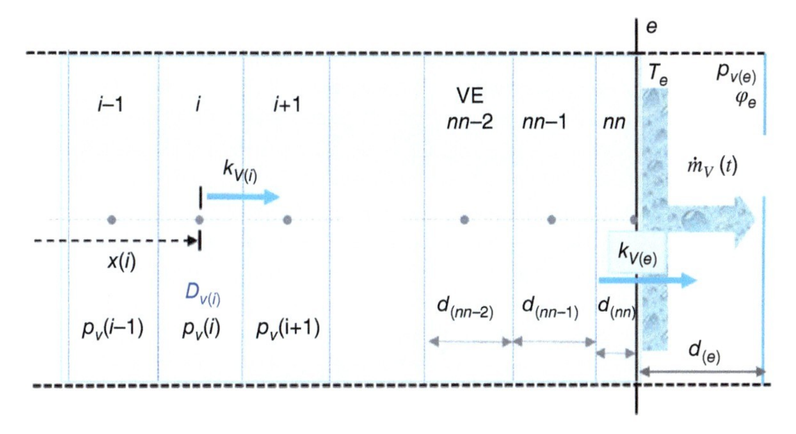

Figure 5.21 Configuration of the volume elements, here related to the evaporation calculations.

The vapor pressures $p_{V(nn)}$ and $p_{V(nn-1)}$ contained therein are obtained from Eq. (5.23). $p_{V(e)}$ is the vapor pressure present at the surface as a boundary condition. $k_{V(nn-1)}$ is obtained from Eq. (5.24).

The timing of **evaporation is largely determined by the vapor transport coefficient $k_{V(e)}$, which captures the vapor transfer at the surface (*e*)** involving vapor transport immediately below the surface as well as the vapor transfer above the surface. Conventionally, $k_{V(e)}$ is modeled based on the Eq. (5.26) (with subsequent consideration of Eq. (5.19)) as

$$k_{V(e)} = \frac{1}{\frac{d \cdot \mu}{\delta_{Air}} + \frac{1}{\beta_e}} = \frac{1}{\frac{d_{(nn)}}{D_{V(nn)}} + \frac{1}{\beta_e}} = \frac{1}{\frac{1}{k_{V(nn)}} + \frac{1}{\beta_e}} \tag{5.30}$$

The comparison calculations show that $k_{V(e)}$ must be independent of the chosen elemental division. This results for the reference material REF, **if** $d_{(nn)} = 0.003$ **[m]** = 3 **[mm] is used as constant value.**

According to the possible boundary conditions of evaporation in practice, numerous experimental investigations were carried out at different external vapor pressures (mainly at RH of 65% and 34%) at different air velocities v_{Air}. Using comparative calculations and experimental evaporation results, we first checked the extent to which the surface vapor transport coefficient in the Eq. (5.30) formulation and the vapor transfer coefficient in Eq. (5.27) could reproduce the measured evaporation curves.

Figure 5.22 shows the evaporation curves calculated using Eq. (5.30) and β_e according to Eq. (5.27) with $\alpha = 1$ and $\alpha = 25$ alternatively compared to the measured evaporation curve.

It can be seen that, on the basis of the above equations, an accurate modeling of evaporation obviously cannot be achieved.

Figure 5.23a,b shows for further understanding evaporation curves, **calculated without influence of a vapor transfer number corresponding to** $\beta_e = \infty$ in comparison to measured evaporation results.

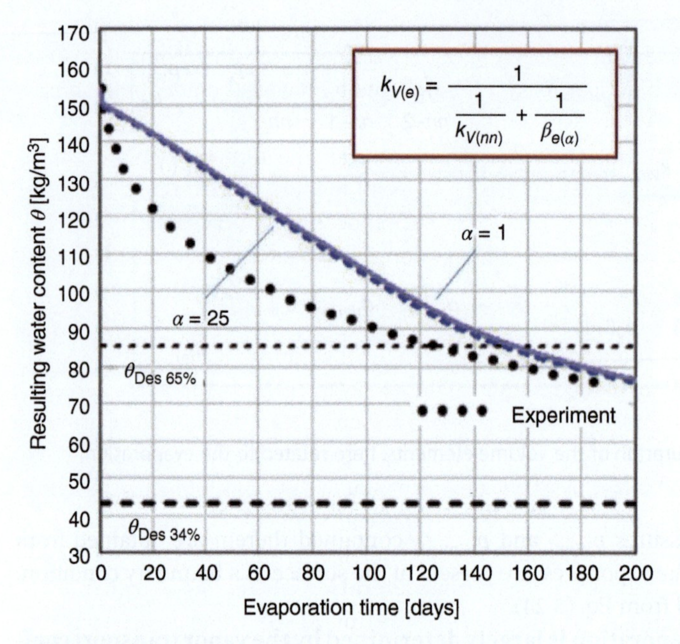

Figure 5.22 Comparison between experimental evaporation curve and curve calculated with Eq. (5.30) for thermal transition coefficients $\alpha = 1$ and $\alpha = 25$. Mortar REF, $L = 72$ [mm], start evaporation after 28d of water storage, evaporation over both face sides, 33.1% RH, 20 °C, $v = 0$ [m/s].

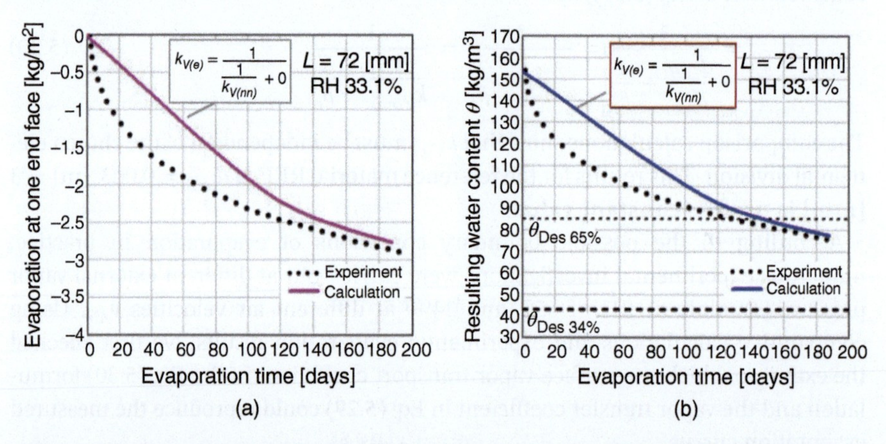

Figure 5.23 Experimental and calculated evaporation curves with boundary conditions according to **Figure 5.22**; the evaporation curves calculated for comparison results solely from the vapor transport fraction out of the boundary volume element (*nn*) without special influence of a vapor transfer coefficient β_e.

Figure 5.23a shows the measured evaporating water amount of one of the two frontal surfaces (dotted curve), i.e. half of the symmetrically proceeding total evaporation. **Figure 5.23b** indicates that the measured total water content of the prisms is decreasing as a result of evaporation.

The experimental evaporation curves shown are mean curves from three independent measurement curves on prisms of mean length $L = 72\,[\mathrm{mm}]$. The prisms were resin-coated on the sides, and evaporation occurred over both face surfaces. The corresponding total measurement duration was 190 days. (Material mortar REF, 28 days water pre-storage, water-saturated at start of storage in closed boxes at 21°C and 33.1% RH over salt solution).

The calculated (colored) curves show the resulting evaporation only from the transport resistance of the material layer close to the surface: Below $\theta = \theta_{gel} = 90\,[\mathrm{kg/m^3}]$ for the test material, the curve progression is determined solely or predominantly by $k_{V(nn)}$. Above $\theta = \theta_{gel}$, the calculated evaporation resulting from $k_{V(nn)}$ alone is too low. This means that, at least in this moisture saturation region of the pore system (close to the material surface), the **vapor transfer at the surface** into the environment may not be formulated as an additional transport resistance, but **as a necessary acceleration of the vapor transport to the outside occurring by a suitable formulation of the vapor transfer in the modeling**. This relationship is obtained from all evaporation experiments, even with different boundary conditions and prism lengths.

5.3.5 Implementation of Modified Vapor Transfer Coefficients

The comparisons show that β by Eq. (5.27) is not very suitable for predictions of the evaporation velocity. This is also due to the fact that the influence of the air velocity cannot be satisfactorily accounted for via the coefficient α, especially at a constant outdoor temperature. Using the low heat transfer coefficient $\alpha = 8$ within β, the calculation of evaporation rates gives several times higher values than measured for a water surface without air movement. **Thus, a different formulation of the vapor transfer coefficient is considered below**.

In addition, it must be taken into account that at high degrees of water saturation of the material, a wet surface resp. a moisture film is present, which is degraded after a short evaporation time resp. is then concentrated in the moisture-transporting pores near the surface.

This means that the vapor transfer number depends on the state of drying or the water content of the pores close to the surface.

A number of studies are available on the question of vapor transfer coefficients via water surfaces at various air velocities.

Poos and Varju [10] have presented a literature review on previously proposed formulations of vapor transfer coefficients over unmoved water surfaces as a function of different air velocities. Poos evaluates the proposed formulations and finds significant differences in the resulting results. Those transfer coefficients, which are formulated independently of the applied vapor pressure difference, give more accurate results, according to Poos.

Recknagel et al. [11] as well as the VDI 2089 Guidelines give for unmoved water surfaces a β-value of $0.5 \cdot 10^{-4}$ [kg/(m$^2 \cdot$ h $\cdot$ Pa)]. For indoor baths $1.5 \cdot 10^{-4}$ [kg/(m$^2\cdot$h$\cdot$Pa)].

Turza and Füri [12] performed new experiments in 2017 on unmoved water surfaces with various relatively low air velocities, the result of which can be prepared in agreement with [11] as:

$$\beta = [0.60 + 0.55 \cdot v_{Air}(\text{m/s})] \cdot 10^{-4}/3600 \quad [\text{kg/(m}^2 \cdot \text{s} \cdot \text{Pa)}] \tag{5.31}$$

This vapor transfer relation was used for the evaporation calculations after water storage of cement mortar and for concrete given here.

The specification of the correct vapor transfer number in each case remains problematic due to the different influences of the boundary conditions, in particular with regard to the influence of the air velocity. Also, the applicability of evaporation test results to other materials with significantly different pore sizes is usually only possible by approximation. Also, the surface roughness of a flat surface covered with a water film may influence the value assumed for a unmoved water surface.

For recalculation of evaporation experiments on ceramic bricks and materials of similar porosity, on the other hand, the following equation is used following Eq. (5.31):

$$\beta = [1.10 + 0.55 \cdot v_{Air}(\text{m/s})] \cdot 10^{-4}/3600 \quad [\text{kg/(m}^2 \cdot \text{s} \cdot \text{Pa)}] \tag{5.32}$$

The correct influence of air velocity v_{Air} [m/s] can be extrapolated from the experiments of [12] at most up to $v_{Air} = 1.0$ [m/s], and at best approximately for higher air velocities.

5.3.6 Realistic Modeling of Drying by Evaporation

A crucial influence on modeling results of drying by evaporation are **the adsorption and desorption isotherms** of the material. From these, the **porosity and water saturation characteristics, respectively, are derived**, which are used for the formulation of the diffusion or transport coefficients and thus for the calculation of the vapor transport. **The characteristic values are shown in Figure 5.24.** In principle, these are the characteristic values already defined in more detail in Section 5.2.1 when modeling the capillary transport fractions.

The aim is now to find a **vapor transport coefficient $k_{V(e)}$ for the transfer at the material surface** which, similar to Eq. (5.30), takes into account the vapor transport in the edge element $VE_{(nn)}$ as well as the vapor transfer β_e and allows a correct modeling resp. a realistic estimation of drying by evaporation.

Evaporation at the porous surface is basically described by the following equation, compare also **Figure 5.21**.

$$\dot{m}_{V(e)} = k_{V(e)} \cdot (p_{V(e)} - p_{V(nn)}) \tag{5.33}$$

The vapor transport coefficient at the material edge or surface(e) $k_{V(e)}$ is now formulated as follows:

$$k_{V(e)} = f_V \cdot \frac{D_{V(nn)}}{d_{(nn)}} + f_\beta \cdot \beta_e \tag{5.34}$$

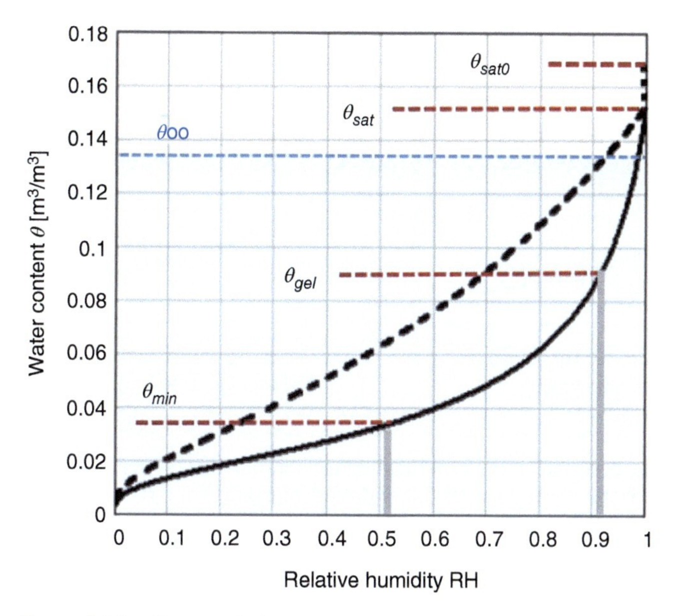

Figure 5.24 Characteristic pore- resp. water saturation parameters of a porous material for modeling drying by evaporation, using the example of the cement mortar REF with the corresponding adsorption and desorption isotherms. θoo, water content at start of evaporation; θ_{sat0}, total saturation including air pores; θ_{sat}, value of the isotherms at 99.8% RH; θ_{gel}, θ value between gel porosity and capillary porosity at ≈ 91.5% RH; θ_{min}, θ value between interlayer/microgel porosity and gel porosity at ≈ 50% RH.

The 1st summand of this equation refers to the pure vapor exchange between the vapor content of the open surface pores and the vapor of the atmosphere. Even if there is no surface film or $\beta_e = 0$, vapor exchange with the atmosphere occurs. This vapor exchange depends, according to Eq. (5.33), not only on the vapor pressure in the surface pores but also on the temperature-dependent vapor pressure of the surrounding air layer. **This moisture exchange mechanism from pure vapor phase takes place inversely in case of moisture absorption by vapor adsorption.**

The 2nd summand in Eq. (5.34) **accounts for the evaporation fraction of a liquid water film on the surface**. This fraction is mainly dependent on the **water** content in the surface pores and is thus only indirectly dependent on the vapor fraction present in these pores. The coefficients f_V and f_β contained in Eq. (5.34) are intended to describe the respective contributions of **the two mechanisms to the transport coefficient** $k_{V(e)}$ during drying by evaporation.

To ensure the independence of the material-side vapor transport fraction from the volume-element mesh or the thickness of the surface element, $d_{(nn)} = 0.003$ [m] is set constant in Eq. (5.34).

Figure 5.25a,b shows the influences of the 1st summand (from vapor transport) and the 2nd summand (from surface evaporation) in Eq. (5.34) on the drying curve calculated from them. The solid curves (pink and blue) are obtained when the diffusion fraction alone acts ($f_\beta = 0,\ f_V = 1$), the dashed lower curves (pink in (a) and

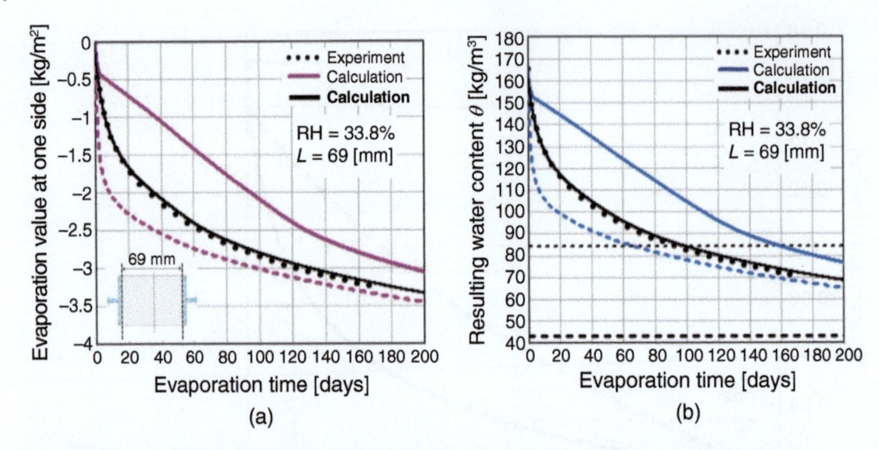

Figure 5.25 Comparison between (a) the experimental evaporation and (b) the corresponding measured water content curves (black dotted) and different calculated curves on the basis of different high influences of the 1st and the 2nd summand of $k_{V(e)}$ in Eq. (5.34). **Mortar REF, prism length** $L = 69$ **[mm]**, starts evaporation after 131d water storage, evaporation over both face sides in an desiccator at 33.8% RH, 20 °C. β_e according to Eq. (5.32) with $v = 0$ [m/s]. **Storage details:** After four months of water storage, sawing off the end faces and halving the 160 [mm] prisms. After a further five days of water storage: specimen removal, drying of the surfaces, coating the longitudinal sides with epoxy resin in the wet state, after one day of curing in protected storage, renewed sawing off of the end faces, then five days more water storage until the start of evaporation. **Parameters used:** $\theta oo = 164, \theta_{sat0} = 165, \theta_{sat} = 151, \theta_{gel} = 95, \theta_{min} = 35$ [kg/m³].

blue in (b)) are obtained when both the diffusion fraction $f_V = 1$ and the evaporation fraction $f_\beta = 1$ are applied together in full (β_e according to Eq. (5.32) with $v = 0$ [m/s]).

From Figure 5.25a, it can be seen that the vapor transport alone (upper curve) results in considerably too long evaporation times or too low drying performance compared to the measured evaporation curve. However, it is not possible to increase this flow from vapor transport, since this flow is limited by the vapor transport within the material. An approximation to the measured evaporation or drying curve is therefore only possible by taking into account an appropriate drying fraction from the surface evaporation of the liquid water from the surface pores or the resulting water film via the coefficient β_e.

The lower dashed line (calculated with $f_V = 1$ and $f_\beta = 1$) gives a too high drying rate and shows that the water content of the surface pores does not lead to a correct surface evaporation when a free water area ($f_\beta = 1$) is assumed; therefore, the surface evaporation has to be adjusted to the pore water content really present.

The relation (5.34) **can therefore be reformulated to**

$$k_{V(e)} = k_{V(nn)} + \beta_e \cdot f(\theta_{(nn)}) \tag{5.35}$$

The control calculations and comparisons with the numerous measurement results show that the surface evaporation fraction is obviously proportional to the volume fraction of the capillary pores ($\theta_{sat} - \theta_{gel}$) and the pore saturation in this pore region. The adjustment to a decreasing water content of the pores at the surface is also made by $f(\theta_{(nn)})$. The effect of the decrease in pore filling is again modeled analogously to the relative decrease of the capillary transport coefficient corresponding to the saturation between θ_{sat} and θ_{gel} and using the exponent *expo*1 of (Eq. (5.1) in Section 5.2.1).

The evaporation measurements show that even below θ_{gel} down to a degree of saturation corresponding to about 90% RH, a small fraction of liquid water evaporation still occurs. This is taken into account in Eq. (5.36) with the characteristic values related to the adsorption isotherm of the material by the reduction factor f_{gel}.

This gives the function $f(\theta_{(nn)}$ **in Eq.** (5.35) **to**

$$f(\theta_{(nn)}) = (\theta_{sat} - \theta_{gel} \cdot f_{gel}) \cdot \left[\frac{\theta_{(nn)} - \theta_{gel} \cdot f_{gel}}{\theta_{sat} - \theta_{gel} \cdot f_{gel}} \right]^{expo1} \tag{5.36}$$

For the cement mortar, REF $exp1 = 3.0$ (compare Section 5.2.1) and $f_{gel} = 0.90$ are used.

From the equation, it can be seen that when the material is capillary water-saturated, corresponding to $\theta_{(nn)} = \theta_{sat}$, the function $f(\theta_{(nn)})$ reflects the capillary water content, so that the value of β_e at the surface is approximately effective on the fraction of area covered with capillary water.

Only when the capillary-water saturation θ_{sat} is exceeded in the direction of θ_{sat0} is a continuous water film assumed, compare the following simplified overview of evaporation modeling.

The vapor diffusion fraction $k_{V(nn)}$ **in Eq.** (5.35) initially (at water saturation) results in zero from the vapor transport calculation and then increases as the water filling of the surface pores decreases. Below a saturation of θ_{gel}, pure vapor diffusion transport predominates with simultaneously vanishing water-film evaporation. **Below** $\theta_{gel} \cdot f_{gel}$, **therefore,** $k_{V(e)} = k_{V(nn)}$ **is set.**

The algorithm for the decisive coefficient $k_{V(e)}$ is formulated as follows depending on the evaporation time resp. of the resulting water saturation degree $\theta_{(nn)}$ and the vapor diffusion fraction $k_{V(nn)} = D_{V(nn)}/d_{(nn)}$ of the material at the edge (*e*) (with $D_{V(nn)}$ from Eq. (5.25)):

IF $(\theta_{(nn)} > \theta_{sat})$ *THEN*

. $k_{V(e)} = \beta_e$

IF $(\theta_{(nn)} \leq \theta_{sat})$ *and* $(\theta_{(nn)} > \theta_{gel} \cdot f_{gel})$ *THEN*

. $f(\theta_{(nn)} = (\theta_{sat} - \theta_{gel} \cdot f_{gel}) \cdot \left[(\theta_{(nn)} - \theta_{gel} \cdot f_{gel})/(\theta_{sat} - \theta_{gel} \cdot f_{gel})\right]^{expo1}$

. $k_{V(e)} = k_{V(nn)} + \beta_e \cdot f(\theta_{(nn)})$

IF $(\theta_{(nn)} \leq \theta_{gel} \cdot f_{gel})$ *THEN*

. $k_{V(e)} = k_{V(nn)}$

The discussed Figure 5.25a,b shows that the explained modeling leads to an almost exact agreement of the calculation (black curve) with the measured values.

Figures 5.26a,b and 5.27a,b represent the evaporation measurements on prisms that had a length of 145 [mm] (after sawing off the faces) for comparison and were

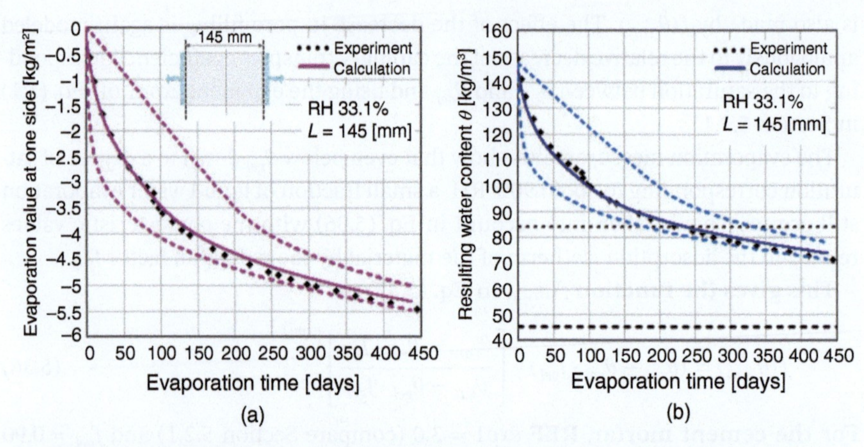

Figure 5.26 Comparison between the experimental evaporation curves (black diamonds) and different calculated curves on the basis of differently high influences of the 1st and the 2nd summand of $k_{V(e)}$ in Eq. (5.34), prism length $L = 145$ [mm]. (a) Evaporation values via one side of the two sided evaporation experiment. (b) Corresponding mass change of the prisms. The prisms were shortened from 145 to 135 [mm] after 90 days of evaporation by cutting off the caps. However, the calculations of evaporation were continued with the original length for 450 days. **Storage details:** Mortar REF, start evaporation after 1100d varying moisture storage stages, evaporation over both face sides in a desiccator at 33.1% RH, 20 °C, β_e according to Eq. (5.32) with $v = 0$ [m/s]. More details of the storage of the prisms before evaporation starts are given in **Figure 5.27a,b**. **Parameters used:** $\theta oo = 147$, $\theta_{sat0} = 165$, $\theta_{sat} = 151$, $\theta_{gel} = 95$, $\theta_{min} = 35$ [kg/m³].

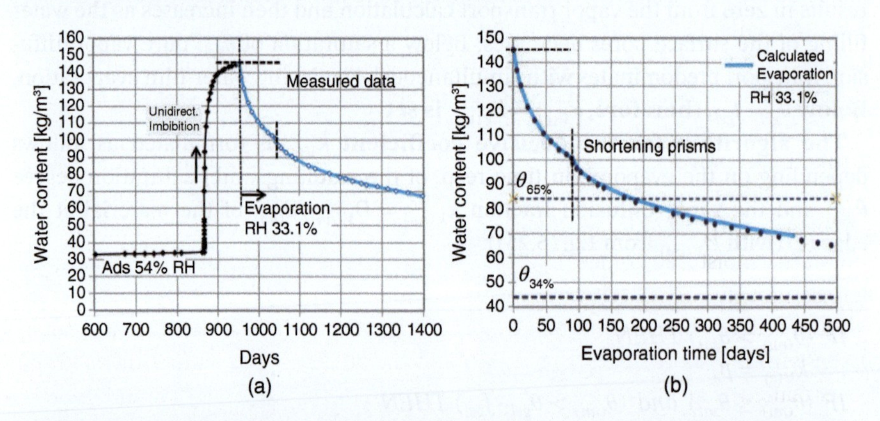

Figure 5.27 **(a)** Varying moisture storage stages of the prism series of **Figure 5.26a,b** before the start of the evaporation experiment and measured evaporation curve drawn in blue (compare **Figure 4.40a**). Total age of the test specimens at evaporation start = 1100 days. **(b)** Comparison of the measured evaporation data with the calculated curves, where in this case the shortening of the prisms was taken into account in the calculation.

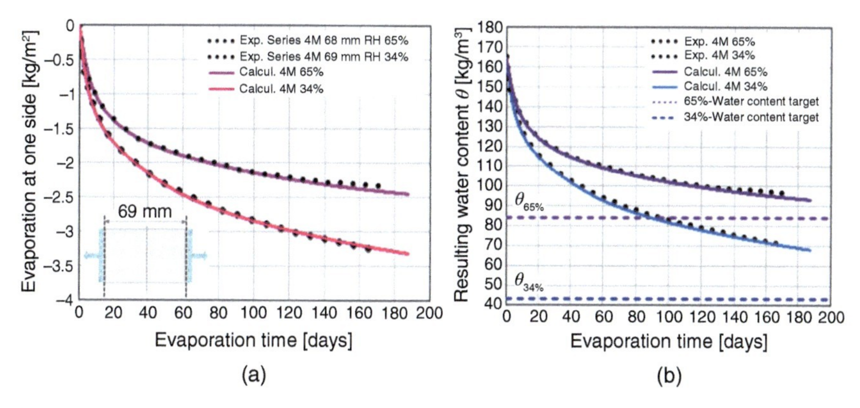

Figure 5.28 Results of two-sided evaporation measurements over the two faces of prism series of material REF after identical prestorage and comparison with evaporation calculation. Evaporation of one series at 65% RH/20 °C, air flow $v = 0.50$ [m/s] in a wind tunnel. 2nd series at 33.8% RH/20 °C in a desiccator, $v = 0$ [m/s]. (a) Evaporation values via one side of the two-sided evaporation experiments. (b) Corresponding mass change of the prisms. Evaporation starts after a total of 4.5 months of water prestorage, halving the 160 [mm] prisms by sawing and cutting off the caps after 4 months. After coating the longitudinal sides in the wet state, we sawed off the end faces again. After additional water storage, the evaporation starts. The following parameters were identical in both calculations: $\theta oo = 163$, $\theta_{sat0} = 165$, $\theta_{sat} = 151$, $\theta_{gel} = 95$, $\theta_{min} = 35$ [kg/m^3]. Only the airboundary conditions were changed.

subjected to various soaking cycles for a long time until the start of the evaporation measurements. Before the start of the evaporation cycle, both faces were ground, and after 90 days of evaporation, both faces were shortened by about 5 [mm] and the experiment continued. The shortening was done to to clarify the possible effects of surface precipitation on evaporation velocity. These relationships will be discussed further in Section 5.3.7. The discernible acceleration of evaporation after the prisms were shortened can also be attributed in part to the now shorter prism length.

Independently of this, however, it can be stated that even after the variable long-term moisture storage and the longer prism length, there is very good agreement between experiment and calculation.

The further course after 90 days shown in Figure 5.27b results in a recalculation with the shortened prism length and the 90-day water saturation as initial water content θoo.

Figure 5.28a,b show the comparison between evaporation measurements and evaporation calculations on identically prestored prism series, which were only exposed to different air-humidity boundary conditions over 170 days.

For more details on the exposure and calculation parameters, please refer to the figure captions. The comparison shows that the calculation or modeling used leads to agreement between experiment and calculation. During the preparation of the test specimen series and during the evaporation test, **care was taken to ensure that no significant deposits or salt precipitation were present on the evaporation surfaces.**

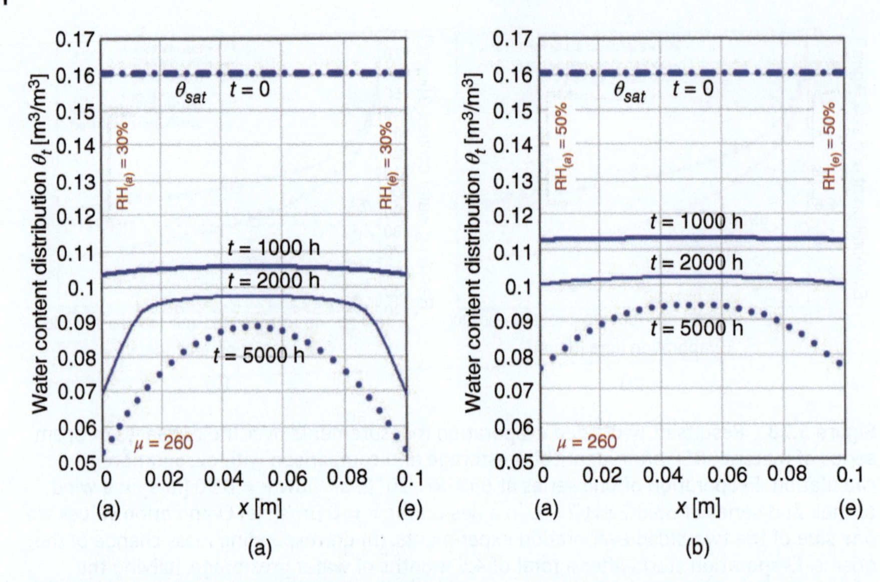

Figure 5.29 Time-dependent water content decrease and associated water content distributions over the cross-section during drying on both sides by evaporation. 10 [cm] thick initially water-saturated concrete walls made of mortar REF. Boundary conditions: Air velocity v = 0.50 [m/s], RH 30% (Figure a) and RH 50% (Figure b). Water content parameters: $\theta oo = 160$, $\theta_{sat0} = 165$, $\theta_{sat} = 151$, $\theta_{gel} = 93$, $\theta_{min} = 35$ [kg/m³].

Figure 5.29a,b are intended to show which water content distributions result as a function of time for initially water-saturated concrete walls made of mortar REF as a result of drying by evaporation. For the calculation, the previously presented relationships were applied. The walls were exposed on both sides to a relative air humidity of 30% RH and 50% RH, respectively, at an air flow $v = 0.50$ [m/s].

It is clear that until the water content of the walls decreases to about $\theta_{gel} \approx 0.09$ [m³/m³] or until the capillary water fraction is completely evaporated, linear moisture distributions across the cross-section result in comparatively fast drying rates.

The investigations have shown that **surface precipitation can lead to a change in the evaporation behavior**, which is shown in Section 5.3.7 on the basis of extensive measurement results and calculations.

5.3.7 Modeling the Influence of Salt Precipitation on the Evaporation Behavior of Concrete Surfaces

5.3.7.1 Measurement Results on the Sources and on the Effect of Salt Precipitation

After storage of cement-bonded test specimens in water or even above water at a high relative humidity of about 98% RH, the formation of a whitish surface deposit is usually observed on the outer surfaces of the test specimens.

Figure 5.30 shows such a sawed surface after water storage on the left and the same surface after removal of the deposit by grinding with corundum on the right.

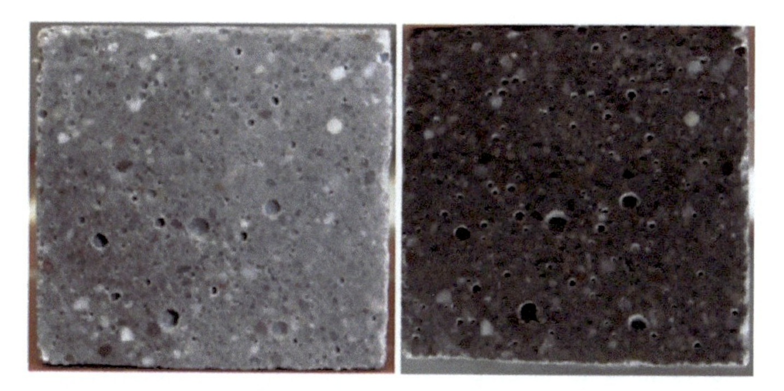

Figure 5.30 Typical images of sawn surfaces after prolonged water storage (left image) and after subsequent grinding of the surface (right image). Approx. 0.2 [mm] of the surface is removed.

Control experiments confirmed that such a deposit can have an influence on the evaporation and drying behavior of the material. The tests showed that such precipitation occurs not only during long prestorage at high water saturation but also in the course of drying from high water saturation. **The measurement results shown in Figures 5.31 and 5.32a,b** demonstrate the effect of such deposits on the evaporation behavior.

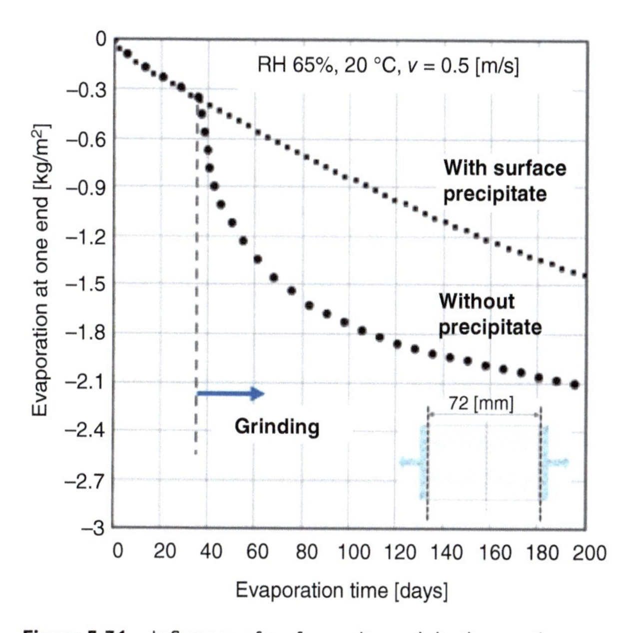

Figure 5.31 Influence of surface salt precipitation on the evaporation behavior of prismatic bodies made of mortar REF. After standard drying of the prisms, the long sides were sealed and the two faces were ground. This was followed by water storage for 210 days. Without further surface treatment, bilateral evaporation was then performed at 65% RH and 20 °C in a wind tunnel at $v = 0.50$ [m/s]. After 36 days of evaporation, the end faces were ground again (compare **Figure 5.30**) and evaporation continued thereafter.

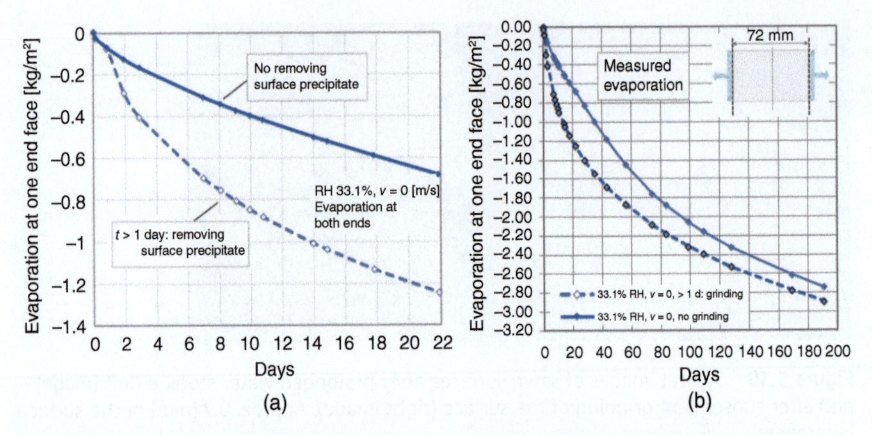

Figure 5.32 Measured evaporation curves on water-stored prisms made of mortar REF. **(a)** Shows the measurement results of the first 22 days, **(b)** the total course of the measurements: **Experiment details:** After 14 days of water storage, the caps of the 160-[mm]-long prisms were sawed off and the prisms were cut in half. After further water storage for 28 days, the sides were sealed, and the prisms were further stored in water. After 49 days, evaporation was started over both faces in a desiccator at 33.1% RH. After one day of evaporation, the faces of one series were ground on after each measurement, while the 2nd series continued to dry without surface treatment for a total of 192 days.

In the experiment on **Figure 5.31**, the end faces of 160 [mm] prisms made of mortar REF were sawed off after 28 days, and the prisms were cut in half. After standard drying (40 °C/1 mbar) for 10 months, the long sides of the prisms were coated, capillary water absorption was performed, and then the prisms were water-saturated by immersion for 180 days.

This was then followed by two-sided evaporation at 65% **RH in the wind tunnel**. After 36 days, the face surfaces were abraded and then abraded again after each subsequent measurement.

The graph shows the dramatic acceleration of evaporation after removal of the deposit. For comparison, the expected behavior without the removal of the deposit was entered, compare Section 5.3.7.2. **For comparison, the evaporation behavior with and without abrasion** of the front surfaces of the prisms was registered after each measurement in two further series after seven weeks of water prestorage when stored in a desiccator for 192 days at 33.1% RH/20 °C. **Figure 5.32a** shows the behavior in the first 22 days, **Figure 5.34b** the measurement results over the total time. Again, there is a clear difference between the curves with and without deposit removal. Due to the lower prestorage times, the differences here are smaller than for the results in **Figure 5.31**.

Infrared ATR analyses were used to investigate the composition of the deposits or salt precipitates on powders obtained by abrading them from their respective surfaces.

The sections of the measured spectra in **Figure 5.33** show that, as expected, the salt precipitates or deposits are calcite formed from $Ca(OH)_2$ in the pore solution and CO_2 dissolved in the surrounding water, or are formed from the $Ca(OH)_2$-containing

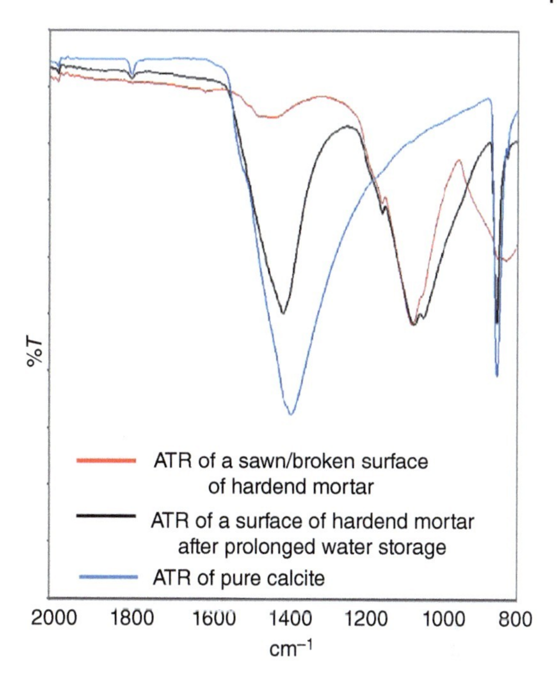

Figure 5.33 IR spectroscopic analysis of the substance removed from the surface by abrasion of a freshly sawn REF prism and of the surface after prolonged water storage, in comparison with the spectrum of pure calcite [TUHH].

water film present on the surface at high water content when exposed to atmospheric carbonic acid.

Ultimately, the formation of such deposits depends on the ambient conditions or on the lime-carbonic acid equilibrium in the area of the surfaces concerned or in the baths in which the test specimens are stored.

When the surfaces are stored in (continuously) exchanged water with an excess dissolved carbonic acid content, existing calcite deposits can be dissolved again by the formation of hydrogen carbonate or not be formed at all.

In the laboratory area, when largely stationary baths are used, deposit formation can usually be assumed.

Further, measurements were performed on sealed stored prisms made of mortar REF. The prisms were stored sealed 28 or 71 days after manufacture until the start of evaporation measurements in the wind tunnel. In each case, the measurement results were compared with evaporation curves calculated according to Section 5.3.6 in the "original condition" without the presence of a deposit.

Figure 5.34a shows the curves obtained at 35% RH, and **Figure 5.34b** shows the curves obtained at 65% RH. Further details can be obtained from the figure captions.

The results show that with sealed storage and the lower water content resulting compared to water prestorage, no deposits were effective at the onset of evaporation.

5.3.7.2 Consideration of Steady-State and Dynamic Salt Precipitation on Evaporation Rate

The preliminary investigations (compare Section 5.3.7.1) have shown that deposits of salt precipitation primarily reduce evaporation from the liquid phase. Below the relative moisture content $\theta_{gel} \cdot f_{gel}$, the influence of vapor diffusion predominates and is

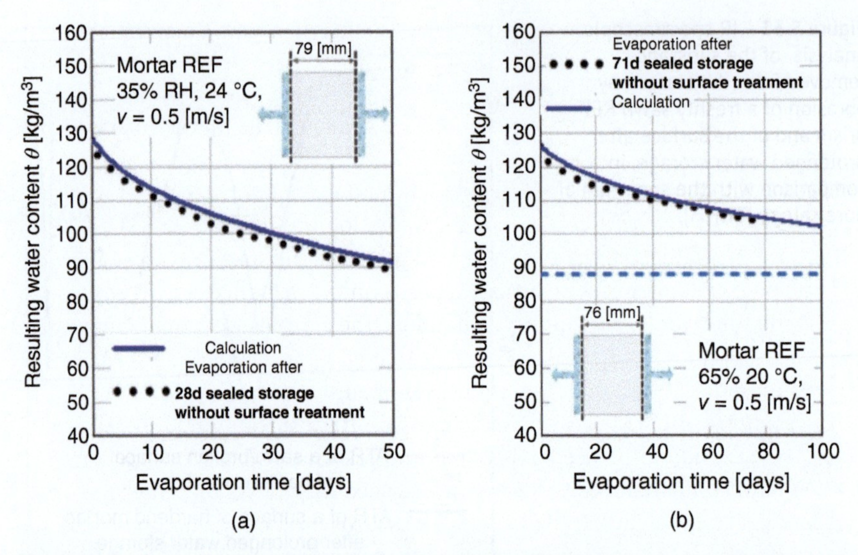

Figure 5.34 Measurement of the two-sided evaporation behavior of test specimens that were **prestored sealed until the start of drying** and comparison with evaporation calculation. The 160 [mm] prisms were prestored/cured in double polyethylene bags after 24 hours of moist prestorage until the start of drying after 28 days (a) or 71 days (b). After sealing the longitudinal sides with aluminum foil, the prisms were divided by clean breaking. Two-sided drying started under different boundary conditions (see details in the figures) for 50 days and 80 days, respectively. One evaporation surface was the original forming area (without formwork oil), the 2nd evaporation surface was the flat fracture surface. The evaporation calculation was performed according to Section 5.3.6 without consideration of a surface covering. A covering influence was not detected.

much less affected by deposit formation. Vapor diffusion then occurs predominantly through the large capillary pores of the pore system, which are not yet closed when a deposit is present.

The reduction in evaporative performance due to a deposit is now accounted for by assigning a reduction factor to both parts of the vapor transport coefficient in the Eq. (5.35):

$$k_{V(e)} = k_{V(nn)} \cdot fsalt_{V(nn)} + \beta_e \cdot f(\theta_{(nn)}) \cdot fsalt_{(e)} \tag{5.37}$$

$fsalt_{(e)}$ is the reduction factor for surface liquid evaporation,

$fsalt_{V(nn)}$ is the reduction factor for the vapor transport fraction.

Both factors have a value of 1.0 in the deposit-free state of the surface.

From the present experimental results, it can be seen that in the case of intensive deposits, the factor $fsalt_{(e)}$ can take the value 0.05, but slowed drying still takes place.

The factor $fsalt_{V(nn)}$, on the other hand, decreases only slightly and ranges between about 0.85 and 1.0.

In the previously unobserved theoretical case of a completely vapor-sealing carbonate deposit with $fsalt_{(e)} = 0$, $fsalt_{V(nn)}$ must also become equal to 0. To ensure this, $fsalt_{V(nn)}$ has been expressed as a function of $fsalt_{(e)}$ with a very low exponent, where $fsalt_{V(nn)}$ only becomes very small when $fsalt_{(e)}$ approaches 0:

$$fsalt_{V(nn)} = fsalt_{(e)}^{\ 0.05} \tag{5.38}$$

$fsalt_{V(nn)}$ hereby becomes 0.96 for e.g. $fsalt_{(e)} = 0.40$ and 0.83 for an intense deposit with e.g. $fsalt_{(e)} = 0.025$.

Figure 5.35a,b shows the measurement and calculation results of a series of specimens of mortar REF tested in the evaporation test after a total of 550 days of prestorage including 180 days of water storage before the start of evaporation.

The side-coated prisms with visible deposits on the face surfaces were first tested in this condition for 36 days in the wind tunnel; the deposits were then removed and the evaporation test continued, with overgrinding of the evaporation surfaces up to a test duration of 100 days after each measurement.

To calculate the evaporation curve in the first 36 via Eqs. (5.37) and (5.38), $fsalt_{(e)} = 0.025$ had to be used. After that, the relevant evaporation fraction in the first 36 days compared to the deposit-free surface was only 2.5%. After removal of the deposits after 36 days, the calculation was also based on the deposit-free condition with $fsalt_{V(nn)}$ and $fsalt_{(e)} = 1.0$, thus calculated according to Section 5.3.6 with Eq. (5.35) and thus obtained agreement with the experimental results.

Figure 5.35a,b further contains an upper dotted curve, which indicates the computationally expected evaporation course under stationary retention of the low initial permeability $fsalt_{(e)} = 0.025$ of the deposit, calculated over the entire evaporation time with Eq. (5.37).

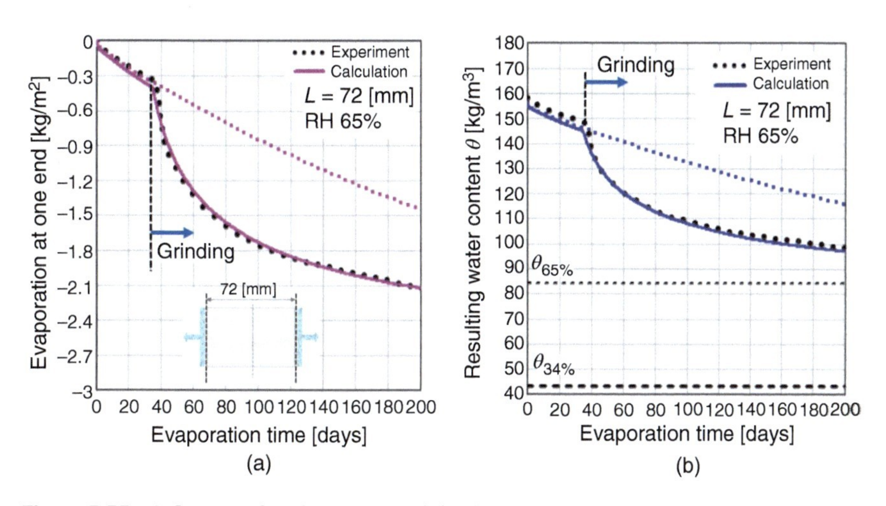

Figure 5.35 Influence of surface salt precipitation and subsequent abrasion by grinding on the evaporation behavior of prismatic bodies made of mortar REF, compare the details to **Figure 5.31**. The drying or evaporation behavior after removal of the covering from the evaporation surfaces was calculated on the basis of the modeling procedure previously presented in Section 5.3.7, and led to a quasi-exact agreement between calculation results and experimental curves. The upper dotted line shows the progress of the evaporation behavior calculated for comparison according to Section 5.3.7.2 when holding constant the covering intensity present after 36 days.

Furthermore, the following parameters were applied to describe the material at the beginning of evaporation: $\theta oo = 156$, $\theta_{sat0} = 165$, $\theta_{sat} = 151$, $\theta_{gel} = 93$, $\theta_{min} = 35$ [kg/m^3].

It can be seen that with intense deposits, there is a significant slowdown but still effective drying of the test specimens

In contrast to the results shown in **Figure 5.35a,b** for evaporation behavior for test specimens that had a deposit at the beginning of evaporation, experiments were also performed on **specimens without an initial deposit under obviously deposit-forming boundary conditions.**

Figure 5.36a,b show such results without initial deposit, determined on 154 [mm] prisms made of mortar REF, which were standard dried after 28 days of water storage and, after coating of the longitudinal sides, were subjected to a capillary water absorption test for 356 days until the test specimens were largely saturated. For each measurement in the water absorption test, the water-absorbing suction surface was ground. In the case of one of the three prisms in the series, an approx. 3 [mm] slice was additionally sawn off the absorbent surface after 26 days. From 356 days onwards, the evaporation test shown in **Figure 5.36a,b** was carried out on one side at the suction surface of the prisms.

The upper curves (with surface deposit) show the measured and calculated evaporation behavior. The lower calculated curves show evaporation without deposit. From the comparison of the curves, it is clear that for these prisms, which were water-saturated at the beginning (without initial deposit), **during evaporation at 65% RH in the wind tunnel, a clear deposit developed**, by which evaporation was greatly reduced.

The dynamic change in the salt deposit and the resulting shape of the evaporation curve were again modeled using Eq. (5.37). For this purpose, the deposit

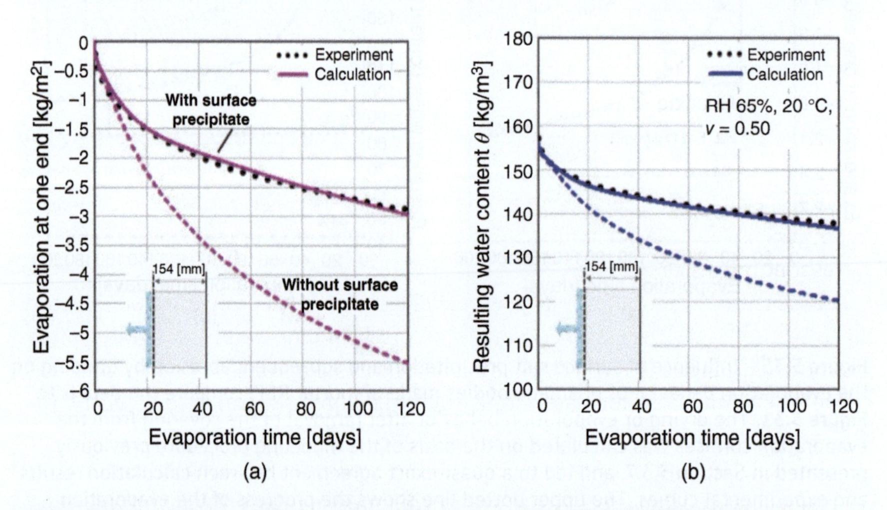

Figure 5.36 Influence of the formation of a surface precipitate on the evaporation behavior when the surface is initially free of coating. Comparison to the behavior calculated without precipitate formation. 154 [mm] prisms made of REF mortar, previously water-saturated for 156 days in a capillary suction test.

permeability $fsalt_{(e)}$ was linearly decreased with evaporation time from 1.0 to 0.10 over 1200 hours and then held constant at this value. Despite the broad agreement between measurement and modeling, the curves indicate that there appears to have been a further slight increase in the deposit after 1200 hours.

Parameters at the beginning of evaporation: $\theta oo = 156$, $\theta_{sat0} = 165$, $\theta_{sat} = 146 = \theta(\varphi = 99.5\%)$ der Adsorptions-Isotherm, $\theta_{gel} = 93$, $\theta_{min} = 35$ [kg/m^3].

These results show that the method shown can be used to model the influence of a dynamic change in the deposit.

If no knowledge of a deposit development is available, the probability of the absence of a deposit or a deposit formation can be inferred based on the chemical/physical environmental conditions and the moisture loading of the material. In the latter case, the deposit permeability value must be estimated on the basis of the previous results.

5.3.7.3 Comparison of Results from One-Sided and Two-Sided Evaporation

For all evaporation results reproduced in Sections 5.3.6–5.3.7.2, the mass loss from evaporation was always presented in relation to one face. Of the two-sided evaporation experiments carried out for the most part, only the evaporation occurring on one face side was shown in each case. For this purpose, the measured total mass losses were divided by 2.

Of further interest is the evaporation occurring on one face side in the one-sided tests compared to the two-sided evaporation tests for test specimens of the same length.

Figure 5.37a shows the measurement results of two series of test specimens of the same length of 46 [mm], made from 160 [mm] prisms of mortar REF and stored in a water bath for five years. Sealing of the longitudinal surfaces was carried out a few days before the start of the evaporation test and subsequent renewed water saturation. By sawing off 5 [mm] of the original end faces and dividing the 160 [mm] prisms to the test specimen length, the evaporation surfaces were in the "original condition" free of deposit.

Figure 5.37a shows the significantly higher water loss in [kg/m^2] in the one-sided evaporation test. The total moisture loss of the test specimens of the two-sided evaporation tests is nevertheless higher because the shown one-sided water loss is present on both sides, and therefore doubling the values of the corresponding upper curve gives the total loss.

The measured evaporation was now calculated without deposit formation, according to Section 5.3.6. The result curves of the corresponding water loss of the samples are shown in **Figure 5.37b**. The coefficient f_{gel} was set to 0.90.

The other parameters for water storage at the beginning of evaporation: $\theta oo = 164$, $\theta_{sat0} = 167$, $\theta_{sat} = 151$, $\theta_{gel} = 93$, $\theta_{min} = 35$ [kg/m^3].

There is agreement in principle between the experiment and the calculation, but the results indicate that **after 40 days of drying, new deposit formation has obviously led to a slowdown in evaporation.**

Another comparison and check of the correctness of the evaporation modeling and the formulation of the moisture transport inside the test specimens is to compare **the**

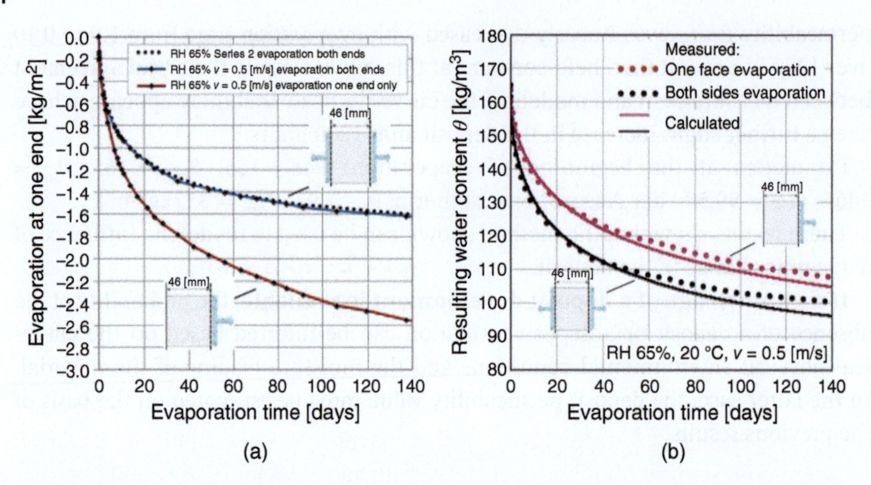

Figure 5.37 Comparison of the one-sided evaporation for two series of test specimens of the same length/thickness (here 46 [mm]) made of five-year-old mortar REF. The drying of one series took place only on one side, the second series dried two-sided for comparison. (a) evaporated mass measured, (b) comparison of resulting water content.

one-sided evaporation of a test specimen of given length with the two-sided evaporation of a test specimen of double length.

The evaporation tests performed show that the evaporation obtained on one side is identical in both cases.

Figure 5.38a,b show the results of such an experiment over 192d and the corresponding calculation results.

For this comparative experiment, REF prisms were used that were water-stored for 49 days to evaporation onset. The required saw cuts to produce the test specimen lengths were made 14 days, and the required side seals were made 28 days after production.

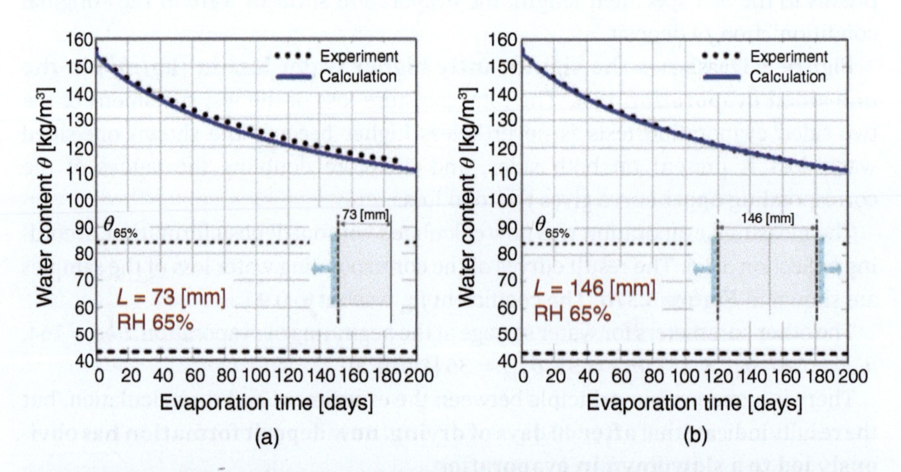

Figure 5.38 Comparison of the one-sided evaporation of a test specimen of given length (73 [mm]) with the two-sided evaporation of a test specimen of double length (146 [mm]) made of mortar REF. In the comparison calculation of both series, the effect of a precipitation with $fsalt_{(e)} = 0.25$ was considered as constant over the test duration.

Shown are the identical total weight losses due to evaporation. In the comparative evaporation calculation, Section 5.3.7.2 was followed with $fsalt_{(e)} = 0.25$, assumed constant over the evaporation time.

One can see the agreement between the curves.

5.4 Realistic Modeling of Drying by Evaporation for Ceramic Bricks, Calcium Silicate Products, and Porous Concrete

Drying by evaporation occurs in principle in the same way for the materials considered in this section as for cement, mortar, or concrete. However, due to the strongly different pore size distributions, a corresponding adjustment of the equations has to be made.

First, the basic equation for the delivered water content is repeated here as Eq. (5.39):

$$\dot{m}_{V(e)} = k_{V(e)} \cdot (p_{V(e)} - p_{V(nn)}) \tag{5.39}$$

The comparative control calculations now showed that for brick materials and materials of comparable porosity ratios, the relationship (5.34) formulated for concrete products can be simplified to Eq. (5.40):

$$k_{V(e)} = f_V \cdot k_{V(nn)} + \beta_e \cdot f(\theta_{(nn)}) \tag{5.40}$$

In the included vapor transport coefficient of the boundary element (nn) $k_{V(nn)} = D_{V(nn)}/d_{(nn)}$ the element width $d_{(nn)} = 0.002$ [m] is set constant to ensure the independence of the coefficient from the boundary-element thickness of the element mesh.

The function $f(\theta_{(nn)})$ gives the relative water content in the near-surface pores. It is defined here as

$$f(\theta_{(nn)}) = \frac{\theta_{(nn)}}{\theta_{sat} - \theta_{lim}} \tag{5.41}$$

θ_{sat} **is the maximum water uptake of the material in the capillary water uptake test** (compare also **Figure 5.24** to the ratios for mortar REF). Many materials absorb additional water during long-term water storage, for which the limit $\theta_{sat0} > \theta_{sat}$ is then reached, which can also be considered in connection with evaporation as shown below.

For the calcium silicate, aerated concrete, and ceramic brick products, all drying tests show a behavior corresponding to **Figure 5.39a,b**, with an extended, **largely linear initial phase (1st Drying section) and a 2nd much slower drying phase (2nd Drying section)** below about 20 – 25% of θ_{sat}.

A discussed problem is to specify an applicable saturation boundary between the 1st and 2nd drying sections. **This limit is denoted here as** θ_{lim}. This will be discussed further below.

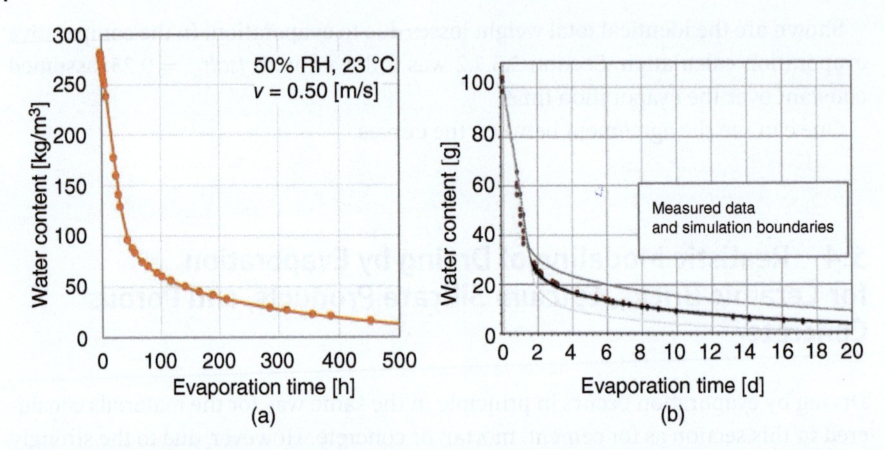

Figure 5.39 Results of one-sided evaporation tests on samples of ceramic brick. (a) Results from Rosa Espinosa/TUHH on brick product Joens. (b) Results from Scheffler and Plagge [13]/John Wiley & Sons.

When drying from a water-saturated state, the water film present on the surface is of paramount importance for the drying speed:

At the beginning of evaporation, when the material is initially saturated with water, a continuous water film is present, whose evaporation can be described solely by the β_e value of a free water surface. As the water saturation $\theta_{(nn)}$ of the surface pores decreases due to drying, the water film becomes smaller. The resulting reduction in evaporative power is accounted for or modeled by the coefficient $f(\theta_{(nn)})$ in the Eqs. (5.40) and (5.42), respectively.

The vapor diffusion fraction in Eq. (5.40) is initially (at water saturation) equal to 0 and then increases with decreasing water filling of the surface pores. In the 2nd drying section below the boundary between sections defined here as **limits**, pure vapor diffusion transport predominates with simultaneously vanishing water-film evaporation.

The Eq. (5.40) is valid only for the 1st drying section. The calculations show that in this equation for the materials considered here in the 1st drying section, the diffusion fraction has a negligible influence in comparison. Therefore, **the equation for the 1st drying section can be further simplified** to:

$$k_{V(e)} = \beta_e \cdot f(\theta_{(nn)}) \tag{5.42}$$

For the control calculations, material data of the brick product Joens were preferably used. The adsorption isotherms determined at our own institute are shown in **Figure 5.40a**, determined on two different batches of the material. **Figure 5.40b**, the derived function J1 of the isotherms is entered as Eq. (4.3) with the associated parameters. These sorption isotherms are necessary for the transport calculations in the material based on the capillary pressure method used.

The calculation of the respective vapor diffusion coefficient was performed in principle as for Eq. (5.25) using Eq. (5.43), but with the adjusted exponent $exp_V = 1.0$ due to the different pore-size distribution of the material considered

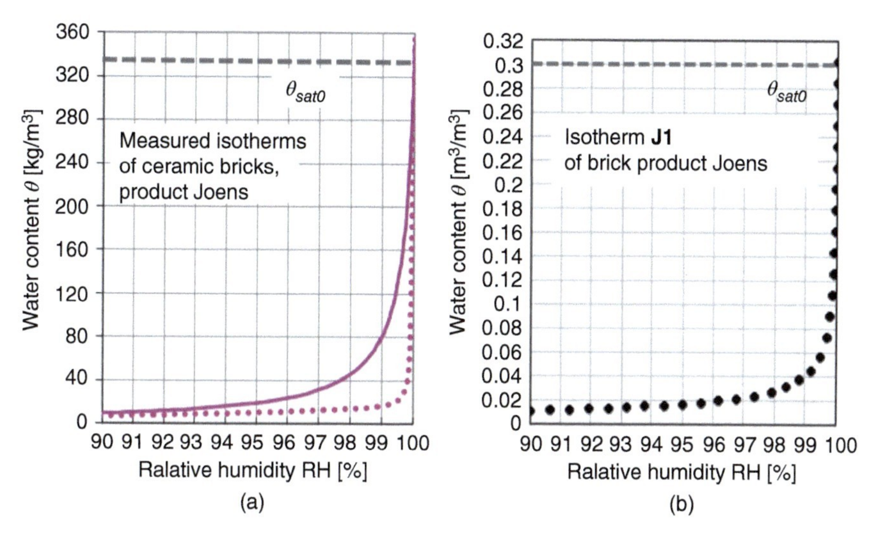

Figure 5.40 (a) Measured adsorption isotherms of two different batches of ceramic bricks, product Joens; data from 0.90 RH to 1.0 RH presented; results from Rosa Espinosa/TUHH. (b) Adsorption isotherm $J1$ as $\theta(\varphi)$ function derived from the measured data of figure (a) function formulation according to Eq. (4.3) with the coefficients $A = 1.00022, a = 0.0058, b = 1.9, c = 2{,}4$.

here. The control calculations show that the size of the exponent has only a minor role (especially in the calculation of drying).

$$D_{V(\theta)} = D_{V(Dry-Cup)} \cdot \left(\frac{\theta_{sat} - \theta}{\theta_{sat} - \theta_{min}} \right)^{1.0} \tag{5.43}$$

θ_{min} corresponds to the θ value of the sorption isotherm of the material under consideration at the upper RH value of 50% corresponding to the dry-cup vapor diffusion experiment.

Further control calculations were performed using the isotherms given in **Figure 5.42a** for ceramic bricks with $\theta_{sat} = 0.202$ [m³/m³].

The other parameters used for the calculations with the Joens product are:

$\theta_{sat} = 0.298$, $\theta oo = 0.295$, $\mu = 13$, max. capillary transport coefficient $k_{sat0} = 3.1 \cdot 10^{-9}$, coefficient for capillary water uptake $ww = 15$ [kg/(m² · h^0.5)], $\theta_{''Gel''} = 0.033$, limit = 0.06 [m³/m³].

In all calculations, the **temperature influence of evaporation enthalpy at the surface and in the test specimen was considered.**

The influence of the wall thickness on the drying rate or the specimen height of cylindrical test specimens during one-sided drying over the face in measuring apparatus is discussed in the relevant literature [13–15], Matiasovsky and Mihalka [16], and again by Zhao et al. in [17].

Figure 5.41a shows the **moisture distributions in a 100 [mm] thick wall of ceramic brick** product Joens calculated (with the previously mentioned material data) of one-sided drying after defined times, starting from the saturation state at time $t = 0$.

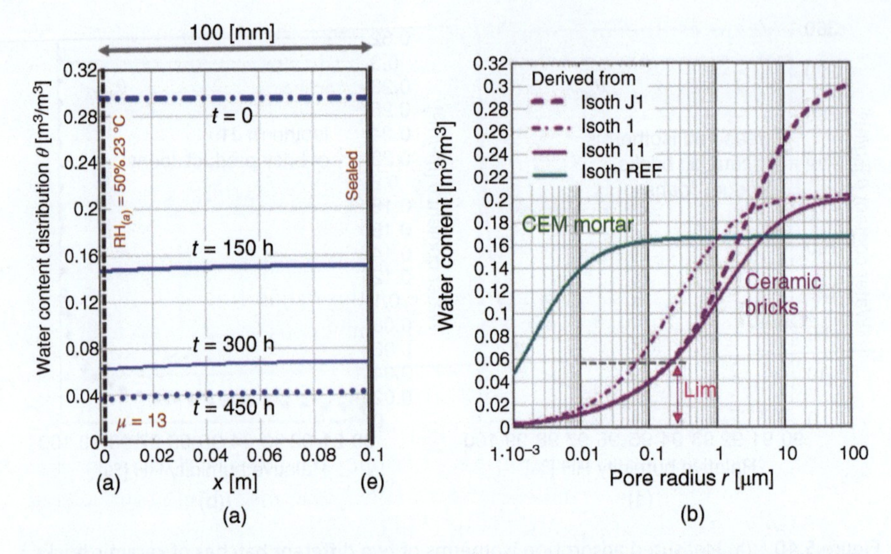

Figure 5.41 **(a)** Calculated water content distributions in a 100 [mm] thick ceramic brick wall as a function of drying time with one-sided drying from a saturated state (product Joens, sorption isotherm J1, 50% RH, 23 °C, $v = 0.40$ [m/s]). **(b)** Picture of the different sorption isotherms used here for ceramic bricks (and CEM mortar for comparison) as a function of the pore radius, also to indicate the $lim_{(Isoth)}$ parameter used in the drying calculation.

The almost constant distribution of the moisture content or pore filling over the cross-section with only a slightly decreasing tendency toward the surface (a) over which the drying takes place is striking. Smaller wall thicknesses result in corresponding distributions with lower saturation at the selected times.

Now consider the times required for different wall thicknesses to dry the wall cross-section by the moisture content $\Delta\theta \cdot d$ (for example, starting from the saturation state or time $t = 0$). It is immediately obvious that for the given drying boundary conditions for half the wall thickness or half the moisture volume $\Delta\theta \cdot d/2$ only half the drying time is required. This indicates that for the materials considered **the drying times, at least in the 1st drying section, must depend largely linearly on the wall thickness or specimen height in the case of face evaporation**, provided that excessive thickness or high elements are not included, where no completely linear behavior can be expected due to the transport processes.

Scheffler and Plagge [15] also find a linear dependence in the 1st drying section on the specimen dimensions via measurements on ceramic bricks and calcium silicate.

For the description and calculation of a complete drying curve, the **transition between the 1st drying section and the 2nd drying section, here called "limit,"** is essential. Crucial to this transition is the specific pore size distribution of these materials, expressed, for example, by the desorption isotherm required for the calculation. Since, for the materials considered, at relative humidities below about 98%, the desorption isotherms are not significantly "higher" than the corresponding adsorption curves and, as a rule, only the latter are measured, these must be adapted or used for the calculation of drying by evaporation.

A look at the isotherm $J1$ **in Figure 5.40b** shows the transition between the capillary pore region and the region responsible for the slower transport in the 2nd drying section (for cement concrete, the gel pore region). This transition is also often referred to as the boundary between the hygroscopic and super-hygroscopic regions. For the adsorption isotherm $J1$, this value would be $RH = 0.99 - 0.995$.

In the case of sorption isotherm 2 in **Figure 5.42a**, this transition is not so clearly definable.

Matiasovsky and Mihalka [16] suggests to use as transition or limit the so-called "hydraulic radius" or the saturation volume below this radius. However, the determination of the hydraulic radius also appears to be ambiguous or not always to provide suitable values.

As an alternative, it is suggested here to derive the Limit "Lim" approximately from the representation of Figure 5.41b. In this figure, the sorption isotherms used are plotted as a function of θ(*radius r*). Initially, the drawn value Lim (defined as lim in the equations) refers to the isotherms J1 and 11 for ceramic bricks with θ_{sat}= 0.30 and 0.20, respectively, and marks the beginning of the curves curvature to the low pore radii and impeded moisture transport, respectively. The extracted value $lim = 0.06$ [m^3/m^3] was used to calculate all the diagrams of drying behavior of the brick material Joens presented in this section.

A simple but good approximation is generally obtained by setting $lim = 0.20 \cdot \theta_{sat}$ **to** $lim = 0.25 \cdot \theta_{sat}$.

Coherently, **the algorithm used for modeling the transport coefficient of evaporation** can be formulated in a simplified way as follows, depending on the

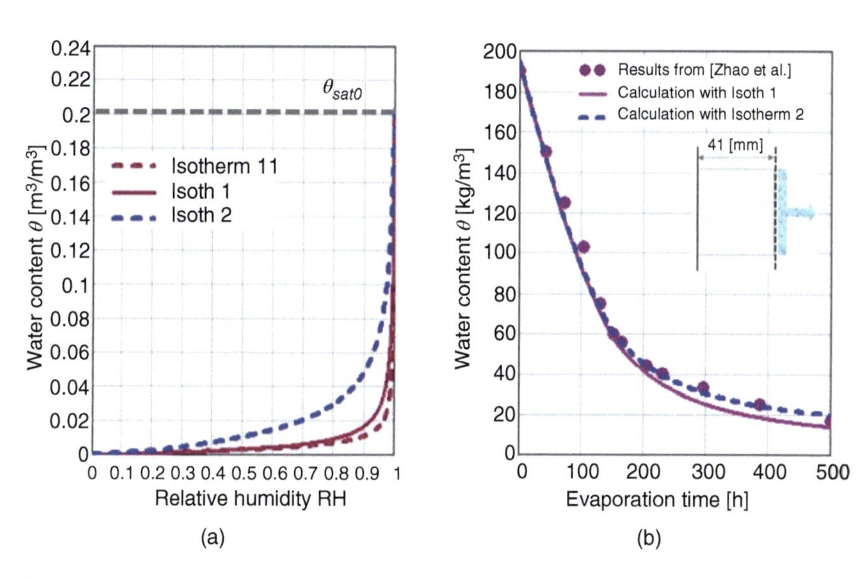

Figure 5.42 **(a)** Sorption isotherms of ceramic bricks with $\theta_{sat0} = 0.20$ [m^3/m^3]; function formulation according to Eq. (4.3) with the coefficients. $A11 = 1.00049$ $a11 = 0.0058$ $b11 = 1.9$ $c11 = 2.4$ for isotherm 11, $A1 = 1.0028$ $a1 = 0.0070$ $b1 = 1.51$ $c1 = 2.2$ for isotherm 1, $A2 = 1.0104$ $a2 = 0.0284$ $b1 = 1.8$ $c2 = 2.8$ for isotherm 2; **(b)** Calculation of evaporation curves using to 40% reduced airflow conditions for fitting to measured values of Zhao et al. [17]/EDP Sciences/CC BY 4.0.

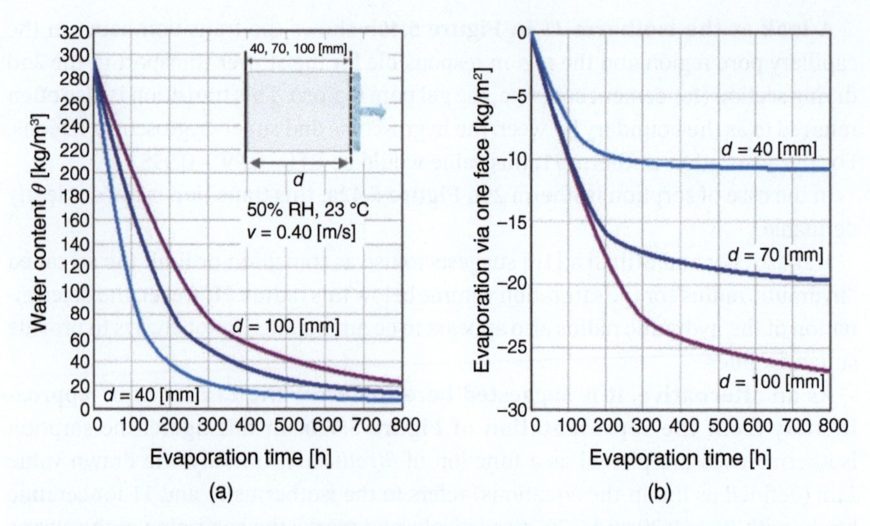

Figure 5.43 Calculated one-sided drying by evaporation of wall panels of different thickness or test specimens of different height, water-saturated at the drying start. **Ceramic brick product Joens**, sorption isotherm J1, parameter Limit = 0.06 [m³/m³], airflow velocity $v = 0.40$ [m/s]. (a) Water content as a function of drying time. (b) Evaporating water mass as a function of specimen thickness and drying time.

evaporation time or the resulting water saturation degree $\theta_{(nn)}$ at the edge (e) of the material:

$IF\ (\theta_{(nn)} > \theta_{sat})\ THEN$

. $k_{V(e)} = \beta_e$

$IF\ (\theta_{(nn)} \leq \theta_{sat})\ and\ (\theta_{(nn)} > lim)\ THEN$

. $f(\theta_{(nn)}) = \theta_{(nn)}/\theta_{sat}$

. $k_{V(e)} = \beta_e \cdot f(\theta_{(nn)})$

$IF\ (\theta_{(nn)} \leq lim)\ THEN$

. $k_{V(e)} = k_{V(nn)} \cdot (\theta_{(nn)}/lim)^2$

The calculation steps show that for any saturations above θ_{sat}, a continuous water film is assumed to evaporate like a free water surface.

The evaporative performance of the water film also depends on the surface roughness. Therefore, when calculating the results of **Figure 5.41a**, **5.43a,b**, and **5.44a,b**, the following β value was used with a higher evaporative power in motionless air:

$$\beta = [1.11 + 0.55 \cdot v_{Air}(\text{m/s})] \cdot 10^{-4}/3600 \quad [\text{kg/(m}^2 \cdot \text{s} \cdot \text{Pa)}] \tag{5.44}$$

Below the transition-water content lim, only the vapor-diffusion fraction is considered and reduced as a quadratic function $(\theta_{(nn)}/lim)^2$.

Figure 5.43a now shows the decreases in water content determined in this way for the ceramic brick material Joens after initial water saturation for wall thicknesses or

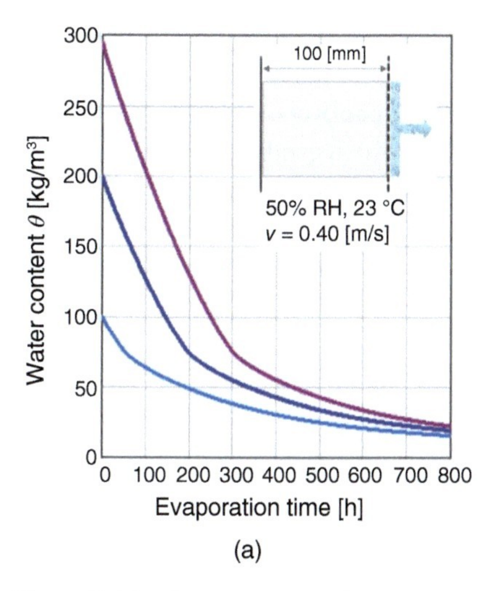

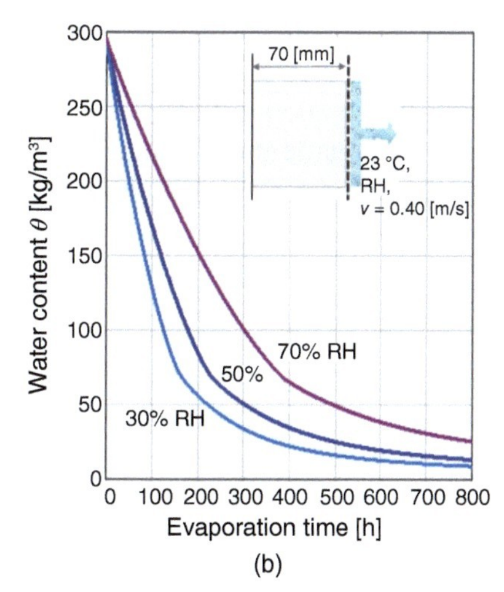

Figure 5.44 Calculated one-sided drying by evaporation of wall panels or test specimens of **ceramic brick product Joens**, sorption isotherm J1, parameter Limit = 0.06 [m³/m³], airflow velocity $v = 0.40$ [m/s]. (a) Course of the water content as a function of drying time after different initial water saturations, wall panel thickness 100 [mm]. (b) Drying curves as a function of different constant air-humidities, wall panel thickness/sample height 70 [mm].

specimen heights of 40, 70 and 100 [mm]. One can see the linear dependence of the water content curves on wall thickness discussed earlier.

The drying time $t(\theta 1)$ of a material layer or wall from saturation degree θ_{sat} to saturation degree $\theta 1$ can be determined with good approach in the 1st drying section for the materials considered here in the following way (without a simulation program).

For this purpose, the Eqs. (5.39) and (5.42) are assumed.

The effect of $f(\theta_{(nn)})$ in drying θ_{sat} to a degree of saturation $\theta 1$ can be expressed in integrated terms as.

$$f1(\theta 1) = \frac{1}{2} \cdot \left[\frac{\theta_{sat} - \theta 1}{\theta_{sat} - \theta_{lim}} \right] \tag{5.45}$$

With $\Delta m(\theta 1) = (\theta_{sat} - \theta 1) \cdot d$ [kg/m²] we get the moisture loss of a [m²] of a considered wall of thickness d [m] when drying from θ_{sat} to $\theta 1$. This moisture loss results from evaporation in time $t(\theta 1)$.

From this follows the required drying time until the saturation degree $\theta 1$ is reached

$$t(\theta 1) = \frac{\Delta m(\theta 1)}{\beta_e \cdot \Delta p_V \cdot [1 - f1(\theta 1)]} \tag{5.46}$$

The equation highlights the linear dependence of drying times on wall thickness, on β_e, and on vapor pressure in the 1st drying section up to θ_{lim}.

The magnitude of Δp_V can be given as follows:

From the sorption isotherm $J1$ (**Figure 5.40b**), it can be seen that for saturation degrees ≥ 60 [kg/m³], $\varphi \geq 99.5\%$ results. The associated vapor pressure $p_{V(nn)} = p_S \cdot 99.5\%$ differs only insignificantly from the saturation pressure and can therefore be assumed approximately constant in the considered saturation range of $\theta_{(nn)}$, at 23°C ≈ 2800 [Pa].

It must be taken into account, however, that cooling occurs in the surface pores as a result of the evaporation enthalpy, which can be approximated to be about 2 [K] or 2 °C. The resulting vapor pressure in the surface pores during the considered evaporation process can then be assumed to be approx. 2500 [Pa] (at initial temperature 23 °C. **The applicability of this equation is verified by the following example:**

Ceramic brick material, wall thickness $d = 0.10$ **[m]**, $\theta_{sat} = 298$ [kg/m³], $\theta_{/lim} = 60$ [kg/m³], $\theta 1 = 75$ [kg/m³], 50% RH/23 °C.

Thus, the vapor pressure of the external atmosphere is 1400 [Pa], and consequently Δp_V is constant ≈2500 − 1400 = 1100 [Pa].

The used transition coefficient at v equal to 0.40 [m/s] is

$\beta_e = (1.1 + 0.55 \cdot 0.40) \cdot 10^{-4} = 1.32 \cdot 10^{-4}$ [kg/(m² · h · Pa)]

From Eq. (5.46) we then obtain

$$t(\theta 1) = \frac{(298 - 75) \cdot 0.10}{1.32 \cdot 10^{-4} \cdot 1100 \cdot \left[1 - 0.5 \cdot \frac{223}{240}\right]} = 292 \text{ [h]} \tag{5.47}$$

This value agrees with the value determined via the simulation program here, compare t in **Figure 5.44a** or **Figure 5.43a** for $\theta 1 = 75$ [kg/m³].

The drying time of the 2nd drying section can be roughly estimated to be five to eight times the drying time of the 1st section.

The total drying time including the 2nd drying section is estimated by Scheffler and Plagge [15] on the basis of a drying coefficient D. They determine this drying coefficient over the duration of the 2nd drying section, calculated and given for ceramic bricks and materials of similar porosity via a simulation program for the standard boundary condition 20 °C and 50% RH:

The drying coefficient is given as $D = 107$ for ceramic bricks. For a given wall thickness d [m], according to [15], the drying time $t = (107 \cdot d)^2$ [days].

From this, for $d = 0.10$ [m]: $t = 114$ days = 2750 hours.

Figure 5.44b shows the drying process as a function of the ambient relative humidity for a wall thickness of 70 [mm]. This dependence can also be predicted up to the transition to the 2nd drying phase from Eq. (5.46).

Figure 5.44a confirms the parallelism of drying curves in the 1st drying section when drying from lower initial saturations.

Zhao et al. [17] investigate the effect of air velocity on evaporation from the faces of cylindrical samples of ceramic brick and calcium silicate products. They use a flow apparatus presented and tested by Scheffler and Plagge [13]. Among other papers, Zhao et al. present measurement results for brick samples, entered in **Figure 5.42b**. However, the paper does not contain any information on the brick material used. A recalculation of these measurement results can be accomplished with the sorption isotherm 2 shown in **Figure 5.42a** with half the value of β_e according to Eq. (5.32) assuming the flow velocity of $v = 1.2$ [m/s] given in the paper.

References

1 Rucker, P. (2008). Modellierung des Feuchte- und Salztransports unter Berücksichtigung der Selbstabdichtung in zementgebundenen Baustoffen. Dissertation. Technische Universität München.

2 Wittmann, F.H., Zhang, P., and Zhao, T. (2011). Damage of concrete under combined actions. *Restoration of Buildings and Monuments* 17: https://doi.org/10.1515/rbm-2011-6469.

3 Zhang, P., Wittmann, F.H., Zhao, T., and Lehmann, E. (2009). Penetration of water into uncracked and cracked steel reinforced concrete elements; visualization by means of neutron radiography + erratum 2013. *Restoration of Buildings and Monuments* 15: 67–76.

4 Zhang, Z. and Angst, U. (2020). A dual-permeability approach to study anomalous moisture transport properties of cement-based materials. *Transport in Porous Media* 135: 59–78.

5 McDonald, P.J., Istok, O., Janota, M. et al. (2020). Sorption, anomalous water transport and dynamic porosity in cement paste: a spatially localised ^{1}H NMR relaxation study and a proposed mechanism A spatially localised ^{1}H NMR relaxation study and a proposed mechanism. *Cement and Concrete Research* 133: 106045.

6 Jooss, M. and Reinhardt, H.W. (2002). Permeability and diffusivity of concrete as function of temperature. *Cement and Concrete Research* 32: 1497–1504.

7 Jooss, M. (2001). Dichtheit von Heißwasser-Langzeitspeichern aus Hochleistungsbeton. Dissertation Stuttgart University (in German).

8 Worch, A. (2002). Überlegungen und Versuche zur Erfassung des Wasserdampfübergangs an Bauteiloberflächen. Dissertation. Technische Universität Dortmund.

9 Matiasovsky, P. and Mihalka, P. (2019). Capillary moisture content – parameter of water suction and drying. *Thermophysics, AIP Conference Proceedings*, Volume 2170.

10 Poós, T. and Varju, E. (2020). Mass transfer coeffcient for water evaporation by theoretical and empirical correlations. *International Journal of Heat and Mass Transfer* 153: 119500.

11 Recknagel, H., Sprenger, E., and Schramek, E.R. (ed.) (2009). *Taschenbuch für Heizung und Klimatechnik*, 74e. Oldenbourg Industrieverlag GmbH.

12 Turza, R. and Füri, B.B. (2017). Experimental measurements of the water evaporation rate of a physical model. *Slovak Journal of Civil Engineering* 25: https://doi.org/10.1515/sjce-2017-0003.

13 Scheffler, G. and Plagge, R. (2005). Bestimmung des Trocknungsverhaltens von Baustoffen unter definierten Randbedingungen. *Bauphysik* 27: 324–330.

14 Scheffler, G.A. (2008). Validierung hygrothermischer Materialmodellierung unter Berücksichtigung der Hysterese der Feuchtespeicherung. Dissertation. Technische Universität Dresden.

15 Scheffler, G.A. and Plagge, R. (2009). Ein Trocknungskoeffizient für Baustoffe. *Bauphysik* 31: 125–138.

16 Matiasovsky, P. and Mihalka, P. (2014). Pore structure parameters and drying rates of building materials. In: *Drying and Wetting of Building Materials and Components* (ed. J.M.P.Q. Delgado), 71–90. Springer Switzerland.
17 Zhao, J., Meissener, F., and Grunewald, J. (2020). Experimental investigation of the drying behaviour of the building materials. *E3S Web of Conferences* 172: 17002.

Bibliography

Adam, Th. (2006). Ein Modell zur Beschreibung der Hydratation von Beton in Abhängigkeit vom Feuchtegehalt. Dissertation. Technische Universität Darmstadt.

Adolphs, J. and Setzer, M.J. (1996). A model to describe adsorption isotherms. *Journal of Colloid and Interface Science* 180: 70–76.

Adolphs, J. and Setzer, M.J. (1996). Energetic classification of adsorption isotherms. *Journal of Colloid and Interface Science* 184: 443–448.

Adolphs, J. and Setzer, M.J. (1998). Description of gas adsorption isotherms on porous and dispersed systems with the excess surface work model. *Journal of Colloid and Interface Science* 207: 349–354.

Ahlgren, L. (1972). Moisture Fixation in Porous Building Materials. *Report 36*. The Lund Inst. of Technology.

Åhs, M.S. (2007). Moisture redistribution in screeded concrete slabs Division of Building Materials. Dissertation. Lund University.

Åhs, M.S. (2008). Sorption scanning curves for hardened cementitious materials. *Construction and Building Materials* 22: 2228–2234.

Aktulga, H., Fogarty, J.C., Pandit, S.A., and Grama, A.Y. (2012). Parallel reactive molecular dynamics: Numerical methods and algorithmic techniques. *Parallel Computing* 38: 245–259.

Ally, J., Kappl, M., Butt, H.J., and Amirfazli, A. (2010). Detachment force of particles from air-liquid interfaces of films and bubbles. *Langmuir* 26: 18135–18143.

Alyafei, N., Al-Menhali, A., and Blunt, M.J. (2016). Experimental and analytical investigation of spontaneous imbibition in water-wet carbonates. *Transport in Porous Media* 115: 189–207.

Andrade, C. and Coppens, E. (2020). Quantifying water permeability and pore size through capillary absorption. *4th International RILEM conference on Microstructure Related Durability of Cementitious Composites (Microdurability2020)*, 485–492.

Andrew, M., Bijeljic, B., and Blunt, M.J. (2014). Pore-scale contact angle measurements at reservoir conditions using X-ray microtomography. *Advances in Water Resources* 68: 24–31.

Antognozzi, M., Humphris, A.D.L., and Miles, M.J. (2001). Observation of molecular layering in a confined water film and study of the layers viscoelastic properties. *Applied Physics Letters* 78 (3): 300–302.

Moisture Storage and Transport in Concrete: Experimental Investigations and Computational Modeling, First Edition. Lutz H. Franke.

Arai, T., Sato, K., Iida, A., and Tomitori, M. (2017). Quasi-stabilized hydration layers on muscovite mica under a thin water film grown from humid air. *Scientific Reports* 7: 4054.

Asay, D.B. and Kim, S.H. (2005). Evolution of the adsorbed water layer structure on silicon oxide at room temperature. *Journal of Physical Chemistry B* 109: 16760–16763.

Auroy, M., Poyet, S., Le Bescop, P., and Torrenti, J.-M. (2013). Impact of carbonation on the durability of cementitious materials: water transport properties characterization. *EPJ Web of Conferences* 56: 1008.

Auroy, M., Poyet, S., Le Bescop, P. et al. (2015). Impact of carbonation on unsaturated water transport properties of cement-based materials. *Cement and Concrete Research* 74: 44–58.

Babaee, M. and Castel, A. (2018). Water vapor sorption isotherms, pore structure, and moisture transport characteristics of alkali-activated and Portland cement-based binders. *Cement and Concrete Research* 113: 99–120.

Badmann, R., Stockhausen, N., and Setzer, M.J. (1981). The statistical thickness and the chemical potential of adsorbed water films. *Journal of Colloid and Interface Science* 82: 534–542.

Barnat-Hunek, D. and Smarzewski, P. (2015). Surface free energie of hydrophobic coatings of hybrid-fiber-reinforced high-performance concrete. *Materiali in Tehnologije / Materials and Technology* 49: 895–902.

Baroghel-Bouny, V. (2007). Water vapour sorption experiments on hardened cementitious materials: Part I: Essential tool for analysis of hygral behaviour and its relation to pore structure. *Cement and Concrete Research* 37: 414–437.

Baroghel-Bouny, V. (2007). Water vapour sorption experiments on hardened cementitious materials. Part II: Essential tool for assessment of transport properties and for durability prediction. *Cement and Concrete Research* 37: 438–454.

Bear, J. and Bachmat, Y. (1990). *Introduction to Modeling of Transport Phenomena in Porous Media*, Theory and Applications of Transport in Porous Media, vol. 4. Springer.

Beaudoin, J. and Odler, I. (2019). Hydration, setting and hardening of Portland cement. *Lea's Chemistry of Cement and Concrete.*

Bentrup, H. (1992). Untersuchungen zur Prüfung der Frostwiderstandsfähigkeit von Ziegeln im Hinblick auf lange Lebensdauer. Dissertation. Hamburg University of Technology TUHH.

Berodier, E. and Scrivener, K. (2015). Evolution of pore structure in blended systems. *Cement and Concrete Research* 73: 25–35.

Bianchi-Janetti, M. (2015). Hygrothermal analysis of building components inclosing air cavities: comparison between different modeling approaches and experimental results. Dissertation. Universität Innsbruck.

Blokhuis, E.M. and Kuipers, J. (2006). Thermodynamic expressions for the Tolman length. *Journal of Chemical Physics* 124: 74701.

Bollmann, K. (2000). Ettringitbildung in nicht wärmebehandelten Betonen. Dissertation. University Weimar (Germany).

van Breugel, K. (1997). *Simulation of Hydration and Formation of Structure in Hardening Cement-Based Materials*, 2e. Delft: Delft University Press.

Brue, F., Liang, Y., Burlion, N. et al. (2010). Influence of temperature and composition upon drying of concretes. In: *Fracture Mechanics of Concrete and Concrete Structures*, 745–750. Seoul: Korea Concrete Institute. ISBN: 978-89-5708-181-5.

Brue, F., Davy, C.A., Skoczylas, F. et al. (2012). Effect of temperature on the water retention properties of two high performance concretes. *Cement and Concrete Research* 42: 384–396.

Brunauer, S., Emmett, P.H., and Teller, E.J. (1938). Adsorption of gases in multimolecular layers. *Journal of the American Chemical Society* 60: 309.

Brunauer, S., Skalny, J., and Bodor, E.E. (1969). Adsorption on nonporous solids. *Journal of Colloid and Interface Science* 30: 546–552.

Bruot, N. and Caupin, F. (2016). Curvature dependence of the liquid-vapor surface tension beyond the Tolman approximation. *Physical Review Letters PRL* 116: 056102.

Bullard, J.W., Jennings, H.M., Livingston, R.A. et al. (2011). Mechanisms of cement hydration. *Cement and Concrete Research* 41: 1208–1223.

de Burgh, J.M. and Foster, S.J. (2017). Influence of temperature on water vapour sorption isotherms and kinetics of hardened cement paste and concrete. *Cement and Concrete Research* 92: 37–55.

de Burgh, J.M., Foster, S.J., and Valipour, H.R. (2016). Prediction of water vapour sorption isotherms and microstructure of hardened Portland cement pastes. *Cement and Concrete Research* 81: 134–150.

Byfors, J. (1980). *Plain Concrete at Early Ages*. Stockholm: Swedish Cement and Concrete Institute.

Cai, J., Perfect, E., Cheng, C.L., and Hu, X. (2014). Generalized modeling of spontaneous imbibition based on Hagen–Poiseuille flow in tortuous capillaries with variably shaped apertures. *Langmuir* 30: 5142–5151.

Calo, A., Domingo, N., Santos, S., and Verdaguer, A. (2015). Revealing water films structure from force reconstruction in dynamic AFM. *Journal of Physical Chemistry C* 119: 8258–8265.

Canut, M.M.C. and Geiker, M.R. (2011). Pore Structure in Blended Cement Pastes. *DTU Civil Engineering Report; No. R-268*. Technical University of Denmark.

Carlier, J.-Ph. and Burlion, M. (2011). Experimental and numerical assessment of the hydrodynamical properties of cementitious materials. *Transport in Porous Media* 86: 87–102.

Carmeliet, J. (2005). Moisture transport in open porous materials.

Carmeliet, J. and Roels, S. (2001). Determination of the isothermal moisture transport properties of porous building materials. *Journal of Thermal Envelope and Building Science* 24: 183–210.

Carmeliet, J. and Roels, S. (2002). Determination of the moisture capacity of porous building materials. *Journal of Building Physics* 25: 209–237.

Carmeliet, J., Descamps, F., and Houvenaghe, G. (1999). A multiscale network model for simulating moisture transfer properties of porous media. *Transport in Porous Media* 35: 67–88.

Carmeliet, J., Hens, H., Roels, S. et al. (2004). Determination of the liquid water diffusivity from transient moisture transfer experiments. *Journal of Thermal Envelope and Building Science* 27: 277–305.

Castrillon, S.R.-V., Giovambattista, N., Aksay, I.A., and Debenedetti, P.G. (2011). Structure and energetics of thin film water. *Journal of Physical Chemistry C* 115: 4624–4635.

Caupin, F., Arvengas, A., Davitt, K. et al. (2012). Exploring water and other liquids at negative pressure. *Journal of Physics: Condensed Matter* 24: 284110.

Chari, M.-N., Shekarchi, M., Sobhani, J., and Chari, M.N. (2016). The effect of temperature on the moisture transfer coefficient of cement-based mortars: an experimental investigation. *Construction and Building Materials* 102: 306–317.

Chen, W., Liu, J., Brue, F. et al. (2012). Water retention and gas relative permeability of two industrial concretes. *Cement and Concrete Research* 42: 1001–1013.

Chen, H., Wyrzykowski, M., Scrivener, K., and Lura, P. (2013). Prediction of self-desiccation in low water-to-cement ratio pastes based on pore structure evolution. *Cement and Concrete Research* 49: 38–47.

Chen, H., Li, Z., Wang, F. et al. (2017). Extrapolation of surface tensions of electrolyte and associating mixtures solutions. *Chemical Engineering Science* 162: 10–20.

Chen, Y., Al-Neshawy, F., and Punkki, J. (2021). Investigation on the effect of entrained air on pore structure in hardened concrete using MIP. *Construction and Building Material* 292: 123441.

Christensen, B.J., Mason, T.O., and Jennings, H.M. (1996). Comparison of measured and calculated permeabilities for hardened cement pastes. *Cement and Concrete Research* 26: 1325–1334.

Churaev, N.V., Setzer, M.J., and Adolphs, J. (1998). Influence of surface wettability on adsorption isotherms of water vapor. *Journal of Colloid and Interface Science* 197: 327–333.

Churaev, N., Starke, G., and Adolphs, J. (2000). Isotherms of capillary condensation influenced by formation of adsorption films. *Journal of Colloid and Interface Science* 221: 246–253.

Cimino, R., Cychosz, K.A., Thommes, M., and Neimark, A.V. (2013). Experimental and theoretical studies of scanning adsorption/desorption isotherms. *Colloids and Surfaces A: Physicochemical and Engineering Aspects* 437: 76–89.

Cohan, H.C. (1938). Sorption hysteresis and the vapor pressure of concave surfaces. *Journal of the American Chemical Society* 433–435.

Cook, R.A. and Hover, K.C. (1999). Mercury porosimetry of hardened cement pastes. *Cement and Concrete Research* 29: 933–943.

Cramer, F. (2016). Mehrfeld-Modell für chemisch-physikalische Alterungsprozesse von Beton. Dissertation. Technische Universität Braunschweig.

Crassous, J., Ciccotti, M., and Charlaix, E. (2011). Capillary force between wetted nanometric contacts and its application to atomic force microscopy. *Langmuir* 27: 3468–3473.

Cwikel, D., Zhao, Q., Liu, C. et al. (2010). Comparing contact angle measurements and surface tension assessments of solid surfaces. *Langmuir* 26 (19): 15289–15294.

D'Orazio, F., Tarczon, J.C., Halperin, W.P. et al. (1989). Application of nuclear magnetic resonance pore structure analysis to porous silica glass. *Journal of Applied Physics* 65: 742.

Dash, J.G. and Peierls, R. (1982). Characteristics of adsorbed films. *Physical Review B* 25: 8.

Davie, C.T., Pearce, C.J., and Kukla, K. (2018). Modelling of transport processes in concrete exposed to elevated temperatures – an alternative formulation for sorption isotherms. *Cement and Concrete Research* 106: 144–154.

De Bruijn, T.J.W., De Jong, W.A., and Van Den Berg, P.J. (1981). Kinetic parameters in Avrami—Erofeev type reactions from isothermal and non-isothermal experiments. *Therrnochirnica Acta* 45: 315–325.

deBoer, J.H., Linsen, B.G., and Osinga, T.J. (1965). Studies on pore systems in catalysis. VI. The universal t-curve. *Journal of Catalysis* 7: 643–648.

Deckelmann, G. (2014). *Handbuch für das Programm ASTRA (Allgemeines Stoff-Transport Simulations-Programm)*. Hamburg University of Technology TUHH.

Deckelmann, G. and Schmidt-Döhl, F. (2015). *Überprüfung von marktüblichen Feuchte- und Temperaturmessgeräten für die Beurteilung der Belegreife von Estrichen*. Inst. für Baustoffe, Bauphysik und Bauchemie. Forschungsbericht. Hamburg University of Technology TUHH.

Deckelmann, G. and Schmidt-Döhl, F. (2018). Das Feuchte-Aufnahme- und Abgabeverhalten zementgebundener Estriche und Konsequenzen für die Bestimmung der KRF. *TKB-Bericht 4*. Köln.

DelRio, F.W., Dunn, M.L., and de Boer, M.P. (2008). Capillary adhesion model for contacting micromachined surfaces. *Scripta Materialia* 59: 916–920.

Derjaguin, B.V. (1939). *A Theory of Capillary Condensation in the Pores of Sorbents and of other Capillary Phenomena Taking into Account the Disjoining Action of Polymolecular Liquid Films*, 46–62. Institute of Colloid Chemistry and Electrochemistry, The Academy of Sciences of the USSR.

Derjaguin, B.V. and Churaeva, N.V. (1992). Polymolecular adsorption and capillary condensation in narrow slit pores. *Progress in Surface Science* 40: 173–191.

Deschner, F., Winnefeld, F., Lothenbach, B. et al. (2012). Hydration of Portland cement with high replacement by siliceous fly ash. *Cement and Concrete Research* 42: 1389–1400.

Diamond, S. (2000). Mercury porosimetry? An inappropriate method for the measurement of pore size distributions in cement-based materials. *Cement and Concrete Research* 30: 1517–1688.

Do, D. (1998). *Adsorption Analysis: Equilibria and Kinetics*, Series on Chemical Engineering, vol. 2, 877 pp. Imperial College Press.

Drouet, E., Poyet, S., and Torrenti, J.-M. (2015). Temperature influence on water transport in hardened cement pastes. *Cement and Concrete Research* 76: 37–50.

Duan, C., Karnik, R., Lu, M.-C., and Majumdar, A. (2012). Evaporation-induced cavitation in nanofluidic channels. *Proceedings of the National Academy of Sciences of the United States of America* 109 (10): 3688–3693.

Dutzer, V., Dridi, W., Poyet, S. et al. (2018). The link between gas diffusion and carbonation in hardened cement pastes. *Cement and Concrete Research* 123: 105795.

Enjilela, R., de J. Cano-Barrita, P.F., Komar, A. et al. (2018). Wet front penetration with unsteady state wicking in mortar studied by Magnetic Resonance Imaging (MRI). *Materials and Structures* 51: 51.

Eriksson, D., Gasch, T., and Ansell, A. (2018). A hygro-thermo-mechanical multiphase model for long-term water absorption into air-entrained concrete. *Transport in Porous Media* 20: 573.

Eslami, F. and Elliott, J.A.W. (2019). Gibbsian thermodynamic study of capillary meniscus depth. *Scientific Reports* 9: 657.

Espinosa, R.-M. (2005). Sorptionsisothermen von Zementstein und Mörtel. Dissertation. Technische Universität Hamburg.

Espinosa, R.M. and Franke, L. (2006). Influence of the age and drying process on pore structure and sorption isotherms of hardened cement paste. *Cement and Concrete Research* 36: 1969–1984.

Espinosa, R.M. and Franke, L. (2006). Inkbottle pore-method: prediction of hygroscopic water content in hardened cement paste at variable climatic conditions. *Cement and Concrete Research* 36: 1954–1968.

Evans, R. (1990). Fluids adsorbed in narrow pores: phase equilibria and structure. *Journal of Physics: Condensed Matter* 2: 8989–9007.

Everett, D.H. (1986). Reporting data on adsorption from solution at the solid/solution interface (Recommendations 1986). *Pure and Applied Chemistry* 58: 967–984.

Extrand, C.W. and Moon, S.I. (2012). Which controls wetting? Contact line versus interfacial area: simple experiments on capillary rise. *Langmuir-Journal of the American Chemical Society* 28: 15629–15633.

Fabbri, A., McGregor, F., Costa, I., and Faria, P. (2017). Effect of temperature on the sorption curves of earthen materials. *Materials and Structures* 50: 253.

Fagerlund, G. (2004). A service life model for internal frost damage in concrete. Division of Building Materials. *Report TVBM*; Vol. 3119. Lund University.

Farah, K., Müller-Plathe, F., and Böhm, M.C. (2012). Classical reactive molecular dynamics implementations: state of the art. *ChemPhysChem* 13: 1127–1151.

Feldman, R.F. and Sereda, P.J. (1968). A model for hydrated Portland cement paste as deduced from sorption-length change and mechanical properties. *Materiaux et Constructions* 1: 509–520.

Feng, C. and Janssen, H. (2016). Hygric properties of porous building materials (II): analysis of temperature influence. *Building and Environment* 99: 107–118.

Feng, C., Janssen, H., Wu, C. et al. (2013). Validating various measures to accelerate the static gravimetric sorption isotherm determination. *Building and Environment* 69: 64–71.

Feng, C., Janssen, H., and Feng, Y. (2013). Effect of temperature on the sorption isotherm and vapor permeability. *The 2nd Central European Symposium on Building Physics 2013*.

Feng, D., Li, X., Wang, X. et al. (2018). Capillary filling of confined water in nanopores: coupling the increased viscosity and slippage. *Chemical Engineering Science* 186: 228–239.

Fenter, P. and Lee, S.S. (2014). Hydration layer structure at solid–water interfaces. *MRS Bulletin* 39: 1056–1061.

Fenter, P., Kerisit, S., Raiteri, P., and Gale, J.D. (2013). Is the calcite–water interface understood? Direct comparisons of molecular dynamics simulations with specular X-ray reflectivity data. *The Journal of Physical Chemistry C* 117: 5028–5042.

Fischer, N., Haerdtl, R., and McDonald, P.J. (2015). Observation of the redistribution of nanoscale water filled porosity in cement based materials during wetting. *Cement and Concrete Research* 68: 148–155.

Fogarty, J., Aktulga, H.M., Grama, A.Y. et al. (2010). A reactive molecular dynamics simulation of the silica-water interface. *Journal of Chemical Physics* 132: 174704.

Franke, L. and Bentrup, H. (1993). Evaluation of the frost resistance of bricks in regard to long service life. II. *ZI International* 46 (9): 528–536.

Franke, L., Deckelmann, G., and Schmidt, H. (2009). Behavior of ultra high performance concrete with respect to chemical attack. *ibausil Conference*, Weimar.

Freymuth, H., Jenisch, R., Klopfer, H. et al. (ed.) (2002). *Lehrbuch der Bauphysik*. Vieweg+Teubner Verlag.

Fukuma, T. (2015). Mechanism of atomic force microscopy imaging of three-dimensional hydration structures at a solid-liquid interface. *Physical Review B* 92: 155412.

Funk, M. (2012). Hysteresis der Feuchtespeicherung in porösen Materialien. Dissertation. Technische Universität Dresden.

Fylak, M.J. (2011). Untersuchungen zum frühen Hydratationsverhalten von Portland- und Portlandkompositzementen. Dissertation. Martin-Luther-Universität Halle-Wittenberg.

Gajewicz, A.M., Gartner, E., Kang, K. et al. (2016). A ^{1}H NMR relaxometry investigation of gel-pore drying shrinkage in cement pastes. *Cement and Concrete Research* 86: 12–19.

Gallage, Ch., Kodikara, J., and Uchimura, T. (2013). Laboratory measurement of hydraulic conductivity functions of two unsaturated sandy soils during drying and wetting processes. *Soils and Foundations (The Japanese Geotechnical Socie)* 53 (3): 417–430.

Gallucci, E., Zhang, X., and Scrivener, K.L. (2013). Effect of temperature on the microstructure of calcium silicate hydrate (C-S-H). *Cement and Concrete Research* 53: 185–195.

Garboczi, E.J. and Bentz, D.P. (1992). Computer simulation of the diffusivity of cement-based materials. *Journal of Materials Science* 27: 2083–2092.

Garrecht, H. (1992). Porenstrukturmodelle für den Feuchtehaushalt von Baustoffen mit und ohne Salzbefrachtung und rechnerische Anwendung auf Mauerwerk. Dissertation. Universität Karlsruhe.

Gätje, B. (2004). Nachträgliche Ermittlung betontechnologischer Parameter in Zementstein, Mörtel und Betonen unbekannter Zusammensetzung. Dissertation. Technische Universität Hamburg.

Gawin, D. and Sanavi, L. (2010). Simulation of cavitation in water saturated porous media considering effects of dissolved air. *Transport in Porous Media* 141–160.

Gawin, D., Lefik, M., and Schrefler, B.A. (2001). ANN approach to sorption hysteresis within a coupled hygro-thermo-mechanical FE analysis. *International Journal for Numerical Methods in Engineering* 50: 299–323.

Geiker, M. and Jacobsen, S. (2019). Self-compacting concrete (SCC). *Developments in the Formulation and Reinforcement of Concrete*, Elsevier.

de Gennes, P.G. (1985). Wetting: statics and dynamics. *Reviews of Modern Physics* 57 (3): 827.

van Genuchten, M.Th. (1980). A closed-form equation for predicting the hydraulic conductivity of unsaturated soils. *Soil Science Society of America Journal* 44: 892–898.

Gertis, T.S.K. (2015). Zur Ermittlung der Sorptionsenthalpie von Baustoffen. *Bauphysik* 37 (2): 71–80.

Gluth, G.J.G. (2011). Die Porenstruktur von Zementstein und seine Eignung zur Gastrennung. Dissertation. Technische Universität Berlin.

Goedeke, H.K. (2018). Lösender Angriff auf zementgebundene Baustoffe - Veränderungen der Porenstruktur und Folgen für Transport- und Korrosionsprozesse. Dissertation. Hamburg University of Technology.

Gómez, I., Sala, J.M., and Millán, J.A. (2007). Characterization of moisture transport properties for lightened clay brick / comparison between two manufacturers. *Journal of Building Physics* 31: 179–194.

Graef, H. and Grube, H. (1986). Einfluss der Zusammensetzung und der Nachbehandlung de Betons auf seine Gasdurchlässigkeit. *Beton* 36 (11+12): 426-9–473-6.

Graue, A. and Fernø, M.A. (2011). Water mixing during spontaneous imbibition at different boundary and wettability conditions. *Journal of Petroleum Science and Engineering* 78: 586–595.

Grosman, A. and Ortega, C. (2005). Nature of capillary condensation and evaporation processes in ordered porous materials. *Langmuir* 21: 10515–10521.

Gruener, S. and Huber, P. (2018). Capillarity-driven oil flow in nanopores: darcy scale analysis of lucas washburn imbibition dynamics. *Transport in Porous Media* 115: 189.

Gruener, S., Hofmann, T., Wallacher, D. et al. (2009). Capillary rise of water in hydrophilic nanopores. *Physical Review E-APS* 1–5.

Gründing, D., Smuda, M., Antritter, T. et al. (2020). A comparative study of transient capillary rise using direct numerical simulations. *Applied Mathematical Modelling* 86: 142–165.

Grunewald, J. (1997). Diffusiver und konvektiver Stoff- und Energietransport in kapillar-porösen Baustoffen. Dissertation. Technische Universität Dresden.

Guizzardi, M., Derome, D., and Carmeliet, J. (2016). Water uptake in clay brick at different temperatures: experiments and numerical simulations. *Journal of Building Physics* 39: 373–389.

Hagentoft, C.-E., Kalagasidis, A.S., Adl-Zarrabi, B. et al. (2004). Assessment method of numerical prediction models for combined heat, air and moisture transfer in building components: benchmarks for one-dimensional cases. *Journal of Thermal Envelope and Building Science* 27: 327–352.

Hall, C. (2006). Anomalous diffusion in unsaturated flow: fact or fiction? *Cement and Concrete Research* 37: 378–385.

Hall, C. and Hoff, W.D. (2012). *Water Transport in Brick, Stone and Concrete*. New York: Spon Press. ISBN: 9-780-41556-4670.

Hall, C. and Hoff, W.D. (2021). *Water Transport in Brick, Stone and Concrete*. CRC Press.

Hansen, K.K. (1986). *Sorption Isotherms. A Catalogue*. Technical University of Denmark. Building Materials Laboratory.

Hansen, T.C. (1986). Physical structure of hardened cement paste. A classical approach. *Materials and Structures* 19: 423–436.
Harkins, W.D. and Brown, F.E. (1919). The determination of surface tension (free surface energy), and the weight of falling drops: the surface tension of water and benzene by the capillary height method. *Journal of the American Chemical Society* 41 (4): 499–524.
Haugen, Å., Fernø, M.A., Mason, G., and Morrow, N.R. (2014). Capillary pressure and relative permeability estimated from a single spontaneous imbibition test. *Journal of Petroleum Science and Engineering* 115: 66–77.
Häupl, P., Homann, M., Kölzow, C. et al. (ed.) (2013). *Lehrbuch der Bauphysik*. Wiesbaden: Springer Fachmedien Wiesbaden.
Hearn, N. (1998). Self-sealing, autogenous healing and continued hydration: what is the difference? *Materials and Structures/Matériaux et Constructions* 31: 563–567.
Hearn, N. and Morley, C.T. (1997). Self-sealing property of concrete - experimental evidence. *Materials and Structures/Matériaux et Constructions* 30: 404–411.
Hedenblad, G. (1993). *Moisture Permeability of Mature Concrete, Cement Mortar and Cement Paste*. Division of Building Materials. Forschungsbericht. Lund Institute of Technology.
Hendrickx, R., Van Balen, K., Van Gemert, D., and Roels, S. (2009). Measuring and modelling water transport from mortar to brick. WTA and München.
Henriksson, U. and Eriksson, J.C. (2004). Thermodynamics of capillary rise: why is the meniscus curved? *Journal of Chemical Education* 81 (1): 150.
Heshmati, M. and Piri, M. (2014). Experimental investigation of dynamic contact angle and capillary rise in tubes with circular and noncircular cross sections. *Langmuir* 30: 14151–14162.
Holm, A.H. (2003). *Baupysik - Feuchteschutz 5*. Vortrag Holzkirchen 25.11.2003. Fraunhofer-Institut für Bauphysik.
Holthausen, R.S. and Raupach, M. (2018). Monitoring the internal swelling in cementitious mortars with single-sided ^{1}H nuclear magnetic resonance. *Cement and Concrete Research* 111: 138–146.
Homann, M. (2013). Feuchtespeicherung. In: *Lehrbuch der Bauphysik* (ed. P. Häupl et al.), 161–195. Wiesbaden: Springer Fachmedien Wiesbaden.
van Honschoten, J.W., Brunets, N., and Tas, N.R. (2010). Capillarity at the nanoscale. *Chemical Society Reviews* 39: 1096–1114.
Hou, D., Li, T., and Wang, P. (2018). Molecular dynamics study on the structure and dynamics of NaCl solution transport in the nanometer channel of CASH Gel. *ACS Sustainable Chemistry & Engineering* 6: 9498–9509.
Houst, Y.F. (1992). Diffusion de gas, carbonatation et retrait de la pâte de ciment durcie. PhD thesis. Ecole Polytechnique Fédérale de Lausanne.
Houst, Y.F. and Wittmann, F.H. (1994). Influence of porosity and water content on the diffusivity of CO_2 and O_2 through hydrated cement paste. *Cement and Concrete Research* 24 (6): 1165–1176.
Hu, Z., Wyrzykowskib, M., Scrivenera, K., and Lura, P. (2018). A novel method to predict internal relative humidity in cementitious materials by ^{1}H NMR. *Cement and Concrete Research* 104: 80–93.

Huang, Q., Jiang, Z., Gu, X. et al. (2015). Numerical simulation of moisture transport in concrete based on a pore size distribution model. *Cement and Concrete Research* 67: 31–43.

Huber, J. (2008). Auswirkungen des Wasserverlustes durch Evaporation in jungem Alter am Beispiel von Straßenbeton. Dissertation. Technische Universität München.

Hundt, J. and Kantelberg, H. (1978). Sorptionsuntersuchungen an Zementstein. *Schriftenreihe DAfStB H.* 297: 25–39.

Hunkeler, F. (2012). Einfluss des CO_2-Gehaltes, der Nach- und Vorbehandlung sowie der Luftfeuchtigkeit auf die Karbonatisierungsgeschwindigkeit von Beton. *Beton- und Stahlbetonbau* 107 (9): 613–624.

Hunkeler, F. and Holtzhauer, K. (1994). Mörtel und Beton: Wassergehalt, Porosität und elektrischer Widerstand. *Schweizer Ingenieur und Architekt* 27/28: 534–541.

Ichikawa, N., Hosokawa, K., and Maeda, R. (2004). Interface motion of capillary-driven flow in rectangular microchannel. *Journal of Colloid and Interface Science* 280: 155–164.

Ioannou, I., Charalambous, C., and Hall, C. (2017). The temperature variation of the water sorptivity of construction materials. *Materials and Structures* 50: 1871.

Ishida, T., Maekawa, K., and Kishi, T. (2007). Enhanced modeling of moisture equilibrium and transport in cementitious materials under arbitrary temperature and relative humidity history. *Cement and Concrete Research* 37: 565–578.

Israelachvili, J.N. (2011). *Intermolecular and Surface Forces*, 3e. Elsevier.

Israelachvili, J., Min, Y., Akbulut, M. et al. (2010). Recent advances in the surface forces apparatus (SFA) technique. *Reports on Progress in Physics* 73: 036601.

Jacobs, F.P. (1994). Permeabilität und Porengefüge zementgebundener Werkstoffe. Dissertation. ETH Zürich.

Jasinska, J. (2011). Particle size and pore structure of nanomaterials. Dissertation. Szczecin.

Jasper, J.J. (1972). The surface tension of pure liquid compounds. *Journal of Physical and Chemical Reference Data* 1: 841.

Jaynes, D.B. (1984). Comparison of soil-water hysteresis models. *Journal of Hydrology* 75: 287–299.

Jennings, H.M. (2000). A model for the microstructure of calcium silicate hydrate in cement paste. *Cement and Concrete Research* 30: 101–116.

Jennings, H.M., Bullard, J.W., Thomas, J.J. et al. (2008). Characterisation and modeling of pores and surfaces in cement paste: correlations to processing and properties. *Journal of Advanced Concrete Technology* 6 (1): 5–29.

Jennings, H.M., Kumar, A., and Gaurav, S. (2015). Quantitative discrimination of the nano-pore-structure of cement paste during drying: new insights from water sorption isotherms. *Cement and Concrete Research* 76: 27–36.

Jensen, O.M. and Hansen, P.F. (2001). Autogenous deformation and RH-change in perspective. *Cement and Concrete Research* 31: 1859–1865.

Jiang, J. and Yuan, Y. (2012). Relationship of moisture content with temperature and relative humidity in concrete. *Magazine of Concrete Research* 65 (11): 685–692.

Jiang, J., Yuan, Y., Zeng, Q., and Mo, T. (2013). Discussion: relationship of moisture content with temperature and relative humidity in concrete. *Magazine of Concrete Research* 65 (24): 1494–1496.

Jiang, Z., Xi, Y., Gu, X. et al. (2019). Modelling of water vapour sorption hysteresis of cement-based materials based on pore size distribution. *Cement and Concrete Research* 115: 8–19.

Johannesson, B. and Nyman, U. (2010). A numerical approach for non-linear moisture flow in porous materials with account to sorption hysteresis. *Transport in Porous Media* 84: 735–754.

Jooss, M. (2001). Dichtheit von Heißwasser-Langzeitspeichern aus Hochleistungsbeton. Dissertation Stuttgart University (in German).

Jooss, M. and Reinhardt, H.W. (2002). Permeability and diffusivity of concrete as function of temperature. *Cement and Concrete Research* 32: 1497–1504.

Julnipitawong, P. (2010). Moisture transport in concrete, experiment investigations and simulation programs. Dissertation. Technische Universität Hamburg.

Kalies, G. (2009). Lecture, Hochschule für Technik und Wirtschaft Dresden.

Kameche, Z.A., Ghomari, F., Choinska, M., and Khelidj, A. (2014). Assessment of liquid water and gas permeabilities of partially saturated ordinary concrete. *Construction and Building Materials* 65: 551–565.

Kargas, G. and Kerkides, P. (2005). Hysteretic curve prediction: comparison of two models. *Transport in Porous Media* 59: 97–113.

Kast, W. and Jokisch, F. (1972). Überlegungen zum Verlauf von Sorptionsisothermen und zur Sorptionskinetik an porösen Feststoffen. *Chemie Ingenieur Technik* 44: 556–563.

Kats, B.M. and Kutarov, V.V. (1993). A three-parameter equation for describing the sorption of water vapor by chemisorptive fibres. *Fibre Chemistry* 25 (5): 374–376.

Kenning, D.B.R., Wen, D.S., Das, K.S., and Wilson, S.K. (2006). Confined growth of a vapour bubble in a capillary tube at initially uniform superheat: experiments and modelling. *International Journal of Heat and Mass Transfer* 49: 23–24.

Kenny, A. and Katz, A. (2020). Cement composition's effect on pore solution composition and on electrochemical behavior of reinforcing steel. Israel Institute for Technology, Haifa, Israel, Reprint.

Kim, S., Kim, D., Kim, J. et al. (2018). Direct evidence for curvature-dependent surface tension in capillary condensation: Kelvin equation at molecular scale. *Physical Review X* 8: 041046.

Kimura, K., Ido, S., Oyabu, N. et al. (2010). Visualizing water molecule distribution by atomic force microscopy. *Journal of Chemical Physics* 132: 194705.

Kinloch, A.J. (1980). Review The science of adhesion Part 1: Surface and interfacia/ aspects. *Journal of Materials Science* 15: 2141–2166.

Knight, A.W., Kalugin, N.G., Coker, E., and Ilgen, A.G. (2019). Water properties under nano-scale confinement. *Scientific Reports* 9: 8246.

Krispel, S. (2009). Einfluss des Luftgehaltes im Frischbeton auf die Luftporenkennwerte. *Zement + Beton* 5: 12.

Kropp, J. (1983). Karbonatisierung und Tansportvorgänge in Zementstein. Dissertation. Universität Stuttgart.

Krus, M. (1996). Moisture transport and storage coefficients of porous mineral building materials: theoretical principles and new test methods. Reprint Dissertation. (Stuttgart University) by Fraunhofer IRB Verlag.

Krus, M. and Künzel, H.M. (1993). Determination of Dw from A-value. *IEA Annex XXIV Report T3-D-39/02*.

Krus, M., Hansen, K.K., and Künzel, H.M. (1997). Porosity and liquid absorption of cement paste. *Materials and Structures/Materiaux et Construction* 30: 394–398.

Kuchin, I.V. and Starov, M. (2016). Hysteresis of the contact angle of a meniscus inside a capillary with smooth, homogeneous solid walls. *Langmuir* 32: 5333–5340.

Kuchin, I.V., Matar, O.K., Craster, R.V., and Starov, V.M. (2014). Influence of the disjoining pressure on the equilibrium interfacial profile in transition zone between a thin film and a capillary meniscus. *Colloids and Interface Science Communications* 1: 18–22.

Kuchin, I.V., Matar, O.K., Crasterc, R.V., and Starov, V.M. (2014). Modeling the effect of surface forces on the equilibrium liquid profile of a capillary meniscus. *Soft Matter* 10: 6024–6037.

Kumar, A., Ketel, S., Vance, K. et al. (2014). Water vapor sorption in cementitious materials—measurement, modeling and interpretation. *Transport in Porous Media* 103: 69–98.

Kumikov, V.K. and Khokonov, Kh.B. (1983). On the measurement of surface free energy and surface tension of solid metals. *Journal of Applied Physics* 54: 1346–1350.

Laliberté, M. (2007). Model for calculating the viscosity of aqueous solutions. *Journal of Chemical and Engineering Data* 52: 321–335.

Laliberté, M. (2009). A model for calculating the heat capacity of aqueous solutions, with updated density and viscosity data. *Journal of Chemical and Engineering Data* 54: 1725–1760.

Langmuir, I. (1940). Monolayers on solids. *Journal of the Chemical Society* 511–543.

Laube, M. (1990). Werkstoff-Modell zur Berechnung von Temperatur-Spannungen in massigen Beton-Bauteilen im jungen Alter. Dissertation. Technische Universität Braunschweig.

Lauth, G.-J. and Kowalczyk, J. (2016). *Einführung in die Physik und Chemie der Grenzflächen und Kolloide. Experimentelle Methoden: Messung der Grenzflächenspannung*. Springer-Spektrum.

Lebeau, M. and Konrad, J.-M. (2010). A new capillary and thin film flow model for predicting the hydraulic conductivity of unsaturated porous media. *Water Resources Research* 46: W12554.

Leech, C.A. (2003). Water movement in unsaturated concrete. Dissertation. University of Queensland.

Lei, W., Rigozzi, M.K., McKenzie, D.R. et al. (2016). The physics of confined flow and its application to water leaks, water permeation and water nanoflows: a review. *Rep. Prog. Phys.* 79: 025901.

Li, J., Li, X., Wu, K. et al. (2016). Water sorption and distribution characteristics in clay and shale: effect of surface force. *Energy Fuels* 30: 8863–8874.

Li, J., Li, X., Wang, X. et al. (2016). Water distribution characteristic and effect on methane adsorption capacity in shale clay. *International Journal of Coal Geology* 159: 135–154.

Li, D., Zhao, W., Hou, D., and Zhao, T. (2017). Molecular dynamics study on the chemical bound, physical adsorbed and ultra-confined water molecules in the nano-pore of calcium silicate hydrate. *Construction and Building Materials* 151: 563–574.

Li, J., Li, X., Wu, K. et al. (2017). Thickness and stability of water film confined inside nanoslits and nanocapillaries of shale and clay. *International Journal of Coal Geology* 179: 253–268.

Lim, E. (2017). A hydrodynamic analysis of thermocapillary convection in evaporating thin liquid films. *International Journal of Heat and Mass Transfer* 108: 1103–1114.

Liu, X. (2009). Transport properties of concrete with lightweight aggregates. Dissertation. University of Singapore.

Liu, Z. (2014). Frost deterioration in concrete due to deicing salt exposure: mechanism, mitigation and conceptual surface scaling model. Dissertation. University of Michigan.

Liu, Z. (2017). Experimental investigation and quantitative calculation of the degree of hydration and products in fly ash-cement mixtures. *Advances in Materials Science and Engineering* 2017: 2437270.

Liu, Z. and Hansen, W. (2016). Effect of hydrophobic surface treatment on freeze-thaw durability of concrete. *Cement and Concrete Composites* 69: 49–60.

Liu, Z., Hansen, W., Wang, F., and Zhang, W. (2018). Simulation of air-void system in hardened concrete using a geometrical model. *Magazine of Concrete Research* 71: 680–689.

Lockington, D., Parlange, J.-Y., and Dux, P. (1999). Sorptivity and the estimation of water penetration into unsaturated concrete. *Materials and Structures/Materiaux et Construction* 32: 342–347.

Lothenbach, B., Winnefeld, F., Alder, C. et al. (2007). Effect of temperature on the pore solution, microstructure and hydration products of Portland cement pastes. *Cement and Concrete Research* 37: 483–491.

Lu, N. (2016). Generalized soil water retention equation for adsorption and capillarity. *Journal of Geotechnical and Geoenvironmental Engineering* 142: 04016051.

Ludwig, R. (2001). Wasser: von Clustern in die Flüssigkeit. *Angewandte Chemie* 113: 1856–1876.

Lunk, P. and Wittmann, F.H. (1996). Feuchtigkeits- und Salztransport in Beton I. Versuchsergebnisse und analytisches Modell. *Werkstoffwissenschaften und Bauinstandsetzen*, 4. Intern., Kolloqu., Esslingen.

Marchand, A., Weijs, J.H., Snoeijer, J.H., and Andreotti, B. (2011). Why is surface tension a force parallel to the interface? *American Journal of Physics* 79: 999–1008.

Maris, H. and Balibar, S. (2000). Negative pressures and cavitation in liquid helium. *Physics Today* 53: 29–34.

Marry, V., Rotenberg, B., and Turq, P. (2008). Structure and dynamics of water at a clay surface from molecular dynamics simulation dynamics simulation. *Physical Chemistry Chemical Physics* 10: 4802–4813.

Martin-Jimenez, D., Chacon, E., Tarazona, P. et al. (2016). Atomically resolved three-dimensional structures of electrolyte aqueous solutions near a solid surface. *Nature Communications* 7: 12164.

Maruyama, I., Rymes, J., Vandamme, M., and Coasne, B. (2018). Cavitation of water in hardened cement paste under short- term desorption measurements. *Materials and Structures* 51: 159.

Maruyama, I., Ohkubo, T., Haji, T., and Kurihara, R. (2019). Dynamic microstructural evolution of hardened cement paste during first drying monitored by ^{1}H NMR relaxometry. *Cement and Concrete Research* 122: 107–117.

Mason, G. and Morrow, N.R. (2013). Developments in spontaneous imbibition and possibilities for future work. *Journal of Petroleum Science and Engineering* 110: 268–293.

Matiasovsky, P. and Mihalka, P. (2014). Pore structure parameters and drying rates of building materials. In: *Drying and Wetting of Building Materials and Components* (ed. J.M.P.Q. Delgado). Springer Switzerland.

Matiasovsky, P. and Mihalka, P. (2019). Capillary moisture content - parameter of water suction and drying. *Thermophysics, AIP Conference Proceedings*, Volume 2170.

Matsuoka, H., Fukui, Sh., and Kato, T. (2002). Nanomeniscus forces in undersaturated vapors: observable limit of macroscopic characteristics. *Langmuir* 18: 6796–6801.

Mattia, D., Starov, V., and Semenov, S. (2012). Thickness, stability and contact angle of liquid films on and inside nanofibres, nanotubes and nanochannels. *Journal of Colloid and Interface Science* 384 (1): 149–156.

McDonald, P.J., Rodin, V., and Valori, A. (2010). Characterisation of intra- and inter-C–S–H gel pore water in white cement based on an analysis of NMR signal amplitudes as a function of water content an analysis of NMR signal amplitudes as a function of water content. *Cement and Concrete Research* 40: 1656–1663.

McDonald, P.J., Istok, O., Janota, M. et al. (2020). Sorption, anomalous water transport and dynamic porosity in cement paste: a spatially localised ^{1}H NMR relaxation study and a proposed mechanism. *Cement and Concrete Research* 133: 106045.

Menzl, G., Gonzalez, M.A., Geiger, P. et al. (2016). Molecular mechanism for cavitation in water under tension. *Proceedings of the National Academy of Sciences of the United States of America* 113 (48): 13582–13587.

Mercury, L., Azaroual, M., Zeyen, H., and Tardy, Y. (2003). Thermodynamic properties of solutions in metastable systems under negative or positive pressures. *Geochimica et Cosmochimica Acta* 67 (10): 1769–1785.

Mette, B. (2014). Experimentelle und numerische Untersuchungen zur Reaktionsführung thermochemischer Energiespeicher. Technische Universität Dresden.

Miller, R.D. (1994). Comment on Paradoxes and realities in unsaturated flow theory. *Water Resources Research* 30 (5): 1623–1624.

Min, W., Johannesson, B., and Geiker, M. (2014). A study of the water vapor sorption isotherms of hardened cement pastes: possible pore structure changes at low relative humidity and the impact of temperature on isotherms. *Cement and Concrete Research* 56: 97–105.

Molendowska, A., Wawrze, J., and Kowalczyk, H. (2020). Development of the measuring techniques for estimating the air void system parameters in concrete using 2D analysis method. *Materials* 13: 482.

Monroe, J., Barry, M., DeStefano, A. et al. (2020). Water structure and properties at hydrophilic and hydrophobic surfaces. *Annual Review of Chemical and Biomolecular Engineering* 11: 523–557.

Morishige, K. and Kittaka, S. (2015). Kinetics of capillary condensation of water in mesoporous carbon: nucleation and meniscus growth. *Journal of Physical Chemistry C* 119: 18287–18292.

Mualem, Y. (1974). A conceptual model of hysteresis. *Water Resources Research* 10: 514–520.

Muller, A.C.A., Scrivener, K.L., Gajewicz, A.M., and McDonald, P.J. (2013). Densification of C–S–H measured by ^{1}H NMR relaxometry. *Journal of Physical Chemistry C* 117: 403–412.

Neumann, T., Bergins, C., and Strauß, K. (2006). Mehrphasige Durchströmung kompressibler und heterogener poröser Medien. *Chemie Ingenieur Technik* 78: 701–708.

Nilsson, L.-O. (2006). Modelling moisture conditions in cementitious materials - some present challenges. *2nd International Symposium on Advances in Concrete through Science and Engineering*, Quebec City, Canada.

Nokken, M.R. and Hooton, R.D. (2008). Using pore parameters to estimate permeability or conductivity of concrete. *Materials and Structures* 41: 1–16.

Papatzani, S., Paine, K., and Calabria-Holley, J. (2015). A comprehensive review of the models on the nanostructure of calcium silicate hydrates. *Construction and Building Materials* 74: 219–234.

Patel, R.A., Phung, Q.T., Seetharam, S.C. et al. (2016). Diffusivity of saturated ordinary Portland cement-based materials: a critical review of experimental and analytical modelling approaches. *Cement and Concrete Research* 90: 52–72.

Pavlik, Z., Žumár, J., Medved, I., and Černý, R. (2012). Water vapor adsorption in porous building materials: experimental measurement and theoretical analysis. *Transport in Porous Media* 91: 939–954.

Pel, L. (1995). Moisture transport in porous building materials. Dissertation. University Eindhoven.

Peng, J., Guo, J., Ma, R., and Jiang, Y. (2022). Water-solid interfaces probed by high-resolution atomic force microscopy. *Surface Science Reports* 71: 100549.

Pérez-Díaz, J.L., Álvarez-Valenzuelab, M.A., and García-Pradab, J.C. (2012). The effect of the partial pressure of water vapor on the surface tension of the liquid water–air interface. *Journal of Colloid and Interface Science* 381: 180–182.

Pham, H.Q., Fredlund, D.G., and Barbour, S.L. (2005). A study of hysteresis models for soil-water characteristic curves. *Canadian Geotechnical Journal* 42: 1548–1568.

Phung, Q.T., Maes, N., De Schutter, G. et al. (2013). Determination of water permeability of cementitious materials using a controlled constant flow method. *Construction and Building Materials* 47: 1488–1496.

Pickett, G. (1945). Modification of the Brunauer-Emmett-Teller theory of multimolecular adsorption. *Contribution from the Portland Cement Association*, Vol. 67, Chicago, Illinois.

Plagge, R. and Teutsch, A. (2003). *Water Retention Transfer Functions of Ceramic Bricks of the Dresden Building Stock*. Institute of Building Climatology, University of Technology Dresden. Institutsmitteilung. Technische Universität Dresden.

Plagge, R., Funk, M., Scheffler, G., and Grunewald, J. (2006). Experimentelle Bestimmung der hygrischen Sorptionsisotherme und des Feuchtetransportes unter instationären Bedingungen. *Bauphysik* 28: 81–87.

Plagge, R., Scheffler, G., and Nicolai, A. (2007). Experimental methods to derive hygrothermal material functions for numerical simulation tools. *Building X Ashrae*.

Poós, T. and Varju, E. (2020). Mass transfer coeffcient for water evaporation by theoretical and empirical correlations. *International Journal of Heat and Mass Transfer* 153: 119500.

Popescu, M.N., Ralston, J., and Sedev, R. (2008). Capillary rise with velocity-dependent dynamic contact angle. *Langmuir* 24: 12710–12716.

Portuguez, E., Alzina, A., Michaud, P. et al. (2017). Evolution of a water pendant droplet: effect of temperature and relative humidity. *Natural Science* 9: 1–20.

Powers, T.C. and Brownyards, T.L. (1948). Studies of the physical properties of hardened cement paste. Research Laboratories of the Portland Cement Association, Chicago Bulletin 22.

Poyet, S. (2009). Experimental investigation of the effect of temperature on the first desorption isotherm of concrete. *Cement and Concrete Research* 39: 1052–1059.

Poyet, S. (2016). Describing the influence of temperature on water retention using van Genuchten equation. *Cement and Concrete Research* 84: 41–47.

Poyet, S. and Charles, S. (2009). Temperature dependence of the sorption isotherms of cement-based materials: heat of sorption and Clausius–Clapeyron formula. *Cement and Concrete Research* 39: 1060–1067.

Qier, W., Rougelot, T., Burlion, N., and Bourbon, X. (2014). Experimental investigation of the first desorption isotherm of a high performance concrete with thin sliced samples. *Construction and building materials* 72: 389–397.

Qin, M., Belarbi, R., Ait-Mokhtar, A., and Nilsson, L.O. (2008). Nonisothermal moisture transport in hygroscopic building materials: modeling for the determination of moisture transport coefficients. *Transport in Porous Media* 72: 255–271.

Radjy, F., Sellevold, E.J., and Hansen, K.K. (2003). *Isosteric Vapour Pressure - Temperature Data for Water Sorption in Hardened Cement Paste*. Trondheim: Report Departement of Civil Engineering MTNU.

Ranaivomanana, H., Verdier, J., Sellier, A., and Bourbon, X. (2011). Toward a better comprehension and modeling of hysteresis cycles in the water sorption / desorption process for cement based materials. *Cement and Concrete Research* 41: 817–827.

Ranaivomanana, H., Verdier, J., Sellier, A., and Bourbon, X. (2013). Prediction of relative permeabilities and water vapor diffusion reduction factor for cement-based materials. *Cement and Concrete Research* 48: 53–63.

Rastogi, M., Müller, A., Haha, M.B., and Scrivener, K.L. (2022). Role of cavitation in drying cementitious materials. *Cement and Concrete Research* 154: 106710.

Recknagel, H., Sprenger, E., and Schramek, E.-R. (ed.) (2009). *Taschenbuch für Heizung und Klimatechnik*, 74e. Oldenbourg Industrieverlag GmbH.

Reichinger, M. (2007). Poröse Silicate mit hierarchischer Porenstruktur: Synthese von mikro-/mesoporösen MCM-41 und MCM-48 Materialien aus zeolithischen Baueinheiten des MFI-Gerüststrukturtyps. Dissertation. Ruhr-Universität Bochum.

Ren, F., Chen, X., Zeng, Q., and Zhou, C. (2022). Effects of pure carbonation on pore structure and water permeability of white cement mortars. *Cement* 9: 100040.

Renard, T. and Guye, P.A. (1907). The rise of liquids in capillary tubes. *Journal de Chimie Physique et de Physico-Chimie Biologique* 5: 81–112.

Ressurecction, A.C., Moldrup, P., Tuller, M. et al. (2011). Relationship between specific surface area and the dry end of the water retention curve for soils with varying clay and organic carbon contents. *Water Resources Research* 47: W06522.

Richards, T.W. and Coombs, L.B. (1915). The surface tension of water, methyl, ethyl and isobutyl alcohols. *Journal of the American Chemical Society*, Contribution from the Kent Chemical Laboratory of the University of Chicago: 1656.

Richardson, I.G. (1997). The structure of the calcium silicate hydrate phases present in hardened pastes of white Portland cement/blast-furnace slag blends. *Journal of Materials Science* 32: 4793–4802.

Richardson, I.G. (1999). The nature of C-S-H in hardened cements. *Cement and Concrete Research* 29: 1131–1147.

Richardson, I.G. (2004). Tobermorite/jennite- and tobermorite/calcium hydroxide-based models for the structure of C-S-H: applicability to hardened pastes of tricalcium silicate, h-dicalcium silicate, Portland cement, and blends of Portland cement with blast-furnace slag, metakaolin, or silica fume. *Cement and Concrete Research* 34: 1733–1777.

Richardson, I.G. and Groves, G.W. (1993). Microstructure and microanalysis of hardened ordinary Portland cement pastes. *Journal of Materials Science* 28: 265–277.

Ridgway, C.J., Schoelkopf, J., Matthews, G.P. et al. (2001). The effects of void geometry and contact angle on the absorption of liquids into porous calcium carbonate structures. *Journal of Colloid and Interface Science* 239: 417–431.

Roels, S. (2000). Modelling unsaturated moisture transport in heteogeneous limestone. Dissertation. Leuven: Katholiek Universiteit Leuven.

Röser, F. (2018). Über die Reaktivität von Betonzusatzstoffen -Ein versuchsbasiertes Hydratationsmodell. Dissertation. Darmstadt: Technische Universität Darmstadt.

Rossen, J.E. (2014). Composition and morphology of C–A–S–H in pastes of alite and cement blended with supplementary cementitious materials. Dissertation. Ecole polytechnique de Lausanne.

Rossen, J.E. and Scrivener, K.L. (2017). Optimization of SEM-EDS to determine the C–A–S–H composition in matured cement paste samples. *Materials Characterization* 123: 294–306.

Rossen, J.E., Lothenbach, B., and Scrivener, K.L. (2015). Composition of C–S–H in pastes with increasing levels of silica fume addition. *Cement and Concrete Research* 75: 14–22.

Rößler, C., Stark, J., Steiniger, F., and Tichelaar, W. (2006). Limited-dose electron microscopy reveals the crystallinity of fibrous C–S–H phases. *Journal of the American Ceramic Society* 89: 627–632.

Rouquerol, F., Rouquerol, F., Llewellyn, P. et al. (2014). *Adsorption by Powders and Porous Solids, Principles, Methodology and Applications*. Elsevier.

Rucker, P. (2008). Modellierung des Feuchte- und Salztransports unter Berücksichtigung der Selbstabdichtung in zementgebundenen Baustoffen. Dissertation. Technische Universität München.

Rucker, P. and Beddoe, R.E. (2007). Transport von drückendem Wasser in Betonbauteilen. *Beton- und Stahlbetonbau* 102: 414–426.

Rucker, P., Beddoe, R.E., and Schießl, P. (2006). Wasser- und Salzhaushalt im Gefüge zementgebundener Baustoffe – Modellierung der auftretenden Mechanismen. *Beton- und Stahlbetonbau* 101 (6): 402–412.

Rucker-Gramm, P. and Beddoe, R.E. (2010). Effect of moisture content of concrete on water uptake. *Cement and Concrete Research* 40: 102–108.

Ruiz-Llamas, A. and Macías-Salinas, R. (2015). Modeling the dynamic viscosity of ionic solutions. *Industrial & Engineering Chemistry Research* 54: 7169–7179.

Saeidpour, M. and Wadsö, L. (2015). Moisture equilibrium of cement based materials containing slag or silica fume and exposed to repeated sorption cycles. *Cement and Concrete Research* 69: 88–95.

Saeidpour, M. and Wadsö, L. (2015). Evidence for anomalous water vapor sorption kinetics in cement based materials. *Cement and Concrete Research* 70: 60–66.

Saeidpour, M. and Wadsö, L. (2016). Moisture diffusion coefficients of mortars in absorption and desorption. *Cement and Concrete Research* 83: 179–187.

Schaan, G., Rybczynski, S., Ritter, M., and Schmidt-Döhl, F. (2019). Transmission electron microscopy investigations of fatigue induced changes in the cement structure of ultra-high performance concrete. *41st International Conference on Cement Microscopy 2019*, San Diego, CA.

Scheffler, G.A. (2008). Validierung hygrothermischer Materialmodellierung unter Berücksichtigung der Hysterese der Feuchtespeicherung. Dissertation. Technische Universität Dresden.

Scheffler, G. and Plagge, R. (2005). Bestimmung des Trocknungsverhaltens von Baustoffen unter definierten Randbedingungen. *Bauphysik* 27: 324–330.

Scheffler, G.A. and Plagge, R. (2009). Ein Trocknungskoeffizient für Baustoffe. *Bauphysik* 31: 125–138.

Scheffler, G.A. and Plagge, R. (2011). Methods for moisture storage and transport property determination of autoclaved aerated concrete. Cement Wapno Beton.

Schießl, P. and Wiegrink, K.H. (2004). *Austrocknungsverhalten von Calciumsulfat-Fließestrichen*, Bericht. Technische Universität München.

Schiller, P., Wahab, M., Bier, Th. et al. (2015). Capillary forces and sorption hysteresis of cement pastes with small slit pores. *Procedia Materials Science* 11: 649–654.

Schiller, P., Wahab, M., Waida, S. et al. (2017). Dislocation model for sorption hysteresis in deformable solids. *Colloids and Surfaces A: Physicochemical and Engineering Aspects* 513: 76–86.

Schimmele, L., Napiórkowski, M., and Dietrich, S. (2007). Conceptual aspects of line tensions. *Journal of Chemical Physics* 127: 164715.

Schlumberger, C. and Thommes, M. (2021). Characterization of hierarchically ordered porous materials by physisorption and mercury porosimetry—a tutorial review. *Advanced Materials Interfaces* 8: 2002181.

Schlüßler, K.H. and Mcedlov-Petrosjan, O.P. (1990). *Der Baustoff Beton*. Berlin: VEB Verlag für Bauwesen.

Schmidt, H. (2010). Korrosionsverhalten von Normalmörtel sowie UHPC - Experimente und numerische Simulation. Dissertation. Technische Universität Hamburg.

Schneider, M. and Goss, K.-U. (2011). Temperature dependence of the water retention curve for dry soils. *Water Resources Research* 47: https://doi.org/10.1029/2010WR009687.

Schulte Holthausen, R., Raupach, M., Merkel, M., and Breit, W. (2020). Auslaugungswiderstand von Betonoberflächen in Trinkwasserbehältern. *Bautechnik* 97 (6): 368–376.

Schulze, G. and Schluender, E.U. (1985). Absorption von einzelnen Gasblasen in vorbeladenem Wasser. *Chemie Ingenieur Technik* 57: 233–235.

Schwotzer, M. (2008). Zur Wechselwirkung zementgebundener Werkstoffe mit Wässern unterschiedlicher Zusammensetzung am Beispiel von Trinkwasserbehälterbeschichtungen. Dissertation. Karlsruhe: Universität Fridericiana zu Karlsruhe.

Schwotzer, M., Scherer, T., and Scherer, A. (2010). Protective or damage promoting effect of calcium carbonate layers on the surface of cement based materials in aqueous environments. *Cement and Concrete Research* 40: 1410–1418.

Setzer, M.J., Duckheim, C., Liebrecht, A., and Kruschwitz, J. (2006). The solid-liquid gel-system of hardened cement paste. *Conference Paper*, www.researchgate.net/publication/301746898 (accessed 2 January 2024).

Shen, W., Li, X., Lu, X. et al. (2018). Experimental study and isotherm models of water vapor adsorption in shale rocks. *Journal of Natural Gas Science and Engineering* 52: 484–491.

Shuangyan, X., Simmons, G.C., Mahadevan, T.S. et al. (2009). Transport of water in small pores. *Langmuir* 25: 5084–5090.

Siebold, A., Walliser, A., Nardin, M. et al. (1997). Capillary rise for thermodynamic characterization of solid particle surface. *Journal of Colloid and Interface Science* 186: 60–70.

Snoeck, D., Velasco, L.F., Mignon, A. et al. (2014). The influence of different drying techniques on the water sorption properties of cement-based materials. *Cement and Concrete Research* 64: 54–62.

Sousa, S.J.G. and de Holanda, J.N.F. (2005). Sintering behavior of porous wall tile bodies during fast single-firing process. *Materials Research* 8: 197–200.

Spragg, R., Villani, Ch., Snyder, K., and Bentz, D. (2013). Factors that influence electrical resistivity measurements in cementitious systems. *Transportation Research Record: Journal of the Transportation Research Board* 232 (1): 90–98.

Stark, J. (2011). Recent advances in the field of cement hydration and microstructure analysis. *Cement and Concrete Research* 41: 666–678.

Stark, J. and Wicht, B. (2000). *Zement und Kalk. Der Baustoff als Werkstoff*. Basel: Birkhäuser.

Starov, V.M. and Velarde, M.G. (2007). *Wetting and Spreading Dynamics*. Boca Raton, FL: CRC Press.

Starov, V.M. and Velarde, M.G. (2009). Surface forces and wetting phenomena. *Journal of Physics: Condensed Matter* 21: 464121.

Striolo, A. (2011). From interfacial water to macroscopic observables: a review. *Adsorption Science & Technology* 29 (3): 211–258.

Sugden, S. (1921). *The Determination of Surface Tension from the Rise in Capillary Tubes*. Downloaded by Hamburg University of Technology.

Tang, X., Ripepi, N., Valentine, K.A. et al. (2017). Water vapor sorption on Marcellus shale: measurement, modeling and thermodynamic analysis. *Fuel* 209: 606–614.

Tarazona, P., Marconi, U.M.B., and Evans, R. (1987). Phase equilibria of fluid interfaces and phase equilibria of fluid interfaces and confined fluids. *Molecular Physics* 60 (3): 573–595.

Tariku, F., Kumaran, K., and Fazio, P. (2010). Transient model for coupled heat, air and moisture transfer through multilayered porous media. *International Journal of Heat and Mass Transfer* 53: 3035–3044.

Taylor, H.F.W. (1997). *Cement chemistry*. London: Thomas Telford Publishing.

Taylor, R., Richardson, I.G., and Brydson, R.M.D. (2010). Composition and microstructure of 20-year-old ordinary Portland cement–ground granulated blast-furnace slag blends containing 0 to 100% slag. *Cement and Concrete Research* 40: 971–983.

Technische Kommission Bauklebstoffe (2018). *Belegreife und Feuchte / Sorptionsisothermen und die Interpretation von KRL-Messungen*. Düsseldorf: Industrieverband Klebstoffe e.V.

Teletzke, G.F., Davis, H.T., and Scriven, L.E. (1988). Wetting hydrodynamics. *Revue de Physique Appliquée* 23: 989–1007.

Thiel, C. and Gehlen, C. (2018). Nuclear Magnetic Resonance and Magnetic Resonance Imaging. State-of-the-Art Report of the RILEM Technical Committee 248-MMB.

Thiéry, M., Faure, P., Morandeau, A. et al. (2011). Effect of carbonation on the microstructure and moisture properties of cement-based materials. *International Conference XII DBMC*, Porto, Portugal.

Thomas, J.J., Biernacki, J., Bullard, J.W. et al. (2011). Modeling and simulation of cement hydration kinetics and microstructure development. *Cement and Concrete Research* 41: 1257–1278.

Thommes, M., Kaneko, K., Neimark, A.V. et al. (2015). Physisorption of gases, with special reference to the evaluation of surface area and pore size distribution (IUPAC Technical Report). *Pure and Applied Chemistry* 87: 160.

Thompson, A.H. (1991). Fractals in rock physics. *Annual Review of Earth and Planetary Sciences* 19: 237–261.

Todd, D.K. (1959). *Ground Water Hydrologie*. London, New York: Wiley.

Tumidajski, P.J. and Lin, B. (1998). On the validity of the Katz-Thompson equation for permeabilities in concrete. *Cement and Concrete Research* 28 (5): 643–647.

Turza, R. and Füri, B.B. (2017). Experimental measurements of the water evaporation rate of a physical model. *Slovak Journal of Civil Engineering* 25: 19–24.

Ungricht, H. (2004). Wasserhaushalt und Chlorideintrag in Beton / Einfluss der Exposition und der Betonzusammensetzung. Dissertation. Technische Hochschule Zürich.

Valeev, A.A. (2014). Simple Kelvin equation applicable in the critical point vicinity. *European Journal of Natural History* 5: 13–14.

Valeev, A.A. and Morozova, E.V. (2017). Simple universal Kelvin equation valid in critical point vicinity and its application to carbon dioxide capillary condensation in mesoporous silica. *Solid State Phenomena* 265: 392–397.

Valori, A., McDonald, P.J., and Scrivener, K.L. (2013). The morphology of C–S–H: Lessons from ^{1}H nuclear magnetic resonance relaxometry. *Cement and Concrete Research* 49: 65–81.

Villagrán-Zaccardi Yury, A., Alderetea, N.M., and De Belieb, N. (2017). Improved model for capillary absorption in cementitious materials: progress over the fourth root of time. *Cement and Concrete Research* 100: 153–165.

Villani, C., Spragg, R., Pour-Ghaz, M., and Weiss, W.J. (2014). The influence of pore solutions properties on drying in cementitious materials. *Journal of the American Ceramic Society* 97: 386–393.

Villmann, B., Slowik, V., Wittmann, F.H. et al. (2014). Time-dependent moisture distributi in drying cement mortars / results of neutron radiography and inverse analysis of drying tests. *Restoration of Buildings and Monuments* 20 (1): 49–62.

Vincent, O., Sessoms, D.A., Huber, E.J. et al. (2014). Drying by cavitation and poroelastic relaxations in porous media with macroscopic pores connected by nanoscale throats pores connected by nanoscale throats. *Physical Review Letters* 113: 134501.

Vincent, O., Marguet, B., and Stroock, A.D. (2017). Imbibition triggered by capillary condensation in nanopores. *Langmuir* 33: 1655–1661.

Wang, F.-C. and Zhao, Y.-P. (2011). Slip boundary conditions based on molecular kinetic theory: the critical shear stress and the energy dissipation at the liquid–solid interface. *Soft Matter* 7: 8628–8634.

Welcome to DataPhysics Instruments (2019). Collection of surface-tension values of liquids and solids. www.dataphysics-instruments.com (accessed 29 December 2023).

Wiegrink, K.-H. (2002). Modellierung des Austrocknungsverhaltens von Calciumsulfat-Fließestrichen und der resultierenden Spannungen und Verformungen. Dissertation. Technische Universität München.

Wiig, E.O. and Juhola, A.J. (1949). The adsorption of water vapor on activated charcoal. *Journal of the American Chemical Society* 372: 561–568.

Wittmann, F.H., Zhang, P., and Zhao, T. (2011). Damage of concrete under combined actions. *Restoration of Buildings and Monuments* 17: 321–330.

Worch, A. (2002). Überlegungen und Versuche zur Erfassung des Wasserdampfübergangs an Bauteiloberflächen. Dissertation. Technische Universität Dortmund.

Wu Ch., Zandavi, S.H., Charles, A., and Ward. (2014). Prediction of the wetting condition from the Zeta adsorption isotherm. *Physical Chemistry Chemical Physics* 16: 25564–25572.

Wu, Q., Rougelot, T., Burlion, N., and Bourbon, X. (eds.) (2014). Experimental study of water retention properties of a high performance concrete with small sliced samples. *5th International Conference on Porous Media and its Applications in Science and Engineering*, Kona, Hawaii.

Wu, M., Johannesson, B., and Geiker, M. (2014). Application of water vapor sorption measurements for porosity characterization of hardened cement pastes. *Construction and Building Materials* 66: 621–633.

Wu, Q., Rougelot, T., Burlion, N., and Bourbon, X. (2015). Representative volume element estimation for desorption isotherm of concrete with sliced samples. *Cement and Concrete Research* 76: 1–9.

Wu, K., Chen, Z., Li, J. et al. (2017). Wettability effect on nanoconfined water flow. *Proceedings of the National Academy of Sciences of the United States of America* 114 (13): 3358–3363.

Wyrzykowski, M., McDonald, P.J., Scrivener, K.L., and Lura, P. (2017). Water redistribution within the microstructure of cementitious materials due to temperature changes studied with ^{1}H NMR. *Journal of Physical Chemistry* C 121: 27950–27962.

Yaghoubian, S., Zandavi, S.H., and Ward, C.A. (2016). From adsorption to condensation: the role of adsorbed molecular clusters. *Physical Chemistry Chemical Physics* 18: 21481–21490.

Yang, F., Lyu, B., and Xu, S. (2021). Water sorption and transport in shales: an experimental and simulation study. *Water Resources Research* 57 (2): e2019WR026888.

Ye, G. (2005). Percolation of capillary pores in hardening cement pastes. *Cement and Concrete Research* 35: 167–176.

Yonghao, F., Zhongli, W., and Yue, Z. (2008). Time-dependent water permeation behavior of concrete under constant hydraulic pressure. *Water Science and Engineering* 1 (4): 61–66.

Yuan, Y. and Lee, T.R. (2013). *Contact Angle and Wetting Properties*, Springer Series in Surface Sciences 51. Surface Science Techniques.

Zandavi, S.H. (2015). Vapours adsorption on non-porous and porous solids: zeta adsorption isotherm approach. Dissertation. University of Toronto.

Zandavi, S.H. and Ward, C.A. (2014). Clusters in the adsorbates of vapours and gases: zeta isotherm approach. *Physical Chemistry Chemical Physics* 16: 10979–10989.

Zandavi, S.H., Charles, A., and Ward. (2015). Characterization of the pore structure and surface properties of shale using the zeta adsorption isotherm approach. *Energy Fuels* 29: 3004–3010.

Zeng, Q. and Xu, S. (2015). Discussion of "Numerical simulation of moisture transport in concrete based on a pore size distribution model". *Cement and Concrete Research* 73: 63–66.

Zeng, Q., Zhang, D.D., Sun, H., and Li, K. (2014). Characterizing pore structure of cement blend pastes using water vapor sorption analysis. *Materials Characterization* 95: 72–84.

Zhang, Z. (2014). Modelling of sorption hysteresis and its effect on moisture transport within cementitious materials. Dissertation. Université Paris Est.

Zhang, Z. and Angst, U. (2020). A dual-permeability approach to study anomalous moisture transport properties of cement-based materials. *Transport in Porous Media* 135: 59–78.

Zhang, Z. and Scherer, G. (2018). Determination of water permeability for cementitious materials with minimized batch effect. Physics.app-ph.

Zhang, P., Wittmann, F.H., Zhao, T., and Lehmann, E. (2009). Penetration of water into uncracked and cracked steel reinforced concrete elements; visualization by means of neutron radiography + erratum 2013. *Restoration of Buildings and Monuments* 15: 67–76.

Zhang, Zh., Thiéry, M., and Baroghel-Bouny, V. (2014). A review and statistical study of existing hysteresis models for cementitious materials. *Cement and Concrete Research* 57: 44–60.

Zhang, Z., Thiery, M., and Baroghel-Bouny, V. (2015). Numerical modelling of moisture transfers with hysteresis within cementitious materials: verification and investigation of the effects of repeated wetting–drying boundary conditions. *Cement and Concrete Research* 68: 10–23.

Zhang, T., Li, X., Sun, Z. et al. (2017). An analytical model for relative permeability in water-wet nanoporous media. *Chemical Engineering Science* 174: 1–12.

Zhang, Y., Ouyang, X., and Yang, Z. (2019). Microstructure-based relative humidity in cementitious system due to self-desiccation. *Materials* 12: 1214.

Zhao, G., Tan, Q., Xiang, L. et al. (2015). Structure and properties of water film adsorbed on mica surfaces. *Journal of Chemical Physics* 143: 104705.

Zhao, J., Meissener, F., and Grunewald, J. (2020). Experimental investigation of the drying behaviour of the building materials. *E3S Web of Conferences* 172: 17002.

Zhou, C. (2014). General solution of hydraulic diffusivity from sorptivity test. *Cement and Concrete Research* 58: 152–160.

Zhou, C. (2014). Predicting water permeability and relative gas permeability of unsaturated cement-based material from hydraulic diffusivity. *Cement and Concrete Research* 58: 143–151.

Zhou, C., Chen, W., Wang, W., and Skoczylas, F. (2016). Indirect assessment of hydraulic diffusivity and permeability for unsaturated cement-based material from sorptivity. *Cement and Concrete Research* 82: 117–129.

Zimmerman, J.R. and Brittin, W.E. (1957). *NMR-Studies in Multiple Phase Systems: Lifetime of a Water Molecule in An Adsorbing Phase On Silica Gel*. Dallas, TX: Magnolia Petroleum Company, Field Research Laboratory.

Index

a

Moisture Storage and Transport in Concrete: Experimental Investigations and Computational Modeling,
First Edition. Lutz H. Franke.

b

c

d

f

h

p